AF839877

Big Data in Predictive Toxicology

Issues in Toxicology

Editor-in-chief:
Diana Anderson, *University of Bradford, UK*

Series editors:
Michael D. Waters, *Michael Waters Consulting, USA*
Timothy C. Marrs, *Edentox Associates, UK*

Editorial advisor:
Alok Dhawan, *CSIR-Indian Institute of Toxicology Research, Lucknow, India*

Titles in the Series:

1: Hair in Toxicology: An Important Bio-monitor
2: Male-mediated Developmental Toxicity
3: Cytochrome P450: Role in the Metabolism and Toxicity of Drugs and other Xenobiotics
4: Bile Acids: Toxicology and Bioactivity
5: The Comet Assay in Toxicology
6: Silver in Healthcare
7: *In Silico* Toxicology: Principles and Applications
8: Environmental Cardiology
9: Biomarkers and Human Biomonitoring, Volume 1: Ongoing Programs and Exposures
10: Biomarkers and Human Biomonitoring, Volume 2: Selected Biomarkers of Current Interest
11: Hormone-disruptive Chemical Contaminants in Food
12: Mammalian Toxicology of Insecticides
13: The Cellular Response to the Genotoxic Insult: The Question of Threshold for Genotoxic Carcinogens
14: Toxicological Effects of Veterinary Medicinal Products in Humans: Volume 1
15: Toxicological Effects of Veterinary Medicinal Products in Humans: Volume 2
16: Aging and Vulnerability to Environmental Chemicals: Age-related Disorders and their Origins in Environmental Exposures
17: Chemical Toxicity Prediction: Category Formation and Read-across
18: The Carcinogenicity of Metals: Human Risk Through Occupational and Environmental Exposure
19: Reducing, Refining and Replacing the Use of Animals in Toxicity Testing
20: Advances in Dermatological Sciences
21: Metabolic Profiling: Disease and Xenobiotics
22: Manganese in Health and Disease
23: Toxicology, Survival and Health Hazards of Combustion Products
24: Masked Mycotoxins in Food: Formation, Occurrence and Toxicological Relevance

25: Aerobiology: The Toxicology of Airborne Pathogens and Toxins
26: Chemical Warfare Toxicology, Volume 1: Fundamental Aspects
27: Chemical Warfare Toxicology, Volume 2: Management of Poisoning
28: Toxicogenomics in Predictive Carcinogenicity
29: Human Stem Cell Toxicology
30: The Comet Assay in Toxicology, 2nd edition
31: Computational Systems Pharmacology and Toxicology
32: Ecotoxicology and Genotoxicology: Non-traditional Terrestrial Models
33: Ecotoxicology and Genotoxicology: Non-traditional Aquatic Models
34: Oxidative Stress and Redox Signalling in Parkinson's Disease
35: Nanotoxicology: Experimental and Computational Perspectives
36: Regulatory Toxicology in the European Union
37: Formaldehyde: Exposure, Toxicity and Health Effects
38: Chemical Health Threats: Assessing and Alerting
39: The Micronucleus Assay in Toxicology
40: Nanoparticle–Protein Corona: Biophysics to Biology
41: Big Data in Predictive Toxicology

How to obtain future titles on publication:
A standing order plan is available for this series. A standing order will bring delivery of each new volume immediately on publication.

For further information please contact:
Book Sales Department, Royal Society of Chemistry, Thomas Graham House, Science Park, Milton Road, Cambridge, CB4 0WF, UK
Telephone: +44 (0)1223 420066, Fax: +44 (0)1223 420247
Email: booksales@rsc.org
Visit our website at www.rsc.org/books

Big Data in Predictive Toxicology

Edited by

Daniel Neagu
University of Bradford, UK
Email: d.neagu@bradford.ac.uk

and

Andrea-Nicole Richarz
European Commission – Joint Research Centre, Italy
Email: andrea.richarz@alumni.tu-berlin.de

ROYAL SOCIETY
OF **CHEMISTRY**

Issues in Toxicology No. 41

Print ISBN: 978-1-78262-298-7
PDF ISBN: 978-1-78262-365-6
EPUB ISBN: 978-1-83916-082-0
Print ISSN: 1757-7179
Electronic ISSN: 1757-7187

A catalogue record for this book is available from the British Library

© The Royal Society of Chemistry 2020

All rights reserved

Apart from fair dealing for the purposes of research for non-commercial purposes or for private study, criticism or review, as permitted under the Copyright, Designs and Patents Act 1988 and the Copyright and Related Rights Regulations 2003, this publication may not be reproduced, stored or transmitted, in any form or by any means, without the prior permission in writing of The Royal Society of Chemistry or the copyright owner, or in the case of reproduction in accordance with the terms of licences issued by the Copyright Licensing Agency in the UK, or in accordance with the terms of the licences issued by the appropriate Reproduction Rights Organization outside the UK. Enquiries concerning reproduction outside the terms stated here should be sent to The Royal Society of Chemistry at the address printed on this page.

Whilst this material has been produced with all due care, The Royal Society of Chemistry cannot be held responsible or liable for its accuracy and completeness, nor for any consequences arising from any errors or the use of the information contained in this publication. The publication of advertisements does not constitute any endorsement by The Royal Society of Chemistry or Authors of any products advertised. The views and opinions advanced by contributors do not necessarily reflect those of The Royal Society of Chemistry which shall not be liable for any resulting loss or damage arising as a result of reliance upon this material.

The Royal Society of Chemistry is a charity, registered in England and Wales, Number 207890, and a company incorporated in England by Royal Charter (Registered No. RC000524), registered office: Burlington House, Piccadilly, London W1J 0BA, UK, Telephone: +44 (0) 20 7437 8656.

For further information see our web site at www.rsc.org

Printed in the United Kingdom by CPI Group (UK) Ltd, Croydon, CR0 4YY, UK

Preface

Big Data science and technologies have won their presence in every section of the digital economy. This was initially motivated by the increasingly efficient and cheaper information gathering and storage solutions, development of non-relational databases and generalisation of accessing internet resources as a service. Such developments, encouraged also by new legislation and industry initiatives, started changing the processes for data acquisition, management, analysis, validation and re-use also of experiments regarding biological and chemical data.

Big Data systems and analytics, as newly established technologies, provide solutions that regard an increasingly fluid number of characteristics that challenge and stretch traditional storage and processing tools and environments, for example, relational databases and off-line machine learning model development. In a traditionalist approach, Big Data is defined by four unanimously acknowledged data features: the *volume*, the *velocity*, the *variety*, and the *veracity* (mentioned by Hilbert[1] as the Big Data 4 Vs) of its available digital content.

The *volume* of Big Data refers to those instances already stored in data warehouses, lakes or less sophisticated databases that have a continuous or intermittent increase of their size. Data management solutions in such cases may not satisfy current challenges and expectations regarding terabytes to exabytes of data available to be processed consistently and efficiently.

The *velocity* of Big Data refers to dynamical attributes of data acquisition, storage, update and transfer mostly in real time, challenged, for example,

Issues in Toxicology No. 41
Big Data in Predictive Toxicology
Edited by Daniel Neagu and Andrea-Nicole Richarz
© The Royal Society of Chemistry 2020
Published by the Royal Society of Chemistry, www.rsc.org

by short times (*e.g.*, milliseconds to seconds) for the database to respond. Regular solutions are offered by well-defined processes but are less prepared to deliver streamed processing and specialised data formats to support realistic and efficient timely services.

The *variety* of Big Data is a consequence of both diversity of sources and provenance of data, and the increased number of multimedia formats that are used to produce, store, exchange and process such data. This is a consequence of successful developments in information gathering, digitalisation, networking, and mobile and communication technologies that allowed a fast democratisation of accessing and exchanging data.

The *veracity* of Big Data refers to the quality, confidence and usefulness of current data resources – features that are requested more and more by regulatory and industry standards, and also acknowledged from the beginning by the data mining and data analytics communities. Data veracity, though, is currently an open field for meta-data capture and processing, consistent data quality measurement and reporting.

Chemical, biological and toxicological data resources show a continuous increase in volume of the number of records and instances recorded, as well as of the number and diversity of the types of data, features and attributes captured or integrated in such repositories. This is due to the current advancements in biology, toxicology, chemistry – as for example in the field of nanomaterials – and related sciences that are producing complex experimental data with increasing digital volumes and increasing speed of data generation, as compared to the traditional experimental toxicological data. A variety of different data – from the chemical and biological side – need to be combined. Thus, more and more sophisticated solutions are required to organise, integrate, analyse and interpret such data and also their digital derivatives (such as statistical analyses, machine learning models), and their further use in multidisciplinary repositories, dashboards and projects. International collaborations and public–private partnership initiatives encourage and further respective research.

Predictive Toxicology is both a multidisciplinary domain and an interdisciplinary field of research and knowledge transfer, critically relying on a broad and qualitatively sound data basis for model development, and as such, it is well suited to take up the challenge and benefit from Big Data technologies. Its multidisciplinarity is an example of how multiple forms of expertise (*e.g.*, biology, chemistry, toxicology, applied statistics, computer science, machine learning) and perspectives are called upon to address toxicology modelling and evaluation challenges. Its interdisciplinarity also combines knowledge and builds on novel disciplines at the crossroads of existing approaches, such as bio-informatics, chemo-informatics, toxico-genomics, applied data sciences, and Big Data.

While Predictive Toxicology has started benefitting from new solutions for large-scale data and model storage, it has also raised additional requirements, specifications and challenges. These regard data gathering, representation, formatting, processing, validation, knowledge extraction and integration from big and diverse toxicology as well as chemistry data, model governance, and visualisation. Therefore, Predictive Toxicology is stepping into the Big Data scene, requiring a diverse and specific effort from different communities of researchers, users and model developers to update and validate sustainable solutions, and take up the opportunities for many applications.

In general, the *volume* and *velocity* of the toxicology and chemistry data resources may not yet create serious technical issues when considered within the context of Big Data, and are the first to offer opportunities for new technologies and progress, such as cloud services and distributed (*e.g.*, Hadoop) processing. More important, though, is currently how the *variety* and *complexity* of these data are managed and systematised, linked together to become a useful information resource, and finally abstracted through large-scale data analysis in the form of models to produce useful pieces of knowledge that can support further decision making. While data organisation is an important topic, knowledge organisation (its storage, accessing and consistent representation both of the data and the resulting predictive models) remains a major issue to be dealt with for any application and institution.

The era of Big Data is already evolving from the relatively normal expectations of traditional measurable features by adding even more challenging characteristics and functionalities,[2] such as *variability, validity, volatility, visibility and value*, aiming for sustainable governance, re-use and automation. While some of these newly added features share common ground with existing ones and require more fundamental research in the definitions of the concepts and measurements of their contributions, nevertheless they contribute towards uncovering the attractiveness and diversity of expectations and requirements from the general public, users and specialists for their area of interest; this also applies to Predictive Toxicology, as described in different chapters of this book.

Big Data in Predictive Toxicology: Overview of the Book Contents

The experience, knowledge, and examples shared by the authors of the chapters are of interest for both Big Data and machine learning researchers as well as for researchers and practitioners in toxicology and chemical safety assessment, to learn about the advances in the field. The content of the book, while sharing features with the topic of Big Data, through the diversity of topics and viewpoints, value of knowledge and variety of approaches, is by no means exhaustive in terms of volume coverage and pace of updates. Its aim is to provide a stepping stone for the acknowledgment of work done to date, for the storage of the current approaches, and to offer an open table for the identification of challenges and for the report and reference of new ideas and contributions.

The book is aimed at giving a broad overview of the area of Big Data in Predictive Toxicology, explaining basic notions and providing examples of applications in twelve chapters ordered in three sections as follows.

Chapter 1 – *Big Data in Predictive Toxicology: Challenges, Opportunities and Perspectives* – gives a systematic introduction of Big Data concepts and features as applied in the context of Predictive Toxicology, from the motivation and relevant characteristics to challenges that need to be addressed to leverage their potential, and opportunities they offer for advancing chemical safety assessment and related urgent issues.

The first section of the book provides in Chapters 2, 3 and 4 the technical basis and a review of types, processing and storage of data in biology, chemistry and toxicology.

Chapter 2 – *Biological Data in the Light of Toxicological Risk Assessment* – reviews the types of toxicological data and studies that are at the basis of predictive model development as well as the toxicological endpoints that are currently information requirements for regulatory toxicological assessment, which Predictive Toxicology is targeting. These comprise *in vivo* and *in vitro* testing data, with advances in high-throughput bioactivity assays and 'omics' technologies building new data resources.

Chapter 3 – *Chemoinformatics Representation of Chemical Structures – A Milestone for Successful Big Data Modelling in Predictive Toxicology* – reviews chemical structure representations in chemoinformatics as a key aspect of linking toxicity information to a chemical in computational toxicology, in the context of storing and searching structural data in and across big databases and its use for predictive modelling.

Chapter 4 – *Organisation of Toxicological Data in Databases* – reviews a variety of approaches and formats for data exchange and database design for organising toxicity, chemistry and related data, and gives an overview of existing toxicological databases as well as examples for integrated approaches. The importance of high-quality databases as support for (*in silico*) safety assessment is highlighted.

The book's second section builds in Chapters 5 and 6 on the available data and data formats by adding the challenges of constructing, using, integrating, storing and sharing data and models as complex, heterogeneous and valid resources for modern Predictive Toxicology applications.

Chapter 5 – *Making Big Data Available: Integrating Technologies for Toxicology Applications* – considers alongside data volume other less studied characteristics of toxicity Big Data: variety and veracity, which add representative challenges in data management and analysis. The solutions presented hereby build on a syntactic level from database integration (Application Programming Interfaces) and ontologies, and on a semantic level from chemo- and bioinformatics methods of data analysis.

Chapter 6 – *Storing and Using Qualitative and Quantitative Structure–Activity Relationships in the Era of Toxicological and Chemical Data Expansion* – presents practices and developments for standardisation of prediction model formats by their systematic representation, organisation and storage. It raises the topic of the need and progress of sustainable representation and fusion of data and models using generalised data governance functionalities.

The third section of the book shows in the contributions contained in Chapters 7 to 12 how the analysis and organisation of different types of toxicological Big Data can leverage building knowledge and support applications in modelling and computational predictions for chemical safety assessment.

Chapter 7 – *Toxicogenomics and Toxicoinformatics: Supporting Systems Biology in the Big Data Era* – reports and exemplifies the current technologies for storage, pre-processing and integration of 'omics' data in order to allow data mining, machine learning and mechanistic analyses revealing new patterns and ways of identification and explanation of adverse effect mechanisms and build knowledge in Systems Toxicology. Examples of toxicoinformatics applications for toxicity prediction are described.

Chapter 8 – *Profiling the Tox21 Chemical Library for Environmental Hazards: Applications in Prioritisation, Predictive Modelling, and Mechanism of Toxicity Characterisation* – highlights the Toxicology for the 21st Century (Tox21) programme, which contributes to Big Data in toxicology with over 120 million data points generated for about 10 000 environmental chemicals and drugs with a high-throughput screening assay battery, and provides examples of *in vivo* toxicity prediction models built from the compounds' activity profiles and how to prioritise compounds for further in-depth toxicological studies.

Chapter 9 – *Big Data Integration and Inference* – reviews ongoing efforts to aggregate toxicological knowledge from different data types and sources in a sustainable knowledge base based on the Adverse Outcome Pathway (AOP) framework, and provides case studies of data integration and inferential analysis for use in Predictive Toxicology from a Big Data volume viewpoint. It also unveils opportunities for mining Big Data sets and producing large libraries of computationally predicted AOPs. The use of such libraries and AOPs in decision support, systems biology applications, testing and risk assessment and model development is discussed.

Chapter 10 – *Chemometrical Analysis of Proteomics Data* – presents case studies for using chemometrical approaches to model data of diverse input formats such as proteomics data and chemical descriptors. The outcome of such an approach relies on opportunities for knowledge extraction from a large amount of data points and using proteomics data analysis to understand toxicity mechanisms and identify biomarkers.

The implication of Big Data in two specific areas of interest in Predictive Toxicology is described in the last two chapters, notably biokinetics, which are at the centre of modern safety assessment as a key for translating *in vitro* and *in silico* toxicity estimates to *in vivo* conditions, and read-across for which there is considerable interest for data gap filling in the regulatory context.

Chapter 11 – *Big Data and Biokinetics* – describes the incorporation of Big Data approaches to improve the 'throughput' of biokinetic modelling, defining a comprehensive physiological/biochemical modelling framework to rapidly predict *in vivo* biokinetics of chemicals, supporting interpretation of high volumes of toxicity data on large numbers of chemicals in an efficient way.

Chapter 12 – *Role of Toxicological Big Data to Support Read-Across for the Assessment of Chemicals* – discusses how the use of Big Data from new methodology approaches supports the justification of similarity in read-across, extending from chemical to biological and toxicological similarity, as well as the mechanistic interpretation, and thus increases the confidence in read-across, essential for regulatory acceptance. Furthermore, Big Data can support read-across approaches in addressing upcoming challenges such as hazard assessment of mixtures and nanomaterials.

We would like to thank all authors and researchers as well as the editors from the Royal Society of Chemistry who contributed to and supported this multi-authored book.

Daniel Neagu
University of Bradford,
Department of Computer Science,
Faculty of Engineering and Informatics,
Bradford, UK
Andrea-Nicole Richarz
European Commission Joint Research Centre,
Ispra, Italy

References

1. M. Hilbert, Big Data for Development: A Review of Promises and Challenges, *Dev. Policy Rev.*, 2016, **34**(1), 135–174.
2. G. Firican, The 10 Vs of Big Data, *TDWI*, 2017, https://tdwi.org/articles/2017/02/08/10-vs-of-big-data.aspx.

Contents

Chapter 1 Big Data in Predictive Toxicology: Challenges, Opportunities and Perspectives 1
Andrea-Nicole Richarz

1.1 Introduction 1
1.2 Big Data in the Area of Predictive Toxicology 2
1.3 The Big Vs of Predictive Toxicology Data 3
1.4 Challenges of Big Data in Predictive Toxicology 6
1.4.1 Need for an Adequate Infrastructure 7
1.4.2 Standardisation and Data Curation 9
1.4.3 Too Big? Identifying the Relevant Data 11
1.4.4 Data Integration Infrastructures 11
1.4.5 Making Sense of the Data 14
1.5 Opportunities Provided by Big Data for Predictive Toxicology 15
1.5.1 More is More – Benefits of a Broader and More Diverse Data Basis 16
1.5.2 Big Data for the Big Picture 18
1.5.3 Creating und Using New Knowledge: Applications for Hot Topics in (Predictive) Toxicology 19
1.6 Conclusions and Perspectives 25
References 27

Issues in Toxicology No. 41
Big Data in Predictive Toxicology
Edited by Daniel Neagu and Andrea-Nicole Richarz
© The Royal Society of Chemistry 2020
Published by the Royal Society of Chemistry, www.rsc.org

Chapter 2 Biological Data in the Light of Toxicological Risk Assessment **38**
Vessela Vitcheva

2.1 Introduction 38
2.2 Data Generated by *In Vivo* Toxicity Testing 39
2.2.1 Acute Toxicity Testing 40
2.2.2 Toxicity Testing after Repeated Dose Administration (Short-term, Sub-chronic and Chronic Toxicity Tests) 42
2.2.3 Toxicity Tests for Specific Endpoints 44
2.3 Data Generated by *In Vitro* Studies 48
2.3.1 *In Vitro* Test Methods for Genotoxicity/ Mutagenicity Testing 48
2.3.2 *In Vitro* Test Methods for Carcinogenicity 51
2.3.3 *In Vitro* and *In Chemico* Test Methods for Skin Sensitisation 51
2.4 Mechanistic Understanding – Adverse Outcome Pathways 52
2.5 Data from New Approach Methodologies 53
2.6 Toxicokinetic Data 54
2.7 Sources of Toxicological Data 55
2.7.1 TOXicology Data NETwork (TOXNET) 55
2.7.2 Registry of Toxic Effects of Chemical Substances (RTECS) 56
2.7.3 Kyoto Encyclopedia of Genes and Genomes (KEGG) 56
2.7.4 Aggregated Computational Toxicology Resource (ACToR) 57
2.7.5 European Chemicals Agency (ECHA) Data 57
2.7.6 COSMOS Next Generation 57
2.7.7 Data from Assessment Reports 58
2.8 Quality of Data 59
2.9 Data Interpretation 61
2.10 Use of Biological Data for Predictive Toxicology 62
References 63

Chapter 3 Chemoinformatics Representation of Chemical Structures – A Milestone for Successful Big Data Modelling in Predictive Toxicology **69**
Nikolay Kochev, Nina Jeliazkova and Ivanka Tsakovska

3.1 Introduction to Chemoinformatics Approaches for Chemical Structure Representation 69
3.2 Structure Representation Characteristics 74

3.3 Constitutional Representations 77
3.4 Topological Representations 78
3.5 Linear Notations 81
3.5.1 SMILES Linear Notation 82
3.5.2 InChI Linear Notation 84
3.5.3 SMARTS Linear Notation 84
3.6 3D and 4D Structure Representations 85
3.6.1 3D Structure Optimisation 87
3.6.2 Conformational Search 89
3.7 Molecular Descriptors, Fingerprints and Hash Codes 91
3.8 Challenging Cases for Structure Representation 95
3.8.1 Handling Aromaticity 97
3.8.2 Unique Structure Representations 98
3.8.3 Tautomerism 99
3.8.4 Miscellaneous Challenges 100
3.9 Chemical Substances 101
3.10 Conclusions 104
References 104

Chapter 4 Organisation of Toxicological Data in Databases 108
David Bower, Kevin Cross and Glenn Myatt

4.1 Introduction 108
4.2 Standards for Exchange and Organising Toxicology Data 111
4.2.1 SD Files 111
4.2.2 ToxML (DB) 113
4.2.3 SEND 120
4.2.4 OECD Harmonised Templates 122
4.2.5 ISA-Tab 123
4.3 Databases and Data Sets of Toxicity Studies 125
4.4 Process Management and Regulatory Compliance Databases 133
4.5 Alternative Approaches Databases 135
4.5.1 ToxBank 135
4.5.2 ToxCast/DSSTox/Tox21 137
4.5.3 PubChem 138
4.5.4 NIEHS/CEBS 138
4.6 Integrated Approaches 138
4.7 Conclusions 144
Appendix 145
References 160

Chapter 5 Making Big Data Available: Integrating Technologies for Toxicology Applications **166**
Nina Jeliazkova and Vedrin Jeliazkov

5.1 Introduction 166
5.2 Data Integration Approaches 168
5.3 Ontologies and Shared Vocabulary 171
5.4 Web Services and Cloud Technology 172
5.5 Standards 172
5.6 Integrative Data Analysis 174
5.7 Data Sharing Infrastructure 174
5.8 Conclusions 175
References 176

Chapter 6 Storing and Using Qualitative and Quantitative Structure–Activity Relationships in the Era of Toxicological and Chemical Data Expansion **185**
Sulev Sild, Geven Piir, Daniel Neagu and Uko Maran

6.1 Introduction 185
6.2 Anatomy of Predictive Models 190
6.2.1 Mathematical Representation 190
6.2.2 Data Set Representation 191
6.2.3 Model (and Data) Provenance and Metadata 192
6.3 Model Governance 193
6.4 Data and File Formats for the Representation of Predictive Models 195
6.4.1 QSAR Model Reporting Format 196
6.4.2 QSAR-ML 198
6.4.3 The Predictive Model Markup Language 199
6.4.4 QsarDB Archive Format 201
6.5 Applications and Solutions for the Storage of Predictive Models 204
6.5.1 Repository Approaches 205
6.5.2 Integrated Modelling Environment Approaches 206
6.6 Conclusions 208
References 209

Chapter 7 Toxicogenomics and Toxicoinformatics: Supporting Systems Biology in the Big Data Era **214**
Terezinha M. Souza, Jos C. S. Kleinjans and Danyel G. J. Jennen

7.1 Introduction to Toxicogenomics and Toxicoinformatics 214
7.2 Current 'Omics Technologies 216

7.2.1 Transcriptomics 216
7.2.2 (Epi)Genomics 217
7.2.3 Proteomics 218
7.2.4 Metabolomics 219
7.3 Data Handling 220
7.3.1 Toxicoinformatics: Standardisation, Storage and Availability of 'Omics Data 220
7.3.2 From Bits to Annotation 223
7.4 Use of 'Omics Data in Predictive Toxicology 226
7.4.1 Comparative Assessments 226
7.4.2 Unsupervised and Supervised Pattern Recognition 226
7.4.3 Connectivity Mapping (CMap) 227
7.4.4 Mechanistic Analysis 228
7.5 Big ('Omics) Data 229
7.6 Toxicogenomics in the Big Data Era: Challenges, Perspectives and Opportunities 232
References 234

Chapter 8 Profiling the Tox21 Chemical Library for Environmental Hazards: Applications in Prioritisation, Predictive Modelling, and Mechanism of Toxicity Characterisation **242**
S. Sakamuru, H. Zhu, M. Xia, A. Simeonov and R. Huang

8.1 Introduction 242
8.2 The Tox21 Process 243
8.2.1 Tox21 Compound Collection 243
8.2.2 Tox21 Robotic Platform 247
8.2.3 Tox21 Screening Process 247
8.2.4 Tox21 Data Analysis Process Including Data Quality Control 248
8.3 Using the Tox21 Big Data Collection for Predictive Modelling 251
8.3.1 Profiling for Compound Mechanism of Toxicity 251
8.3.2 Modelling Tox21 Data for *In Vivo* Toxicity Prediction 252
8.4 The Tox21 Data Challenge – New Methods for Data Modelling in the Big Data Era 257
8.5 Conclusion 259
Acknowledgements 259
References 260

Chapter 9 Big Data Integration and Inference **264**
Karen H. Watanabe-Sailor, Hristo Aladjov, Shannon M. Bell, Lyle Burgoon, Wan-Yun Cheng, Rory Conolly, Stephen W. Edwards, Nàtalia Garcia-Reyero, Michael L. Mayo, Anthony Schroeder, Clemens Wittwehr and Edward J. Perkins

9.1 Introduction: New Toxicology Paradigm and Challenges of Big Data Integration and Application 265
9.2 Structuring Knowledge to Support Data Mining, Modelling, and Decision-making 268
9.2.1 Development of a Globally Accessible Knowledge Base for Data Integration 268
9.2.2 Use of Structured Knowledge 269
9.2.3 Supporting Decision-making 273
9.3 Integration of High-throughput Screening Assays and Curated Databases for Understanding Chemical–Bioactivity Relationships 274
9.3.1 Data Generated from HTS Assays 274
9.3.2 Publicly Available Databases 275
9.3.3 Current Approaches for Integrating HTS Data and Databases and Computational Modelling to Predict *In Vivo* Toxicity From *In Vitro* Data 278
9.4 Systems Biology: *De Novo* Network Inference of Chemical, Gene, Protein, and Metabolite Relationships Underlying Toxicity 279
9.4.1 Network Inference 280
9.4.2 Steady-state Networks 280
9.4.3 Dynamic Networks 281
9.4.4 Networks and Inference of Toxicological Effects 282
9.5 Computationally Predicted and Putative AOP Generation 283
9.5.1 Computational AOP Generation 284
9.5.2 From Computationally Predicted AOPs to Putative AOPs 287
9.6 Use of AOPs to Support Testing and Risk Assessment 287
9.6.1 Identification of Sufficient Key Events in AOP Networks 287
9.6.2 Case Study: Steatosis 289
9.7 Translating Experimental Data into Model Structure: Integration of High-throughput Datasets or Expert Elicitation? 290

9.7.1 Mechanistic Modelling Methods 291
9.7.2 Case Study: Aromatase Inhibition Model Development 292
9.8 Conclusions 295
9.9 Future Directions 296
References 297

Chapter 10 Chemometrical Analysis of Proteomics Data 307
Marjan Vračko

10.1 Introduction 307
10.2 Chemometrical Methods for Data Analysis 310
10.2.1 Measure of Similarity 310
10.2.2 Clustering and Classification 311
10.2.3 Genetic Algorithm 313
10.2.4 Chemical Descriptors and Bio-descriptors 313
10.3 Application Examples of Chemometrical Analyses of Proteomics Data 315
10.3.1 Clustering and Classification of Proteomics Data 315
10.3.2 Genetic Algorithm Applied to Proteomics Data 317
10.3.3 Proteomics as Bio-descriptors 320
10.3.4 Biomarkers 324
10.4 Conclusions 326
References 326

Chapter 11 Big Data and Biokinetics 331
Miyoung Yoon, Gina Song, Harvey Clewell and Bas Blaauboer

11.1 Introduction 331
11.2 Simple Biokinetic Modelling Approaches for Supporting the Use for Large Data Amounts 333
11.2.1 Biokinetic Approach for Supporting High-throughput Testing 333
11.2.2 Biokinetic Approach for Predicting Acute Toxicity 336
11.2.3 Biokinetics Supporting Chemical Grouping 339
11.2.4 Minimalist Biokinetics Approach 340
11.3 Big Data Facilitating the Parameterisation of Biokinetic Modelling to Support the Use of Big Data for Risk Assessment 342
11.3.1 Example of an Approach for Rapid PBPK Modelling: PLETHEM 343
11.3.2 Building *In silico* Humans 344
11.3.3 Obtaining Chemical Specific Parameters 346

11.4 Conclusion 348
List of Abbreviations 348
Appendix: Examples of Life Stage Physiological Parameters for Human Populations 349
Height 349
Weight 350
Linking the Height and Weight Growth Curves 351
Body Fat Volume 352
References 353

Chapter 12 Role of Toxicological Big Data to Support Read-across for the Assessment of Chemicals 359
Mark T. D. Cronin and Andrea-Nicole Richarz

12.1 Introduction 359
12.2 Role of Toxicological Big Data to Inform and Support Read-across 364
12.2.1 Supporting a Read-across Justification 364
12.2.2 Investigation of Toxicological Mechanisms of Action 365
12.2.3 Biological Read-across 366
12.2.4 The Use of Omics Technologies for Read-across 368
12.2.5 Making Sense of Big Data Compilations to Form the Bridge With Chemical Structure 370
12.2.6 Read-across for Nanomaterials 372
12.3 The Future – Needs for the Use of Big Data for Read-across 373
12.4 Conclusions 374
References 375

Subject Index 385

CHAPTER 1

Big Data in Predictive Toxicology: Challenges, Opportunities and Perspectives

ANDREA-NICOLE RICHARZ*

European Commission, Joint Research Centre (JRC), Ispra, Italy
*E-mail: andrea.richarz@alumni.tu-berlin.de

1.1 Introduction

Predictive toxicology and model development rely heavily on data to draw upon and have historically suffered from the paucity of available and good quality datasets. As recently as 20 years ago, obtaining even relatively limited datasets of chemical structures and experimental bioactivities involved the laborious and manual collection, curation and compilation of data. Historically, the foremost problem in predictive toxicity modelling has been the scarcity of data, in addition to that of data quality. The data were produced manually, and thus slowly, by measuring physico–chemical properties and activities experimentally in the laboratory. As a result, models have been built based on the same training sets, and data were only available for a limited number of properties and endpoints.

With the advance in laboratory automation and the paradigm change towards 21st Century Toxicology, focussing on adverse effect pathways and

Issues in Toxicology No. 41
Big Data in Predictive Toxicology
Edited by Daniel Neagu and Andrea-Nicole Richarz
© The Royal Society of Chemistry 2020
Published by the Royal Society of Chemistry, www.rsc.org

elucidating mechanisms of action, the availability of data has changed and, overall, improved. For example, initiatives such as ToxCast and Tox21 have produced publicly accessible bioactivity data for a large number of chemicals for many endpoints with high throughput/high content assays. Furthermore, omics technologies for screening changes in genomes, proteomes, metabolomes *etc.* have produced many data. Biomonitoring and epidemiological data are also increasingly available.

Thus, in a short time period, the situation for predictive toxicology has changed dramatically from a lack of data hampering model development to that of "data overload". It can be argued that big data in predictive toxicology have not yet reached the dimension of other sectors, such as large-scale real-world data generated in the health sector.[1,2] However, these data encompass similar characteristics and related challenges and represent big data for this specific field.

This introductory chapter looks at the definition of big data in the context of predictive toxicology and the challenges and opportunities big data present in this field.

1.2 Big Data in the Area of Predictive Toxicology

Big data is a "buzz word" in many areas of science and technology and without doubt have changed everyday lives in many areas of society, for example, with the "internet of things", by offering new opportunities to integrate and use information.

Given the ubiquity of the term, what are the "big data" as relates to predictive toxicology? As compared to other fields, the data taken into account for predictive toxicology are not yet generated in real time for instant analysis. In the closely related sector of health care, for example, 24-hour data produced by monitoring functions in patients are already a reality. There are many initiatives in the area of drug discovery or medicine and public health that address the leverage of big data. For example, the European Union Innovative Medicines Initiative (IMI) programme Big Data for Better Outcomes (BD4BO[3]) supports the use of big data to improve health care. This programme assists by developing platforms for integrating and analysing diverse big data sets. In the area of drug discovery, possibilities to accelerate medicinal chemistry by using big data and artificial intelligence are being investigated.[4] The European Lead Factory,[5] a collaborative public–private partnership, has established the first European Compound Library and the first European Screening Centre, comprising up to 500 000 novel compounds. The EU Integrated Training Network BIGCHEM[6] project provides education in large chemical data analysis for drug discovery, to be able to process and use the growing amount of biomedical data in chemistry and life sciences.[7] The EU ExCAPE project[8] applies the power of supercomputers and machine learning to accelerate drug discovery.

It must be said that compared to the use of big data in other fields (*e.g.*, finance, economics, *etc.*), the data in toxicology are still relatively small scale, but with regard to what went before, they have broken the mould and can be considered big data. Typical examples of the types of big data in the context of toxicology are high throughput/high content screening data and data generated with omics technologies, gene arrays *etc.* Compared to "absolute" or "static" toxicity data, in addition to dose–response curves, resolution of responses in time is becoming a major area of investigation, including toxicokinetic measurements – or models – to take into account, for example, the rate of absorption or metabolism. As a consequence, more data are being measured. This is particularly true for the area of high throughput kinetics. Furthermore, as discussed in the 2017 US National Academies of Sciences, Engineering, and Medicine (NAS) report on "Using 21st Century Science to Improve Risk-Related Evaluations",[9] the use of epidemiological data will play an important role in chemical risk assessment in the future. Monitoring data, either environmental monitoring or human biomonitoring, will also contribute to risk assessment and predictive toxicology, and form an important part of the increasing big data in the area.

Similar challenges in terms of compiling, storing, processing, analysing and integrating these big data apply as for other areas and types of big data. Their quality, adequacy and appropriateness for specific purposes need to be investigated. They do, however, offer opportunities to tackle problems which were previously out of reach, and make completely new applications possible, to support toxicology evaluation and chemical safety assessment.

1.3 The Big Vs of Predictive Toxicology Data

The characteristic attributes of big data, the "Big Vs", have been discussed frequently in the literature,[10] starting from the three major V attributes that define the characteristics of big data, *i.e.*, Volume, Velocity, Variety. Currently, there are seven to up to 10 or more Vs described for big data,[11–14] which point at other attributes of big data and issues that need to be taken into account when dealing with big data; these vary slightly according to situation and use. The additional Vs include: Veracity, Variability, Validity, Visibility, Visualisation, Volatility, Vulnerability and Value (see Table 1.1). This section addresses how these characteristics of big data apply to the field of (predictive) toxicology; the 11th V, Vulnerability, is not considered in this context.

Volume: one of the three major defining characteristics of big data is the amount of data generated. In the field of toxicology, the large datasets produced are currently high throughput/content screening (HTS/HCS) assay data, high throughput kinetics, and omics technology read-outs; it is highly probable that in the future the use of large epidemiological and (bio)monitoring data sets will increase. Examples of large data sets and databases

Table 1.1 Ten Vs of big data and their applicability in predictive toxicology; related opportunities and challenges.

Big V	Applicability in predictive toxicology	Opportunities	Challenges
Volume	High number of data generated, *e.g.*, high content screening (HCS), omics test read-outs, epidemiological and (bio)monitoring data	Broader data basis for modelling and elucidation of modes of action and pathways	Storing and processing large amounts of data; finding the relevant data in the flood of information; limits of capacities for data curation
Velocity	Speed of data generation increased, *e.g.*, high throughput screening (HTS)	More rapid generation of data to fill gaps; ability to generate time-dependent data	Speed of data generation overtaking speed of storage, analysis and processing capacities
Variety	Many different types of data, *e.g.*, chemical structures, results from a variety of assays, omics, time-dependent kinetics *etc.*	Different types of information that can be combined to get the full picture	Integration of different types of data. Representation and informatics processing of chemical structures is a challenge, especially if 3D transformations of structures have to be computed for many chemicals. Comparability of data from varied sources might not always be given
Veracity	Data quality, accuracy and reliability; uncertainties requiring data curation and evaluation of data quality	Large amounts of data might statistically compensate for inaccuracy in individual data when integrated	Data curation and evaluation of data quality for large amount of data
Variability	Intrinsic variability of biological data, *e.g.*, inter- and intra- individual variations, genetic, population variations	Availability of a large amount of data might enable taking into account variations of parameters with the models built, enabling better prediction of population variations	Processing large amounts of variable data
Validity	Validity of the data for a specific application, *e.g.*, for prediction of toxicity (for a specific endpoint)	Generation of many (types) of data possible in a targeted way, tailored for the specific prediction and toxicity assessment goal	Finding/choosing the relevant data for the specific application
Visibility	Data sharing, access to data sources leading to centralised databases and repositories	Large data sets more visible than small disparate data sets, preferred storage in centralised repositories or linked *via* hubs/portals	Making the data available and visible in an appropriate way
Visualisation	Representation of the data content	Supports and facilitates making sense of the varied and complex data sets, also supporting the organisation of the data	Visualisation of complex data difficult, methods to improve representations of the data content in a clear way are needed
Volatility	Data access might cease, or repositories disappear, which affects the sustainability of data resources	N/A	Appropriate storage and sustainability concept necessary
Value	Adequacy and usefulness for predictive toxicology and hazard/risk assessment, depending on specific risk assessment goal	Availability of many (types) of data as a broad data basis to understand mechanisms and pathways, in order to build informed predictive models	Extracting and distilling the knowledge from the large amount of data

making these data available are ToxCast/Tox21 [15–18] data for high throughput screening data – screening thousands of chemicals in about 70 assays covering over 125 processes in the organism, over 120 million data points have been produced so far[19] (see Chapter 8); ChEMBL[20] from the European Bioinformatics Institute (EMBL-EBI) for chemical, bioactivity and genomic data for bioactive molecules with drug-like properties;[21] and TG-GATEs[22] for large-scale toxicogenomics data[23] (see Chapters 4 and 7). The EU research initiative HBM4EU,[24] a joint effort of 28 countries, the European Environment Agency and the European Commission, aims to advance human biomonitoring and use of the data for evaluating the actual exposure to chemicals and impact on human health. Data will be made accessible *via* the Information Platform for Chemical Monitoring IPCHEM,[25] the European Commission's reference access point for chemical occurrence data collected and managed in Europe.

Velocity: the speed of generation of toxicity data has increased significantly compared to historic data, by using automated assays, in particular high-throughput screening. Omics technologies also produce large datasets in a short time, compared to traditional assays. In addition, toxicology is moving away from static data to time-resolved recording of bioactivities. (Bio)monitoring data will gain increasing importance for risk assessment; these might also be recorded in real-time, for example, as daily monitoring and thus rapidly produce large amounts of data.

Variety: big data are also particularly characterised by their variety. Data relevant for predictive toxicology encompass many different types of data, such as results from a variety of assays in different formats, omics, time-dependent kinetics, metabolism information, and most importantly in chemical assessment: chemical structures. These are linked both to the unique identity of the compounds as well as specific activities or effects related to specific structural features ((quantitative) structure–activity relationships ((Q)SARs)). The informatics processing of chemical structures is a particular challenge (see Chapter 3). Furthermore, there is also a variety of data sources, resulting in datasets being dispersed and often not connected electronically, and the comparability of data produced in different laboratories might not always be given.

Veracity: this relates to the accuracy of the data and the reliability of the methods used to generate them. Data quality should always be evaluated; data may be prone to certain uncertainties. Data curation is needed in order to provide a reliable dataset. Data should ideally be generated according to Good Laboratory Practice (GLP) or Good *In Vitro* Method Practices (GIVIMP)[26] and transparently documented. Whilst Klimisch scores have traditionally been used to assign quality scores to a study,[27] there are currently other methods to evaluate study quality or potential bias.[28,29]

Variability: biological data possess an intrinsic variability: inter-individual variations, intra-individual variations of parameters, for example, depending on the time of the day, genetic and population variations. There is also variability due to different assay conditions or measurement errors. In the case of nanomaterials, there is an inherent distribution of the properties between nanoparticles.

Validity: validity encompasses, amongst many other aspects, both the reliability and relevance of the data for a specific application, *e.g.*, the prediction of toxicity for a specific endpoint and risk assessment goal.[30] In toxicological assessments, there is a formal process of test validation for *in vitro* assays, and validation of new methodologies, including predictive computational models, is currently being discussed. In general, the target chemical needs to be in the applicability domain of the method used. For example, some classic toxicity assays are not applicable to nanomaterials.[31] Therefore, the validity of the data produced needs to be verified.

Visibility: datasets should be visible to potential users. Data sharing and free access to data is important for data use in predictive toxicology (if not in-house use from companies), in order to allow best use, for example for model building. Ideally, data should be made available in centralised databases and repositories, and/or different data sources should be interconnected to facilitate integration and to allow the full potential of use of the information.

Visualisation: in order to facilitate making sense of the varied and complex data sets, methods to improve representations of the data content in a clear way are needed, also supporting the organisation of the data.

Volatility: access to data might cease or websites with repositories disappear over time, as is the case, for example, at the end of research projects or companies going out of business. In order to provide continued access to the data, a sustainability plan should guarantee long-term archiving and access to the data resources.

Value: this relates to the adequacy and usefulness of the data for predictive toxicology and hazard or risk assessment, depending on the specific use and assessment goal. For example, data can be adequate for a number of purposes including: producing models or profilers useful for screening or flagging a potential hazard; allowing quantitative predictions for decision making; or for filling data gaps complying with regulatory information requirements. Value also relates to financial considerations, *i.e.*, the cost, time taken and animal use, or the value of providing a valid alternative to an animal study, as well as ensuring the safety of the human population and environment.

1.4 Challenges of Big Data in Predictive Toxicology

After modellers have been requesting more data for decades, the situation seems to have turned into the question of whether there are now too many data. There might be a risk of being overwhelmed by the flood of data and losing sight of the objective of the hazard or risk assessment to be undertaken. It will be increasingly necessary to identify the relevant data for the specific application to avoid being lost in "noise". Moreover, it is also true that quantity cannot replace quality, and a thorough evaluation of the data and their adequacy for the purpose needs to be carried out, which is increasingly difficult with increasing volume, variety and speed of generation of the data.

The 2017 NAS report highlights the big data challenge for risk-related evaluations, stating that

> *The 21st century science with its diverse, complex, and very large datasets, however, poses challenges related to analysis, interpretation, and integration of data and evidence for risk assessment. In fact, the technology has evolved far faster than the approaches for those activities.*[9]

In this section, a number of the challenges that big data pose are highlighted – some are general issues encountered when dealing with data, magnified on a bigger scale, others specifically concern challenges of big data in the field of (predictive) toxicology.

The major challenges to take into account when considering the use of big data in predictive toxicology are represented in the scheme shown in Figure 1.1 and also summarised in relation to the Big Vs in Table 1.1. They concern sharing, accessibility, and processing of the data, data quality and comparability, interoperability of the data resources and integration of varied types of data, as well as identifying the relevant data needed for the intended purpose.

1.4.1 Need for an Adequate Infrastructure

First of all, data need to be available, which is the case for a variety of types of data with increasing speed and volume. In addition to data from high-throughput and omics technologies, large compilations of bioscience data, large datasets from monitoring and epidemiology are increasingly available. Moreover, new sources of data such as the mining of the grey literature and social media, *e.g.*, Twitter, might be used more in the future, as well as more connections to the pharmaceutical and biomedical fields.

Potential users need to know about the existence of the data in the first place. However, broad visibility of research project data is not always given and, in the past, they were mostly scattered, individual data collections. For significantly-sized datasets generated in large-scale endeavours, visibility is generally higher, for example the ToxCast/Tox21 data highlighted through different channels. Ideally, data should be linked to major, well-known and well-trusted data repositories or portals.

The data should also be easily, publicly and freely accessible, which is not the case for in-house data, and has not always been for research project data. In the future, public crowd-sourcing data collection activities could play a role, as well as the use of new types of information from public sources, such as from mining social media and the world wide web. Questions of

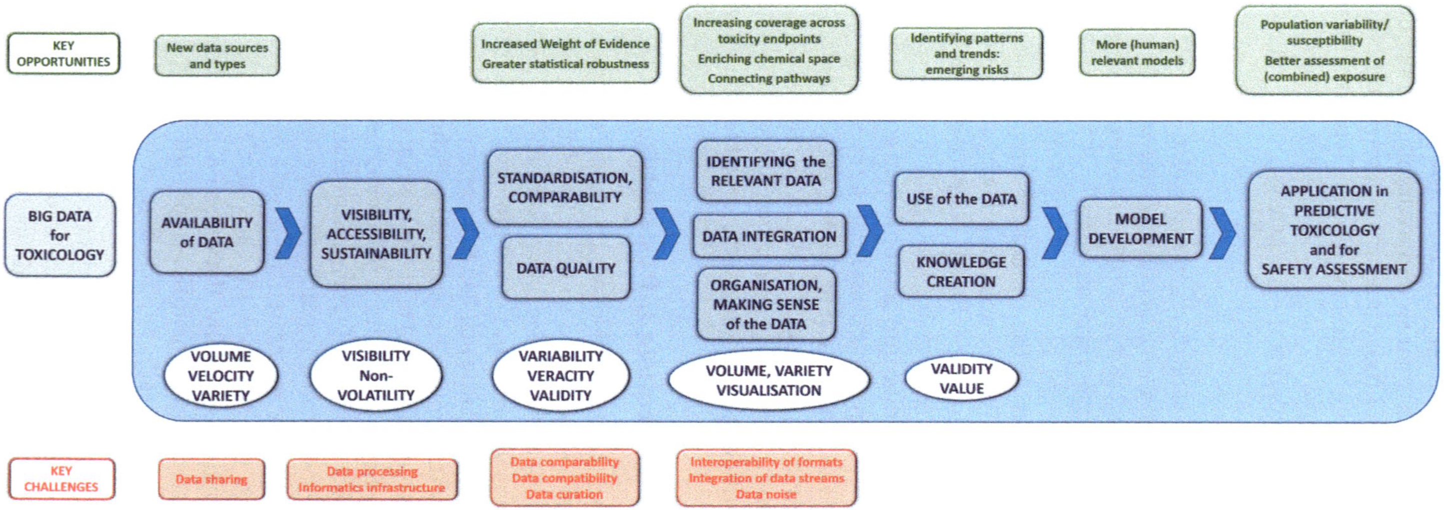

Figure 1.1 Key steps of big data in predictive toxicology from generation to application, related challenges and opportunities.

intellectual property rights and licensing will need to be solved. It is also of upmost importance that the data are reliably accessible, and respective measures to achieve this should be taken. Again, linking to established repositories with defined strategies for sustainability and long-time archiving will be helpful. Generally, data sharing and access to data resources has significantly increased as part of the big data era.

As is the case for big data in general, handling large volumes of data poses challenges regarding storage, retrieval, sharing and processing of the data, with the speed of data generation overtaking the speed of storage, analysis and processing capacities. In particular, representation and informatics processing of chemical structures are a challenge, especially if 3D transformations have to be computed for a large amount of chemicals (see Chapter 3). An adequate informatics infrastructure is essential. Technical advances have been made and have enabled the big data era, but have specifically to be adapted to the type of data used in the area of (predictive) toxicology (see Chapters 2, 7 and 5). This field has to further adapt itself to the new realities, possibilities and availabilities of data and what they can offer.

1.4.2 Standardisation and Data Curation

Big data sets for predictive toxicology comprise a high degree of both variety and variability of the data. This concerns intrinsic variability of biological data, variability related to different measurement procedures used by different laboratories, as well as measurement errors. In addition, there are errors in recording or transferring of study results in general. Thus, the data sets are prone to some form of uncertainties. To minimise these uncertainties, there is a need to increase data quality as well as compatibility and comparability of the data. Efforts should be undertaken to standardise formats, agreeing on the minimum information required to be recorded for experimental data, and linking to standardised ontologies. Standard test guidelines and standard operating procedures should be used to enable the compatibility and comparability of the results and data from varied sources, *e.g.*, when measured in different laboratories. This would also allow for the connection or merging of several datasets. Such guidelines generally contribute to assuring a higher quality of the data. Good quality data are essential for building of models that give accurate predictions of toxicity.

An example of the standardisation of tests are the standard Test Guidelines (TGs), which have been adopted by the Organisation for Economic Co-operation and Development (OECD)[32] and are regularly updated. Guidelines exist for most established toxicological *in vivo* and *in vitro* tests (see Chapter 2). The OECD Harmonised Templates (OHTs) are standardised formats supporting the processing of the results in databases, for example, by means of the IUCLID software.[33] Most templates correspond to specific OECD Test Guidelines. OECD Guidance Document 211 aims to harmonise the description of non-guideline (*in vitro*) test methods to allow evaluation and comparison of the results.[34] ToxCast assays and results are described in this format. For other "new" types of data such as omics data, development of reporting frameworks is ongoing in several international initiatives, to increase the transparent and standardised recording and reporting of omics assay results. These are, for example: the summary of Minimum Information about a Microarray Experiment for Toxicogenomics (MIAME; US National Research Council),[35] the development of a Transcriptomics Reporting Framework (TRF) at OECD, and the ECETOC MEtabolomics standaRds Initiative in Toxicology (MERIT).[36,37] At the laboratory level, Standard Operating Procedures (SOPs) with detailed descriptions of the test procedures to be performed, facilitate the generation of reliable and reproducible data.

A particular issue of data quality for experimental chemical toxicity data is the correct identity and identification of the chemical used for testing and being referred to. The measured bioactivity should be linked to an unambiguous chemical structure. Data curation of data sets is needed for creation of high-quality databases.[38]

Although these requirements for generating, processing and curating data apply to data in general, they are more complex and challenging when considering big data sets of a large variety of data types to be investigated together in view of their use for predictive toxicology; there are capacity limits for manual curation. Therefore, for the evaluation and curation of large data sets, including correct assignment and processing of chemical structures, automated procedures can be used. For example, a workflow has been developed by the US EPA for, amongst others, normalising chemical structures and resolving chemical ID conflicts. A significant effort was also undertaken to curate, update and resolve conflicts in the over 15 000 chemicals in the KOWWIN™ data file (underlying the model in the US EPA Estimation Programme Interface (EPI) Suite™ software[39]) to improve (re)building of the models. Quality flags were added to the datasets.[40,41] A publicly available KNIME[42] workflow was developed to process Distributed Structure-Searchable Toxicity (DSSTox) Database files (which includes about 22 000 chemical structures) to create "QSAR-ready" structures, for example standardising tautomers and removing stereochemistry.

The large data streams of ToxCast and Tox21 undergo rigorous quality assurance procedures (see Chapter 8). These relate to both a quality check to assure the chemical's identity and purity, tracking sample problems such as limited solubility or volatility, which can impact on the assay result,[43] as well as checking the bioactivity data. The automated processing of the results, necessary because of the large volume of data, may have produced possible false positive or negative hit calls, and annotations with different flags are added in an additional processing step.[44] Another example of a data curation process for a large data repository is the ChEMBL database, which has been performed by automated and manual steps, in order to increase accessibility, comparability as well as usefulness of the data, especially also for modellers, by adding curated annotations.[45–47]

1.4.3 Too Big? Identifying the Relevant Data

The wealth of data brings new opportunities for predictive toxicology, both in terms of amount of information as well as more diverse types of information. However, it will also become increasingly difficult to grasp all existing data in the flood of information. As stated at a NAS workshop, big data can be described as "more data than you know what to do with".[48] It is key that users need to be able to find the information that is actually relevant for the purpose and toxicological question they are looking at. This might seem like needing to find a needle in a haystack, for example, which gene changes are connected with the endpoint in question and will be predictive for an adverse health outcome? With more data becoming available, the same is true for more noise, or spurious correlations being made with unrelated data. It will be necessary to identify the appropriate, adequate and relevant data and distil out the adequate information – which might depend on the context of the prediction to make or the toxicological question to answer. Thus, the challenge has shifted from finding any useful data at all to needing to identify and select the relevant data for the specific application, and possibly try to find the sweet spot in the amount (and types) of data that are sufficient to support predictive modelling and safety assessment.

There are different approaches to find data and information: either broad data – or text – mining, in an attempt to "fish" for all possibly relevant data, increasingly narrowing down the search results in view of the respective application;[49,50] or a targeted search in all possible data resources to answer a specific question or fill a data gap.

Text mining for extracting information from large amounts of data is combined with complementary approaches and analytic tools in order to leverage the information and enable knowledge discovery. It has been applied, for example, in the field of toxicogenomics to discover information not yet available in structured databases, *e.g.*, to identify early predictors of toxicity.[50] Another example is the development of automatic data mining methods to extract relevant bioassays – ranked by means of a newly developed scoring system – from publicly available bioassays databases. For instance, data from 739 000 assays in PubChem have been mined to generate response profiles to evaluate a set toxicity for a large group of nearly 5000 target compounds.[49,51]

1.4.4 Data Integration Infrastructures

The data sets and repositories in toxicology comprise a variety of different types of data: *in vitro* or *in chemico* assay results, *in silico* predictions, gene arrays and omics read-outs for different toxicological endpoints (see Chapters 2 and 7), epidemiological, monitoring and exposure data. In addition, chemical structures may be stored in different representations (see Chapter 3), along with further substance and methodological details. All this information needs to be processed informatically and data set structures need to be harmonised to allow integration of the different data streams in large repositories (see Chapter 5).

There are different technical approaches related to data integration:

- Creating large repositories and feeding in available data after curation and standardisation
- Connecting data resources directly by enabling technical compatibility and interoperability of data resources and data formats
- Portals allowing the search of data repositories in parallel
- Overarching data infrastructures.

The interoperability between data resources is enabled by Application Programming Interfaces (APIs). Connection of data from different data repositories, which might use different terms for the same content, also relies on a mapping to ontologies for harmonisation and comparability (see Chapters 4 and 5). An example is the ChEMBL database, which has been integrated with other data resources using open standards and ontologies, after mapping to the Resource Description Framework (RDF).[52] Similarly, the annotation of data with metadata is crucial for interpretation, comparability and connection of data across data sources.

Integration of structurally different data is required, such as structured data from curated databases and unstructured content such as full text from the scientific literature. The Semantic Enrichment of the Scientific Literature (SESL) project has used approaches to connect these two types of data *via* semantic web standards.[53] Similarly, the Open PHACTS project created an "open pharmacological space" using semantic web standards.[54] There are some other examples of approaches from the field of drug discovery, for example connecting, mining and sharing heterogeneous biological and chemical data from low-throughput and big data sources in a community-based platform linking informatics and social networks.[55] An analysis of the concordance of pharmaceutical toxicity in different species, with over 1 600 000 recorded adverse events, combining text mining and statistical analysis, showed the advantage of a large body of analytical results, but, on the other hand, the challenge of lacking controlled vocabulary and non-aligned ontologies between data sources.[56]

Portals or warehouses can encompass many databases as an overarching "one stop" reference point, and allow for properties of a target chemical to be searched for simultaneously across all linked databases. For example, the OECD eChemPortal[57] currently comprises 34 data collections that can be queried in a single search. The data resources in eChemPortal include Canada's Existing Substances Assessment Repository (CESAR), the

European Chemicals Agency's Dissemination portal with information on REACH registered chemicals, the Organisation for Economic Co-operation and Development Existing Chemicals Database, the Data Bank of Environmental Properties of Chemicals, the Information Platform for Chemical Monitoring IPCHEM. The US EPA Aggregated Computational Toxicology Resource (ACToR),[58,59] as a warehouse for US EPA web applications, allowed the search for thousands of toxicity testing results, hazard, exposure and risk assessment data for over 500 000 chemicals, by aggregating over a thousand public sources. ACToR is now integrated in the iCSS CompTox Chemistry Dashboard,[60,61] which also links to external resources such as PubChem, ChemSpider and DrugBank. The European Commission's Joint Research Centre provides the ChemAgora Portal,[62] which allows the on-the-fly search of connected repositories, including ChEMBL, ToxNet and the AOP Wiki. The Data Infrastructure for Chemical Safety Assessment (diXa) resource has been created as a central data repository and warehouse for toxicogenomics data, connected to a portal linking to chemical, molecular and phenotype data.[63] ELIXIR is an intergovernmental organisation coordinating a "distributed infrastructure for life-science information".[64] Part of its activities is the ELIXIR Data Platform, which identifies Core Data Resources and Deposition Databases.

Other initiatives and approaches contribute to facilitating the handling, analysis and integration of big data resources. For example, the EU ExCAPE project uses supercomputers and machine learning technologies to accelerate drug discovery. The project's ExcapeDB[65] is an integrated large-scale dataset facilitating big data analysis and predictive model building in chemogenomics. BIOGPS is a semiautomated approach to navigate biological space and identify structurally similar protein binding sites. It automatically prepares protein structure data, identifies and aligns binding sites, and compares GRID Molecular Interactions Fields.[66]

The Online Chemical Modelling Environment (OCHEM)[67] aims to connect data resources of experimental measurements with a modelling framework, as a "crowd-sourcing" platform where users can share and thus connect results and cheminformatics tools in order to facilitate QSAR model development.[68] The National Institutes of Health (NIH) Big Data to Knowledge (BD2K) initiative,[69] launched in 2012, aims to support the broad use of biomedical big data by making them Findable, Accessible, Interoperable, and Reusable (FAIR), as well as facilitating discovery and new knowledge. Supported by the BD2K Initiative, the Center for Expanded Data Annotation and Retrieval (CEDAR)[70] was established in 2014 to further the development and use of biomedical metadata, with a view to enhancing data interpretation and discovery.[71] CEDAR is working on community-driven standards for representing biomedical metadata, focussing on templates for describing different types of experiments, as well as a machine-readable repository for these standards.

In general, in the big data era, new technologies such as clouds have opened up new opportunities to store, share and handle data as well as virtually connecting disparate data sources. New cheminformatics/bioinformatics technologies are available to integrate the different types of data available in different formats across the connected repositories, *i.e.*, to apply an integrative data analysis (see Chapter 5).

Apart from the informatics and technical issues, a major challenge in integrating toxicological data and results for predictive toxicology and chemical safety assessment lies in the integration of the data content, assessing the weight of the different data, possible contradictions, their overall relevance and contribution to the predictive question, as well as their uncertainties or possible bias.

1.4.5 Making Sense of the Data

From large-scale datasets and repositories, connected or separate sources, to official sources or data brought together in the scientific communities *via* crowd-sourcing, there are many different data, data types and data formats available. It has been said that "We are drowning in information, while starving for wisdom",[72] a situation that has been amplified in the big data era. It is essential to make sense of the data in order to be able to use them successfully and build new knowledge.

Setting aside the technical interoperability of datasets and formats, the next challenge is the structuring of the data, to help the potential user find a way through the data jungle. Ideally, data should be presented in a form that helps the user to make sense of them. In general, visualisation of large compound collections can, for example, be achieved by projecting them in a low dimensional space, to reduce the complexity and allow visual and intuitive analysis.[7,73–75] Techniques for data reduction and comparison include Principal Component Analysis (PCA), Generative Topographic Mapping (GTM), K-means clustering and Bayesian methods, amongst many others. Visualisation of large collections of diverse and complex data is *per se* more challenging. Visualisation can be multidimensional and in particular needs to include a temporal dimension for time-dependent data. The US EPA iCSS CompTox Chemistry Dashboard is also an example of visualisation of toxicity/bioactivity data. It serves as a graphical user interface for the user to search for information either from a chemical, including drawing chemical structures, or an assay viewpoint. The results of the query and biological activity for assay combinations are presented in the form of graphs or charts.

Another important aspect is the adequate annotation of the data, by introducing sufficient metadata describing the context of the data generation, the data format *etc.* and, in general, a description of the data source and data collection, helping to understand the nature of the data. An example is the ToxCast Manual,[76] linking to a list of chemical tests, assay descriptions and protocols, and other related user guidance and documentation.[77] In view of harmonisation, the assay descriptions are summarised in the format recommended in the OECD Guidance Document 211 [34] for describing non-guideline *in vitro* test methods.

In terms of rationalising and visualising the relationships between different biological activities, reactions with biological molecules and the adverse effects that may result, the concept of pathways, and in particular the Adverse Outcome Pathway (AOP) framework, has been established in the last decade.[78] The AOP Knowledge Base[79] and AOP Wiki[80] have been introduced as practical tools to collect and bring mechanistic data and knowledge together in a crowd-sourcing effort. In addition, the Effectopedia[81] application aims to collect qualitative and quantitative data that are related to the (key) events in the pathways, as a means of organising and integrating different assay results and data types related to the same biological events. As such, it envisions guiding the user who is attempting to navigate the information.

The challenges described in this section need to be addressed and are prerequisites for making the best use of the wealth of data for supporting human health or environmental safety assessment. Once the relevant data are located and initially processed, they need to be analysed, evaluated, interpreted and integrated to answer set problems or create knowledge. The question is whether the new plethora of data are adequate for use in predictive toxicology and risk assessment and whether they can address some of the most urgent problems. Opportunities and examples for the use of big data in predictive toxicology are discussed in the next section.

1.5 Opportunities Provided by Big Data for Predictive Toxicology

Biological and toxicological big data, as new and diverse data streams, offer many opportunities in the context of chemical risk assessment, by virtue of the big data characteristics related to their volume, velocity, variety or overall value (see Figure 1.1). This section discusses how these big data contribute to forming a comprehensive overview of chemical hazards and risks, as well as increasing and creating new knowledge in toxicology and providing support for the building of better predictive models.

Some issues in predictive toxicology and chemical risk assessment cannot be adequately addressed at the current time due to the complexity of the problem. This relates, for example, to safety assessment and computational models for predicting nanomaterial toxicity[82] as well as complex chemical mixtures or exposure scenarios, such as aggregated exposure or "real-life" co-exposure from different sectors of chemical uses.[83] Specific aspects and examples of how big data could be applied to address these "hot topics" in predictive toxicology for chemical risk assessment are discussed in this section.

1.5.1 More is More – Benefits of a Broader and More Diverse Data Basis

The 2017 NAS report on "Using 21st Century Science to Improve Risk-Related Evaluations",[9] which followed up on the reports on the vision of 21st Century toxicity testing[84] as well as exposure science in the 21st Century,[85] acknowledges that various data streams from exposure science, toxicology and epidemiology have not been taken into account sufficiently. It also emphasises the value of these new types and volumes of data for risk assessment.[9] The importance of exposure information is highlighted in particular. More and more diverse information from new data sources is becoming available. This broader data basis, also obtained by connecting different large data sources, gives a broader picture, and enhances the weight of evidence, providing more certainty. In addition, new types of information not previously taken into consideration can provide new perspectives for the same issues of chemical risk assessment. Therefore, a transdisciplinary approach, exploring new and increasingly diverse types of data beyond classic toxicology, is key to leverage the new information available to enhance (predictive) toxicology.

New technologies have provided new means to monitor the impact of chemicals on biological systems and in the environment, for example mobile and remote sensors to trace pollutant concentrations[86–88] or "crowd-sourced" monitoring data.[89] Biomonitoring initiatives contribute to increasing exposure information: the HBM4EU project is coordinating human biomonitoring in Europe and centralised data collection and sharing. IPCHEM is key for this endeavour and generally for enhancing access to monitoring data, connecting separate data sets and making them explorable in a single search by chemical or location. Epidemiological studies are able to provide the link, possibly confirming causality, between exposure and health effects by biologically plausible statistical associations.[90] Furthermore, data and knowledge from close disciplines such as pharmaceutical research and public health should be increasingly considered to support toxicology and chemical safety assessment.[2]

New sources and types of information can also be found by mining grey literature and other sources in the world wide web, information on bioactive chemicals from patents,[91] as well as social media for communication and consumer behaviour information. Similar techniques are already being used, for example, in the area of pharmacovigilance,[92–94] and could systematically be extended to identify and capture early signs of adverse effects from chemicals. Furthermore, additional information on intended or non-intended actual uses might be obtained from these channels and contribute to a realistic picture of exposure to a substance.

It is generally acknowledged that "quantity is not quality". However, it is sometimes considered that possible problems with data quality deficiencies might be overruled by a sufficiently large amount of data, *i.e.*, that shortcomings of lower quality or wrong data points might be overcome by the sheer number of – mostly correct – data points. If large-scale data sets are available, ideally from different sources, connecting these data and identifying the overall trends can compensate some "inaccurate" data points. As part of this process, outliers can be identified and ignored in model development. Overall, an increasing number of data increases the weight of evidence which leads to less uncertainty.

In order to obtain a "critical mass" of data, *i.e.*, a big data set of sufficient size, several data sources can be compiled or integrated in a single overarching repository, or otherwise connected. This again shows that data sharing is essential. Crowd-sourcing efforts can contribute to obtaining the necessary number and types of data. As such, models for predictive toxicology can be created from data even if scattered over several systems or repositories, as long as they can be technically, and semantically, connected (see also Chapter 5). Boyles *et al.* described a vision for computationally enabled semantic toxicology linked to the concept of ontologies.[95]

Thus, big data offer the opportunity to build models on a broader data basis and with greater statistical robustness. Furthermore, with new types of data becoming available, such as data from human cell lines, targeted high-throughput data generation with assays relevant to the human organism, large-scale human biomonitoring data and epidemiologic data on exposure and adverse health effects, there is an opportunity to build more specific and human relevant predictive models. This corresponds to the aim of chemical safety assessment, which is to secure safety for human health for the population – as well as an intact environment. Classic toxicological assessments using animal tests were used as a surrogate. However, predictive toxicology should not restrict itself to simply reproducing the results of toxicity tests in surrogate species *etc.*, but contribute with relevant and valid predictions to the overall aim of safety assessment and decision making.

The new wealth of data also calls for new approaches for modelling to use their full potential, opening up new opportunities for model development at the same time. Deep Learning approaches, machine learning algorithms based on networks of artificial neural networks, used in other areas such as face or speech recognition or in the detection of fraud and suggested for use in pharmaceutical research,[96] have been introduced specifically for toxicity prediction and virtual screening.[97] Since Deep Learning is able to analyse and learn from large amounts of uncategorised data, it is very well suited to deal with big data and their data variety.[98] In fact, Deep Learning approaches were used in the winning entries for the Tox21 Grand Data Challenge for building predictive models based on the Tox21 big data to predict toxicity of nuclear receptor and stress response pathways. This was one of the first applications of Deep Learning methods to computational toxicology[99] (see Chapter 8). The deep nets automatically learned features resembling well-established toxicophores.[100] In general, the application of artificial intelligence approaches in predictive toxicology is being explored. This goes beyond the concept of machine learning, using the wealth of information now available in order to distil and simulate human expert knowledge and decisions.[101]

1.5.2 Big Data for the Big Picture

A European Commission Communication on data, information and knowledge management defined that knowledge "is acquired through analysis and aggregation of data and information, supported by expert opinion, skills and expertise".[102] The availability of a wealth of information and a broad data basis of a variety of types of data from different disciplines thus offers the opportunity to create new knowledge from the analysis and integration of these data, giving the full picture of chemical toxicity assessment from different perspectives.

This "big picture" that the big data provide is a holistic view of the area, enabled through fitting together the many "pieces of the puzzle" of a variety of information. These would not be apparent if big data were analysed piecemeal. The integration of the diverse data streams of biological/toxicological big data, which might not have been considered together in the past, enables new insights. Since the whole is greater than the sum of its parts, the variety of the toxicological big data pool plays an important role, contributing different perspectives and enabling the identification of inferences and distilling of new insights in toxicology and the effect of chemicals, and thus contributing to the understanding of their mechanisms of action – a prerequisite for building predictive models. In particular, it is the large-scale nature of the data resources that allows recognition of overarching patterns and general trends that would be lost and remain unidentified if only smaller datasets were investigated. Bespoke pattern recognition technologies can contribute to the analysis of these datasets. Examples for "big pictures" supported by toxicological big data include the ensemble of toxicological pathways and their perturbations as well as the universe of chemicals as considered for chemical safety assessment.

Biological big data have enormous potential to contribute to the continuing elucidation of AOPs and related knowledge disovery.[78,103] Firstly, screening the data and identifying patterns and connections can identify new pathways, or recognise so far undiscovered connections between AOPs forming AOP networks.[104] Secondly, big data can help to enrich these AOP networks with quantitative bioactivity data to establish quantitative AOPs with dose–response relationships. Conversely, AOPs can be seen as a framework useful to anchor the huge amount of increasingly available big data, contributing to their comprehension in a toxicological context. As such, AOPs can serve as a reference point to connect different data streams and integrate their information content (see Section 1.4.5).

The biotechnology revolution and its omics technologies have made it possible, in a systems toxicology approach,[105] to monitor many cellular pathways in parallel[106] and indicate perturbations of pathways if these are known. A workshop has explored in particular how omics data can further support the

AOP framework for chemical risk assessment, identify molecular initiating events as well as provide evidence for key events at different organisational levels,[107] contributing to a data-driven shift in the assessment of large numbers of chemicals. Adverse outcomes can be multifactorial in nature, *i.e.*, be linked to multiple mechanisms or connected to different pathways and initiating events,[9] even over a long period of time. For example, a combination of effects of individual, non-carcinogenic chemicals acting *via* different mechanisms on different pathways and target levels in the organism can lead to carcinogenesis,[108,109] according to the hallmarks of cancer.[110] Therefore, it is important, and within reach with new big data streams, to cover the big picture of pathways and consequences of their perturbation and synergies of dissimilar processes.[108]

The increasing amount of data results in an increasing coverage across different toxicity endpoints, thus enabling us to see the full picture of properties and hazards of a substance, enabling us to make connections between endpoints and the leveraging of knowledge from one toxic effect to others. In this way, development of predictive toxicology models and expert systems can draw on a broad basis and a holistic view of the various interconnected effects that a toxicant can have in the organism. Discovering new knowledge by cross-linking unconnected areas has been successfully applied in the domain of drug discovery. Studies have shown, for example, that drugs with similar safety profiles have similar polypharmacologies.[111] Finding such cross-links within the pool of toxicological big data could be applied in an analogous way in the field of toxicology and chemical safety assessment.

The large universe of information available through chemical, biological, bioactivity and toxicological big data also provides the opportunity to contribute to the big picture of chemicals and assist the overall mapping of the chemical universe. Big data may allow the connection of disconnected areas of information, filling hitherto existing data gaps and further enriching the chemical space. ToxCast/Tox21, for example, contributes to the chemical landscape by targeted *in vitro* bioactivity data generation, and enables strategic data mining and predictive toxicity model development.[112,113] In terms of computational modelling, the mapping of the chemical space also contributes to the definition of applicability domains for model development and application.

1.5.3 Creating und Using New Knowledge: Applications for Hot Topics in (Predictive) Toxicology

The new data streams and new knowledge drawn from toxicological big data might be able to help address the major challenges and support current trends of (predictive) toxicology in chemical risk assessment. These are areas where toxicological big data can add most value.

1.5.3.1 Prediction of Emerging Risks

In chemical risk assessment and management, even when all precautions are taken and evaluations are performed according to the best current knowledge, the emergence of some unexpected hazard or new risks can never be excluded. The International Risk Governance Council (IRGC) defines emerging risks as "new risks or familiar risks that become apparent in new or unfamiliar conditions".[114] There is also often a reference to so-called black swan events, which refer to unexpected events that were not foreseen or envisioned as possible at all. Flage and Aven state that for both types of events "knowledge is the key concept".[115]

Risk assessment is confronted with known unknowns and unknown unknowns; it might either be known that there is a potential problem, but there are not sufficient or adequate data available to evaluate the situation, or the nature of the risk has not been encountered before at all. Emerging risks can be newly created risks that are caused by new substances or new technologies (such as nanoforms of the same already known bulk material with different properties) or by new types of applications; existing issues which are only identified or understood by means of new knowledge (for example, the relationship and impact between biological pathways); an increased incidence of an issue due to changed habits or – qualitatively or quantitatively – new types of exposure (for example, hazardous compounds from electronic waste unexpectedly appearing in toy articles due to recycled plastics).

Detecting and recognising "early" warning signals, *e.g.*, of adverse effects, is crucial; however, they might not always be apparent with the standard information and processes available. The concept of gathering, centralising or connecting as much information as possible was already realised in networks such as the network of reference laboratories, research centres and related organisations for the monitoring and biomonitoring of emerging environmental substances (NORMAN[116]) or the MODERNET network (Monitoring trends in Occupational Diseases and tracing new and Emerging Risks in a NETwork[117]).

However, the wealth of available data now offers new and so far unavailable opportunities to detect and amplify signals, and to identify patterns in the overall "noise" of the big data. An example from the clinical sciences: the innovative real-time quantification of the arterial waveform shape and variability, with a mathematical method adequate for handling long non-stationary data streams, allowed the prediction of an emerging health problem (sepsis) with high sensitivity at an early stage, which was not possible with the classic clinical measurement of the pulse, *i.e.*, small parts of the signal.[118] Toxicological big data streams provide the opportunity for similar applications in future. With more data streams being available, it is possible to link different types of information, making connections, and discovering trends and patterns that were not visible and would have been missed otherwise. In other words, big data related to chemical properties and toxicological information can create new knowledge that supports the identification of emerging risks, and even opens up the possibility to predict otherwise unforeseen events.

As discussed, new data streams can be explored for toxicological applications, such as epidemiological data, biological monitoring data or mining of "signals" in the world wide web or social media. Mining and analysing social media data is already common practice in the field of pharmacovigilance for detecting emerging adverse drug effects.[92,93] Another example is untargeted environmental metabolomics to detect environmental stressors (xenobiotic substances) and elucidate mechanisms, when the mode of action is so far unknown.[119]

1.5.3.2 Complex Exposure Scenarios of Combined Exposure to Multiple Chemicals

The assessment of the effect of chemicals is getting more challenging with increasingly complex exposure scenarios. Exposure to one chemical alone can already occur *via* several sources or routes (aggregated exposure), and exposure is getting more complex for the combined exposure to multiple chemicals *via* different sources and routes. This is the real-world situation, *i.e.*, humans and ecosystems are exposed to a quasi-infinite number of different combinations of chemicals *via* sources such as food, consumer products as well as environmental chemicals or in an occupational context.

The assessment of chemical mixtures and combined exposure is a challenge in chemical risk assessment, as well as for predictive toxicology.[120] Exposure to multiple chemicals, which do not cause harm individually, can lead to adverse effects in combination.[83,121,122] Even if the toxicity of all individual components is known or can be predicted, as well as the relative concentrations, the determination of the effect of the combined exposure is not always straightforward. Interactions might occur between chemicals and lead to synergistic or antagonistic effects. In addition, a time aspect needs to be taken into account, *i.e.*, the exposure to several chemicals which might interact to disturb a biological pathway could be sequential and even years apart, but would need to be considered together. It is impossible to physically test all possible combinations of chemicals considering the huge number of possible combinations of compounds and their concentrations. Computational prediction methods will support the assessment of mixtures, but need an adequate data basis.

There is an opportunity to combine information from disparate sources and exposure from different media and chemical sectors (*e.g.*, collected in the framework of different legislations) at different times, and integrate the information in order to provide an overall big data resource closer to the diverse real-life exposure to chemicals/chemical mixtures and for assessing co-exposures. Data can be mined to find trends and elucidate chemical interaction mechanisms, for a better assessment of mixtures.

Chemical monitoring is a step towards better knowledge of the actual exposure (numbers of compounds and their concentrations). Biomonitoring data, including real-time epidemiological monitoring, can contribute towards understanding causal connections between diseases and – sometimes also geographically specific – exposure to chemicals; epidemiological evidence, for example, has linked air pollution to cancer and neurodevelopment effects.[9] They also provide insights into co-exposure patterns.[123] Exposure data are increasingly becoming available, on a big data scale. Many initiatives are measuring, compiling, or predicting exposure data. Examples are biomonitoring initiatives such as HBM4EU, databases such as IPCHEM or the US EPA project to compile an exposure data landscape for manufactured chemicals.[124,125] ExpoCast, the US EPAs exposure-based chemical prioritisation programme, is providing high-throughput exposure predictions for thousands of chemicals, looking at multiple routes of exposure.[126] This exposure information is integrated with ToxCast data, also exploring the use of big data techniques, including data mining, for example for defining consumer product use profiles.[127]

Furthermore, technological and methodological advances in the big data era, such as the use of crowd-sourced monitoring and epidemiological data from mobile personal devices, provide an increasing amount of new exposure data sources and thus new opportunities for further development of the emerging discipline of computational exposure science[128] and predictive models from big data. Predictive toxicology approaches simulating the different combinations of exposure to chemicals and related probabilities, as well as possible interactions, will benefit from the new large-scale data resources and the possible application of artificial intelligence approaches[101] to gain new knowledge.

Overall, big data offer opportunities, which have so far not been available, to address the complex exposure scenarios of combined exposure to multiple chemicals; their potential is waiting to be fully exploited in the future.

1.5.3.3 Addressing Population Variability and Specific Susceptibility

Big data resources also cover intrinsic variabilities of measured properties and have the advantage of being sufficiently large and complex to represent and address population and genome variabilities and related different susceptibilities to chemical agents. Genetic data have already established a link between genetic variants (polymorphisms) with diseases or adverse effects of toxicants or drugs, recognising the importance of taking into account this type of information on different sensitivities to chemical agents and disease susceptibilities.[106]

Variability in response to chemical stressors can also be due to factors such as life stage and sex or influence of external factors or other, non-chemical, stressors. Epidemiological and exposure, such as biomonitoring, data can contribute towards identifying these influences and susceptibilities.[9] Omics data also contribute to elucidating species sensitivity.[107]

These new approaches and generated large-scale data will contribute to regulatory toxicology practice in the future. The US National Center for Biotechnology Information at the National Library of Medicine at the National Institutes of Health, for example, has a selection of resources available related to human variability, especially genotype to phenotype associations. They include the database of single nucleotide polymorphisms and estimates of their occurrence within the population (dbSNP), the database of Genotypes and Phenotypes (dbGaP), the GTEx database on Genotype-Tissue Expression and a Phenotype-Genotype Integrator (PheGenI).[129]

1.5.3.4 Grouping Chemicals and Read-across for Chemical Safety Assessment

Being faced with a very large amount – tens of thousands – of compounds that need to be safety assessed, and which might appear as a chemical big data entity in itself, grouping approaches seem to be the only viable solution for their evaluation in an acceptable time frame. This strategy goes beyond mere computational screening with QSAR models, although *in silico* models can be part of the category formation. Grouping will make it possible to bin the chemicals and flag and prioritise them for further assessment, where applicable. Furthermore, evaluation of several to many chemicals at the same time is possible, in order to enable a faster and more efficient evaluation compared to single chemical evaluation. Such an approach of organising the chemical space for prioritisation and chemical safety assessment is being pursued, for example, by the Canadian Authorities within the context of the Chemicals Management Plan.[130] The European Chemicals Agency (ECHA) also strategically maps the chemical universe to address substances of concern.[131] The grouping of chemicals with similar properties also facilitates the substitution of chemicals. In this sense, grouping and read-across is used to solve the problem of chemical big data.

On the other hand, this type of group formation relies on the variety of biological/toxicological property data connected with the chemical substances to find trends, identify and justify the similarity hypothesis. The recent availability of large-scale repositories of bioactivity data, such as PubChem[132] and ChEMBL, or other data that support mechanistic understanding, such as information from omics technologies[133] (see Chapter 7), now offers the opportunity to provide more and better evidence for the justification of similarity between chemicals. Thus, biological big data provide crucial support to read-across approaches (see also Chapter 12). Different types of data cover different parts of the biological/toxicological space. More evidence gives more confidence overall in similarity argumentation and read-across reasoning.

1.5.3.5 Predictive Nanotoxicology

The interactions of nanomaterials with biological molecules are still not fully understood, small differences in properties being able to make significant differences in effects, and overall nanotoxicology is struggling with its complexity.[134] Nanomaterials have a high inherent complexity: there is a distribution of sizes and variability of properties, structural complexity, as well as different possible coatings, surface corona *etc.* There is, thus, an almost infinite number of potential combinations and variations in the materials, which – together with a lack of representative samples, as well as standard operation procedures – makes it difficult to define the physico-chemical characteristics and reliably measure toxic effects. Standard test guidelines are not always applicable to nanomaterials and there is also a high variability of the nanomaterials depending on the experimental conditions.[82,135] There might be many data on nanomaterials available; however, the issues described and the related insufficient quality of the data are also impairing the development of computational models to predict nanotoxicity.[82,136] In particular, the correlation of potential adverse effects and the properties of the nanomaterials that are causing them are not fully elucidated yet.

Big data could help to identify property–activity relationships by providing larger numbers of data points to base the analyses upon and to be able to identify trends. In order to support the use of big data in nanotoxicology, means of processing and analysing large-scale datasets produced by high-throughput screening have been developed. For example, a web-based platform for HTS data analyses tools (HDAT) was created as a publicly available computational nanoinformatics infrastructure for statistical analysis of nanomaterial toxicity data, offering, amongst others, self-organising map (SOM)-based clustering analysis as well as possibilities to visualise raw or processed data.[137] Furthermore, natural language processing techniques are available to mine nanomaterial-related information.[138]

In another line of research related to nanomaterial discovery, multivariate data analytics were applied to identify nanoparticle prototypes and archetypes and their relevant properties.[139] The method helps to identify representative types of nanostructure in the big data sets, which can then be further studied to understand the structure/property–effect relationships. The authors highlight that creating value from data is a separate endeavour from creating the data in the first place, in particular considering the uncertainties and variabilities of big data, and conclude that this type of data-driven search will have its own value as a nanoscience discipline.[139] Similarly, it was emphasised that in order to make value of the large amount of compiled nanomaterial data and research efforts,[140] it is necessary to integrate top-down decision analytic tools to identify the relevant information and bottom-up data management approaches for visualising these data, in order not to "waste" efforts on collecting and processing information irrelevant for decision-making.[141] The use of multicriteria decision analysis (MCDA), value of information (VOI), weight of evidence (WoE) and portfolio decision analysis (PDA) in order to integrate and interpret nanomaterial research data in a way to distil out the information relevant to decision-making needs was discussed in a recent paper by Bates *et al.*[141]

1.6 Conclusions and Perspectives

The quantity of big data in toxicology is growing, and whilst that may not be at the same level as seen in other fields, they are profoundly transforming the possibilities for predicting chemical hazards and undertaking risk assessment, opening up many new opportunities along the way. The era of big toxicological data coincides with the developing paradigm of Toxicology in the 21st Century, which is moving from testing and observation of apical effects to a mechanistic toxicology. The new direction in toxicology is based on an understanding – and continuing elucidation – of the pathways leading to adverse outcomes, in the framework of systems toxicology, and focusses more on the human relevance of the assessments. As such, this new toxicology is immensely data hungry and requires support from the related targeted screening of chemicals as well as investigation of pathway perturbations. This is only truly possible with advances in technologies, and associated big data, such as the development of high-throughput and high-content bioassays as well as omics technologies measuring changes at the protein and genetic level *etc.* These new technologies have produced large numbers and a high variety of data points at a significantly higher speed than classic experimental toxicology, corresponding to the three basic characteristics of big data. Moving from static measurement of effects to time-resolved data has also contributed towards increasing the volume of information, in particular taking into consideration high-throughput kinetics.

These "new methodology" data are complemented by an overall large legacy of historical data in toxicology, which is becoming increasingly available. A major challenge is that the historical data are not always available in a structured way such as in easily accessible databases, but rather in isolated archives and in the form of reports, as well as an abundant quantity of scientific journal publications. This unstructured information needs to be extracted to add it to the pool of – easily accessible and usable – toxicological data. Importantly, toxicology can draw from the "bigger" data streams from closely related fields: pharmaceutical sciences and drug discovery, health sciences, as well as epidemiology. (Bio)monitoring data, for example, or data from large cohort epidemiological investigations are valuable if not essential additions for risk assessment at the population level. They will improve the connection between chemical hazards and diseases, as well as going a long way towards integrating exposure. Knowledge of exposure to realistic concentrations and combinations of chemical substances is fundamental to risk assessment and essential as a basis for decision making with the goal of protecting human health and the environment. Another new type of data can bring additional value to toxicology and decision making: information extracted from the internet or social media, which might bring, for example, knowledge of the actual uses of substances or potential first warning signs about adverse effects, as has been shown for pharmaceuticals. In the future, predictive toxicology should integrate the experience from related disciplines even more, embrace the approaches and draw upon the data streams used in these fields to allow further cross-fertilisation. It has to be a truly interdisciplinary approach. Big data in (predictive) toxicology could move to another level.

These developments in big data sources and availability have also advanced computational toxicology, by providing comprehensive data as an input for modelling, the volume of data points also providing an improved statistical robustness, as well as by strengthening the mechanistic basis of the models to improve the predictions, their (human) relevance and credibility for human safety assessment.

Big data have thus brought significant advances, but also constitute significant challenges to toxicology. The challenges are from both the technical side relating to the integration and interpretation of their content, as well as the overall challenge not to be overwhelmed by the quantity of data and being able to determine their meaning. Advances in data sciences have made the handling, processing, storing, integrating and analysing of these data possible such that data sciences have become an integral part of modern toxicology. The lines between experimental testing and computational modelling have generally become blurred with the analysis and interpretation of certain types of data, for example from omics, requiring routine use of bespoke computational algorithms. In this sense, predictive toxicology is seen here not only as the application of computational predictive modelling. Predictive toxicology used to be understood mainly as computational models predicting the hazard of a chemical from a chemical structure utilising structure–activity relationships or being based on statistical correlations between a number of physico-chemical properties and toxic effects. However, many methods in chemical safety assessment make predictions about the impact of a chemical on human health (or the environment and wildlife), including *in chemico* and *in vitro* assays and even animal studies. These are models used to predict the outcome in humans. In this sense, predictive toxicology has been and will continue to be extended to many other aspects in chemical safety assessment.

Another change in chemical safety assessment is the emphasis on risk assessment as compared to the evaluation of intrinsic hazards of chemicals, thus putting more focus on the exposure aspect, both exposure of individuals or populations as well as actual internal exposure of target sites in the organism. Therefore, exposure science has grown to be an essential part of risk assessment, and data streams from (bio)monitoring or epidemiological research are an integral part of risk assessment in the 21st

Century. Real-life exposure to chemicals, *i.e.*, combined exposure to multiple chemicals from many different sources over a long time, as well as the different susceptibilities in the population, also changing with the life stage, need to be addressed.

The velocity of big data is an important aspect and will not only support the increase in the pace of chemical safety assessment, but will also enable predictive toxicology to foresee forthcoming risks. These emerging risks may appear with new substances or new uses of chemicals. Through their broad coverage, big data seem to be best suited for the early identification of emerging risks; this could be a major scope of predictive toxicology in the future.

Big data are now an accepted reality in toxicology and chemical risk assessment. They are being increasingly well characterised and understood, especially as the data science from other fields (where big data are more advanced) develops. They provide numerous opportunities to change the face of this area of science and improve the prospects for chemical risk assessment. There will be even more advanced methods and technologies to integrate the data and make sense of them, finding trends and connections; artificial intelligence will undoubtedly increase possible applications and opportunities. However, in the end, it comes down to how the wealth of data, new methods and new knowledge gained can actually be used for risk assessment and achieving the goal of the protection of human health and the environment. Whilst there is excitement and optimism about the use of big data, in particular for addressing major current challenges in chemical assessment, there is still, however, some way to go before they reach their full potential in regulatory risk assessment. As such, big data support efforts in new toxicological assessment and will be translated to routine chemical evaluation in a regulatory context in the future.

References

1. G. Singh, D. Schulthess, N. Hughes, B. Vannieuwenhuyse and D. Kalra, Real world big data for clinical research and drug development, *Drug Discovery Today*, 2018, **23**(3), 652–660.
2. J. G. Diaz Ochoa and F. Weil, From personalization to patient centered systems toxicology and pharmacology, *Comput. Toxicol.*, 2019, **11**, 14–22.
3. http://bd4bo.eu.
4. E. J. Griffen, A. G. Dossetter, A. G. Leach and S. Montague, Can we accelerate medicinal chemistry by augmenting the chemist with Big Data and artificial intelligence?, *Drug Discovery Today*, 2018, **23**(7), 1373–1384.
5. http://www.europeanleadfactory.eu.
6. http://bigchem.eu.
7. I. V. Tetko, O. Engkvist, U. Koch, J.-L. Reymond and H. Chen, BIGCHEM: Challenges and opportunities for big data analysis in chemistry, *Mol. Inf.*, 2016, **35**, 615–621.

8. http://excape-h2020.eu/index.php.
9. National Academies of Sciences, Engineering, and Medicine, *Using 21st Century Science to Improve Risk-Related Evaluations*, Washington, DC, The National Academies Press, 2017.
10. M. Hilbert, Big Data for development: A review of promises and challenges, *Dev. Policy Rev.*, 2016, **34**(1), 135–174.
11. M. van Rijmenam, Why the 3V's are not sufficient to describe big data, blog post available at https://datafloq.com/read/3vs-sufficient-describe-big-data/166, 2013.
12. M. Ali-ud-din Khan, M. Fahim Uddin and N. Gupta, Seven V's of Big Data - Understanding Big Data to extract value, Proceedings of 2014 Zone 1 Conference of the American Society for Engineering Education (ASEE Zone 1).
13. R. Livingstone, The 7 Vs of Big Data – and by they are important for you and your business, blog post available at https://livingstoneadvisory.com/2013/06/vs-big-data/.
14. G. Firican, The 10 Vs of Big Data, TDWI, 2017, https://tdwi.org/articles/2017/02/08/10-vs-of-big-data.aspx.
15. D. J. Dix, K. A. Houck, M. T. Martin, A. M. Richard, R. W. Setzer and R. J. Kavlock, The ToxCast program for prioritizing toxicity testing of environmental chemicals, *Toxicol. Sci.*, 2007, **95**, 5–12.
16. R. Kavlock, K. Chandler, K. Houck, S. Hunter, R. Judson, N. Kleinstreuer, T. Knudsen, M. Martin, S. Padilla, D. Reif, A. Richard, D. Rotroff, N. Sipes and D. Dix, Update on EPA's ToxCast Program: Providing high throughput decision support tools for chemical risk management, *Chem. Res. Toxicol.*, 2012, **25**, 1287–1302.
17. C. W. Schmidt, TOX21: New dimensions of toxicity testing, *Environ. Health Perspect.*, 2009, **117**, A348–A353.
18. R. R. Tice, C. P. Austin, R. J. Kavlock and J. R. Bucher, Improving the human hazard characterization of chemicals: A Tox21 update, *Environ. Health Perspect.*, 2013, **171**, 756–765.
19. U.S. Environmental Protection Agency, United States Federal Government TOX21 collaboration advancing toxicology to improve environmental health and pharmaceutical safety, 2019, Tox21 factsheet https://tox21.gov/wp-content/uploads/2019/04/Tox21_FactSheet_Apr2019.pdf.
20. https://www.ebi.ac.uk/chembl.
21. A. Gaulton, A. Hersey, M. Nowotka, A. P. Bento, J. Chambers, D. Mendez, P. Mutowo, F. Atkinson, L. J. Bellis, E. Cibrián-Uhalte, M. Davies, N. Dedman, A. Karlsson, M. P. Magariños, J. P. Overington, G. Papadatos, I. Smit and A. R. Leach, The ChEMBL database in 2017, *Nucleic Acids Res.*, 2017, **45**(D1), D945–D954.
22. Toxicogenomics Project-Genomics Assisted Toxicity Evaluation system, https://toxico.nibiohn.go.jp/english.
23. Y. Igarashi, N. Nakatsu, T. Yamashita, A. Ono, Y. Ohno, T. Urushidani and H. Yamada, Open TG-GATEs: a large-scale toxicogenomics database, *Nucleic Acids Res.*, 2015, **43**(Database issue), D921–D927.

24. https://www.hbm4eu.eu.
25. Information Platform for Chemical Monitoring, https://ipchem.jrc.ec.europa.eu/RDSIdiscovery/ipchem/index.html.
26. Organisation for Economic Co-operation and Development (OECD), *Guidance Document on Good In Vitro Method Practices (GIVIMP)*, Series on Testing and Assessment No. 286, ENV/JM/MONO(2018)19, 2018.
27. H. J. Klimisch, M. Andreae and U. Tillmann, A systematic approach for evaluating the quality of experimental toxicological and ecotoxicological data, *Regul. Toxicol. Pharmacol.*, 1997, **25**, 1–5.
28. L. Molander, M. Ågerstrand, A. Beronius, A. Hanberg and C. Rudén, Science in risk assessment and policy (SciRAP): an online resource for evaluating and reporting in vivo (Eco)Toxicity studies, *Hum. Ecol. Risk Assess. Int. J.*, 2014, **21**, 753–762.
29. C. R. Hooijmans, M. M. Rovers, R. B. M. de Vries, M. Leenaars, M. Ritskes-Hoitinga and M. W. Langendam, SYRCLE's risk of bias tool for animal studies, *BMC Med. Res. Methodol.*, 2014, **14**, 43.
30. Organisation for Economic Co-operation and Development (OECD), *Guidance Document on the validation and international acceptance of new or updated test methods for hazard assessment*, Series on Testing and Assessment No. 34, ENV/JM/MONO(2005)14, 2005.
31. Organisation for Economic Co-operation and Development (OECD), *Genotoxicity of manufactured nanomaterials: report of the OECD expert meeting*, Series on the safety of manufactured nanomaterials no. 43, ENV/JM/MONO(2014)34, 2014.
32. https://www.oecd.org/env/ehs/testing/oecdguidelinesforthetestingofchemicals.htm.
33. https://iuclid6.echa.europa.eu.
34. Organisation for Economic Co-operation and Development (OECD), *Guidance Document for describing non-guideline in vitro test methods*, Series on Testing and Assessment No. 211, ENV/JM/MONO(2014)35, 2014.
35. A. Brazma, P. Hingamp, J. Quackenbush, G. Sherlock, P. Spellman, C. Stoeckert, J. Aach, W. Ansorge, C. A. Ball, H. C. Causton, T. Gaasterland, P. Glenisson, F. C. Holstege, I. F. Kim, V. Markowitz, J. C. Matese, H. Parkinson, A. Robinson, U. Sarkans, S. Schulze-Kremer, J. Stewart, R. Taylor, J. Vilo and M. Vingron, Minimum information about a microarray experiment (MIAME)-toward standards for microarray data, *Nat. Genet.*, 2001, **29**, 365–371.
36. http://www.ecetoc.org/topics/standardisation-metabolomics-assays-regulatory-toxicology.
37. H. Kamp, R. Beger, J.-L. Dorne, T. Ebbels, D. Ekman, C. Guillou, A. Goetz, G. Loizou, P. Leonards, B. Van Ravenzwaay, S. Sperber, M. Viant and T. Walk, MEtabolomics standaRds Initiative in Toxicology (MERIT), *Toxicol. Lett.*, 2018, **295**(Supplement 1), S214.
38. A. J. Williams and S. Ekins, A quality alert and call for improved curation of public chemistry databases, *Drug Discovery Today*, 2011, **16**(17/18), 747–750.

39. https://www.epa.gov/tsca-screening-tools/epi-suitetm-estimation-program-interface.
40. K. Mansouri, C. Grulke, A. Richard, R. Judson and A. Williams, An automated curation procedure for addressing chemical errors and inconsistencies in public datasets used in QSAR modelling, *SAR QSAR Environ. Res.*, 2016, **27**(11), 911–937.
41. C. Grulke, I. Thillainadarajah, P. Brown, A. Williams and A. Richard, *Mapping of chemical identifiers to DSSTox to enable data integration in the US-EPA CompTox Chemicals Dashboard*, Presented at American Chemical Society Spring Meeting, Orlando, FL, March 31-April 04, 2019, https://doi.org/10.23645/epacomptox.8089133.
42. http://www.KNIME.org.
43. US Environmental Protection Agency, ToxCast chemical inventory: data management & data quality considerations, https://www.epa.gov/sites/production/files/2015-08/documents/toxcast_chemicals_qa_qc_management_141204.pdf.
44. N. Ryan, A User's guide for accessing and interpreting-ToxCast-data, https://lri.americanchemistry.com/Users-Guide-for-Accessing-and-Interpreting-ToxCast-Data.pdf, 2017.
45. A. Gaulton, L. J. Bellis, A. P. Bento, J. Chambers, M. Davies, A. Hersey, Y. Light, S. McGlinchey, D. Michalovich, B. Al-Lazikani and J. P. Overington, ChEMBL: a large-scale bioactivity database for drug discovery, *Nucleic Acids Res.*, 2012, **40**(Database issue), D1100–D1107.
46. A. P. Bento, A. Gaulton, A. Hersey, L. J. Bellis, J. Chambers, M. Davies, F. A. Krüger, Y. Light, L. Mak, S. McGlinchey, M. Nowotka, G. Papadatos, R. Santos and J. P. Overington, The ChEMBL bioactivity database: an update, *Nucleic Acids Res.*, 2014, **42**(Database issue), D1083–D1090.
47. G. Papadatos, A. Gaulton, A. Hersey and J. P. Overington, Activity, assay and target data curation and quality in the ChEMBL database, *J. Comput.-Aided Mol. Des.*, 2015, **29**(9), 885–896.
48. National Academies of Sciences, Engineering, and Medicine, *Data Science: Opportunities to Transform Chemical Sciences and Engineering: Proceedings of a Workshop in Brief*, The National Academies Press, Washington, DC, 2018.
49. J. Zhang, J.-H. Hsieh and H. Zhu, Profiling animal toxicants by automatically mining public bioassay data: A big data approach for computational toxicology, *PLoS One*, 2014, **9**(6), e99863.
50. K. M. Hettne, A. Boorsma, D. A. van Dartel, J. J. Goeman, E. de Jong, A. H. Piersma, R. H. Stierum, J. C. Kleinjans and J. A. Kors, Next-generation text mining mediated generation of chemical-response-specific gene sets for interpretation of gene expression data, *BMC Med. Genomics*, 2013, **6**, 2.
51. H. Zhu, J. Zhang, M. T. Kim, A. Boison, A. Sedykh and K. Moran, Big data in chemical toxicity research: the use of high-throughput screening assays to identify potential toxicants, *Chem. Res. Toxicol.*, 2014, **27**, 1643–1651.

52. E. L. Willighagen, A. Waagmeester, O. Spjuth, P. Ansell, A. J. Williams, V. Tkachenko, J. Hastings, B. Chen and D. J. Wild, The ChEMBL database as linked open data, *J. Cheminf.*, 2013, **5**, 23.
53. I. Harrow, W. Filsell, P. Woollard, I. Dix, M. Braxenthaler, R. Gedye, D. Hoole, R. Kidd, J. Wilson and D. Rebholz-Schuhmann, Towards virtual knowledge broker services for semantic integration of life science literature and data sources, *Drug Discovery Today*, 2013, **18**(9/10), 428–434.
54. A. J. Williams, L. Harland, P. Groth, S. Pettifer, C. Chichester, E. L. Willighagen, C. T. Evelo, N. Blomberg, G. Ecker, C. Goble and B. Mons, Open PHACTS: semantic interoperability for drug discovery, *Drug Discovery Today*, 2012, **17**(21/22), 1188–1198.
55. M. Hohman, K. Gregory, K. Chibale, P. J. Smith, S. Ekins and B. Bunin, Novel web-based tools combining chemistry informatics, biology and social networks for drug discovery, *Drug Discovery Today*, 2009, **14**, 261–270.
56. M. Clark and T. Steger-Hartmann, A big data approach to the concordance of the toxicity of pharmaceuticals in animals and humans, *Regul. Toxicol. Pharmacol.*, 2018, **96**, 94–105.
57. http://www.echemportal.org, participating databases: http://www.echemportal.org/echemportal/page.action?pageID=2.
58. https://actor.epa.gov.
59. R. S. Judson, M. T. Martin, P. Egeghy, S. Gangwal, D. M. Reif, P. Kothiya, M. Wolf, T. Cathey, T. Transue, D. Smith, J. Vail, A. Frame, S. Mosher, E. A. Cohen Hubal and A. M. Richard, Aggregating data for computational toxicology applications: The U.S., "Environmental Protection Agency (EPA) Aggregated Computational Toxicology Resource (ACToR) System", *Int. J. Mol. Sci.*, 2012, **13**(2), 1805–1831.
60. https://comptox.epa.gov/dashboard, https://comptox.epa.gov.
61. A. J. Williams, C. M. Grulke, J. Edwards, A. D. McEachran, K. Mansouri, N. C. Baker, G. Patlewicz, I. Shah, J. F. Wambaugh, R. S. Judson and A. M. Richard, The CompTox Chemistry Dashboard: a community data resource for environmental chemistry, *J. Cheminf.*, 2017, **9**, 61.
62. chemagora.jrc.ec.europa.eu, connected third party resources: https://chemagora.jrc.ec.europa.eu/chemagora/thirdPartyData.htm.
63. D. M. Hendrickx, H. J. W. L. Aerts, F. Caiment, D. Clark, T. M. D. Ebbels, C. T. Evelo, H. Gmuender, D. G. A. J. Hebels, R. Herwig, J. Hescheler, D. G. J. Jennen, M. J. A. Jetten, S. Kanterakis, H. C. Keun, V. Matser, J. P. Overington, E. Pilicheva, U. Sarkans, M. P. Segura-Lepe, I. Sotiriadou, T. Wittenberger, C. Wittwehr, A. Zanzi and J. C. S. Kleinjans, diXa: a data infrastructure for chemical safety assessment, *Bioinformatics*, 2015, **31**(9), 1505–1507.
64. https://elixir-europe.org.
65. https://zenodo.org/record/173258#.XRiiKnduJyx.
66. L. Siragusa, S. Cross, M. Baroni, L. Goracci and G. Cruciani, BioGPS: Navigating biological space to predict polypharmacology, off-targeting, and selectivity, *Proteins*, 2015, **83**(3), 517–532.

67. https://ochem.eu.
68. I. Sushko, S. Novotarskyi, R. Körner, A. K. Pandey, M. Rupp, W. Teetz, S. Brandmaier, A. Abdelaziz, V. V. Prokopenko, V. Y. Tanchuk, R. Todeschini, A. Varnek, G. Marcou, P. Ertl, V. Potemkin, M. Grishina, J. Gasteiger, C. Schwab, I. I. Baskin, V. A. Palyulin, E. V. Radchenko, W. J. Welsh, V. Kholodovych, D. Chekmarev, A. Cherkasov, J. Aires-de-Sousa, Q. Y. Zhang, A. Bender, F. Nigsch, L. Patiny, A. Williams, V. Tkachenko and I. V. Tetko, Online chemical modeling environment (OCHEM): web platform for data storage, model development and publishing of chemical information, *J. Comput.-Aided Mol. Des.*, 2011, **25**, 533–554.
69. https://commonfund.nih.gov/bd2k.
70. https://metadatacenter.org/#about.
71. M. A. Musen, C. A. Bean, K. H. Cheung, M. Dumontier, K. A. Durante, O. Gevaert, A. Gonzalez-Beltran, P. Khatri, S. H. Kleinstein, M. J. O'Connor, Y. Pouliot, P. Rocca-Serra, S. A. Sansone, J. A. Wiser and CEDAR team, The center for expanded data annotation and retrieval, *J. Am. Med. Inform. Assoc.*, 2015, **22**(6), 1148–1152.
72. E. O. Wilson, Consilience: the unity of knowledge, *Vintage Books*, Random House, New York, 1998.
73. H. A. Gaspar, I. I. Baskin, G. Marcou, D. Horvath and A. Varnek, Chemical data visualization and analysis with incremental generative topographic mapping: big data challenge, *J. Chem. Inf. Model.*, 2015, **55**(1), 84–94.
74. S. J. Lusher, R. McGuire, R. C. van Schaik, C. D. Nicholson and J. de Vlieg, Data-driven medicinal chemistry in the era of big data, *Drug Discovery Today*, 2014, **19**(7), 859–868.
75. T. J. Howe, G. Mahieu, P. Marichal, T. Tabruyn and P. Vugts, Data reduction and representation in drug discovery, *Drug Discovery Today*, 2007, **12**(1–2), 45–53.
76. US EPA, ToxCast Manual, https://www.epa.gov/sites/production/files/2018-04/documents/toxcastownermanual4252018.pdf.
77. ToxCast Manual, https://www.epa.gov/chemical-research/exploring-toxcast-data-downloadable-data.
78. G. T. Ankley, R. S. Bennett, R. J. Erickson, D. J. Hoff, M. W. Hornung, R. D. Johnson, D. R. Mount, J. W. Nichols, C. L. Russom, P. K. Schmieder, J. A. Serrrano, J. E. Tietge and D. L. Villeneuve, Adverse outcome pathways: A conceptual framework to support ecotoxicology research and risk assessment, *Environ. Toxicol. Chem.*, 2010, **29**, 730–741.
79. https://aopkb.oecd.org.
80. https://aopwiki.org.
81. https://www.effectopedia.org.
82. A. Worth, K. Aschberger, D. Asturiol, J. Bessems, K. Gerloff, R. Graepel, E. Joossens, L. Lamon, T. Palosaari and A.-N. Richarz, Evaluation of the availability and applicability of computational approaches in the safety assessment of nanomaterials, *EUR 28617*, Publications Office of the European Union, Luxembourg, 2017.

83. S. K. Bopp, A. Kienzler, A.-N. Richarz, S. C. van der Linden, A. Paini, N. Parissis and A. P. Worth, Regulatory assessment and risk management of chemical mixtures: challenges and ways forward, *Crit. Rev. Toxicol.*, 2019, **49**(2), 174–189.
84. NAS 2007, *Toxicity Testing in the 21st Century: A Vision and a Strategy*, National Academies Press, Washington DC, 2007, DOI: 10.17226/11970.
85. NAS 2012, *Exposure Science in the 21st Century: A Vision and a Strategy*, National Academies Press, Washington DC, 2012, DOI: 10.17226/13507.
86. T. A. Burke, W. E. Cascio, D. L. Costa, K. Deener, T. D. Fontaine, F. A. Fulk, L. E. Jackson, W. R. Munns Jr, J. Orme-Zavaleta, M. W. Slimak and V. G. Zartarian, Rethinking environmental protection: Meeting the challenges of a changing world, *Environ. Health Perspect.*, 2017, **125**(3), A43–A49.
87. E. G. Snyder, T. H. Watkins, P. A. Solomon, E. D. Thoma, R. W. Williams, G. S. Hagler, D. Shelow, D. A. Hindin, V. J. Kilaru and P. W. Preuss, The changing paradigm of air pollution monitoring, *Environ. Sci. Technol.*, 2013, **47**(20), 11369–11377.
88. R. M. Hoff and S. A. Christopher, Remote sensing of particulate pollution from space: have we reached the promised land?, *J. Air Waste Manage. Assoc.*, 2009, **59**(6), 645–675.
89. J. L. Dickinson, B. Zuckerberg and D. N. Bonter, Citizen science as an ecological research tool: challenges and benefits, *Annu. Rev. Ecol. Evol. Syst.*, 2010, **41**, 149–172.
90. R. Albertini, M. Bird, N. Doerrer, L. Needham, S. Robison, L. Sheldon and H. Zenic, The use of biomonitoring data in exposure and human health risk assessment, *Environ. Health Perspect.*, 2006, **114**, 1755–1762.
91. C. Southan, Expanding opportunities for mining bioactive chemistry from patents, *Drug Discovery Today: Technol.*, 2015, **14**, 3–9.
92. A. Sarker, R. Ginn, A. Nikfarjam, K. O'Connor, K. Smith, S. Jayaraman, T. Upadhaya and G. Gonzalez, Utilizing social media data for pharmacovigilance: A review, *J. Biomed. Inf.*, 2015, **54**, 202–212.
93. I. Korkontzelos, A. Nikfarjam, M. Shardlow, A. Sarker, S. Ananiadou and G. H. Gonzalez, Analysis of the effect of sentiment analysis on extracting adverse drug reactions from tweets and forum posts, *J. Biomed. Inf.*, 2016, **62**, 148–158.
94. G. Eysenbach, Infodemiology and infoveillance: framework for an emerging set of public health informatics methods to analyze search, communication and publication behavior on the Internet, *J. Med. Internet Res.*, 2009, **11**, e11.
95. R. R. Boyles, A. E. Thessen, A. Waldrop and M. A. Haendel, Ontology-based data integration for advancing toxicological knowledge, *Curr. Opin. Toxicol.*, 2019, **16**, 67–74.
96. S. Ekins, The next era: Deep Learning in pharmaceutical research, *Pharm. Res.*, 2016, **33**(11), 2594–2603.

97. T. Unterthiner, A. Mayr, G. Klambauer, M. Steijaert, J. Wegner, H. Ceulemans and S. Hochreiter, Deep Learning as an opportunity in virtual screening, Conference Neural Information Processing Systems Foundation (NIPS 2014), Workshop on Machine Learning for Clinical Data Analysis, Healthcare and Genomics, Montreal, Canada, December 8–13, 2014, http://www.bioinf.jku.at/publications/2014/NIPS2014a.pdf.
98. M. M. Najafabadi, F. Villanustre, T. M. Khoshgoftaar, N. Seliya, R. Wald and E. Muharemagic, Deep learning applications and challenges in big data analytics, *J. Big Data*, 2015, **2**, 1, DOI: 10.1186/s40537-014-0007-7.
99. A. Mayr, G. Klambauer, T. Unterthiner and S. Hochreiter, DeepTox: Toxicity prediction using Deep Learning, *Front. Environ. Sci.*, 2016, **3**, 80.
100. T. Unterthiner, A. Mayr, G. Klambauer and S. Hochreiter, *Toxicity Prediction Using Deep Learning*, 2015, arXiv:1503.01445 [stat.ML].
101. C. Wittwehr, P. Blomsted, J.-P. Gosling, T. Peltola, B. Raffael, A. Richarz, M. Sienkiewicz, P. Whaley, A. Worth and M. Whelan, Artificial Intelligence for chemical risk assessment, *Comput. Toxicol.*, submitted.
102. European Commission, Communication to the Commission, Data, information and knowledge management at the European Commission, 2016, SWD(2016) 333 final.
103. M. Vinken, Omics-based input and output in the development and use of adverse outcome pathways, *Curr. Opin. Toxicol.*, 2019, **18**, 8–12.
104. C. Wittwehr, H. Aladjov, G. T. Ankley, H. J. Byrne, J. de Knecht, E. Henzie, G. Klambauer, B. Landesmann, M. Luijten, C. MacKay, G. Maxwell, M. E. Meek, A. Paini, E. Perkins, T. Sobanski, D. Villeneuve, K. M. Waters and M. Whelan, How Adverse Outcome Pathways can aid the development and use of computational prediction models for regulatory toxicology, *Toxicol. Sci.*, 2017, **155**(2), 326–336.
105. S. J. Sturla, A. R. Boobis, R. E. FitzGerald, J. Hoeng, R. J. Kavlock, K. Schirmer, M. Whelan, M. F. Wilks and M. C. Peitsch, Systems Toxicology: From basic research to risk assessment, *Chem. Res. Toxicol.*, 2014, **27**, 314–329.
106. J. T. MacGregor, The future of regulatory toxicology: impact of the biotechnology revolution, *Toxicol. Sci.*, 2003, **75**(2), 236–248.
107. E. K. Brockmeier, G. Hodges, T. H. Hutchinson, E. Butler, M. Hecker, K. E. Tollefsen, N. Garcia-Reyero, P. Kille, D. Becker, K. Chipman, J. Colbourne, T. W. Collette, A. Cossins, M. Cronin, P. Graystock, S. Gutsell, D. Knapen, I. Katsiadaki, A. Lange, S. Marshall, S. F. Owen, E. J. Perkins, S. Plaistow, A. Schroeder, D. Taylor, M. Viant, G. Ankley and F. Falciani, The role of omics in the application of Adverse Outcome Pathways for chemical risk assessment, *Toxicol. Sci.*, 2017, **158**, 252–262.

108. W. H. Goodson, *et al.*, Assessing the carcinogenic potential of low-dose exposures to chemical mixtures in the environment: the challenge ahead, *Carcinogenesis*, 2015, **36**(Suppl. 1), S254–S296.
109. S. Bopp, A. Kienzler, S. van der Linden, L. Lamon, A. Paini, N. Parissis, A. Richarz, J. Triebe and A. Worth, Review of case studies on the human and environmental risk assessment of chemical mixtures, JRC Technical Report, EUR 27968, Publications Office of the European Union, 2016.
110. D. Hanahan and R. A. Weinberg, Hallmarks of cancer: the next generation, *Cell*, 2011, **144**, 646–674.
111. R. Garcia-Serna, D. Vidal, N. Remez and J. Mestres, Large-scale predictive drug safety: From structural alerts to biological mechanisms, *Chem. Res. Toxicol.*, 2015, **28**(10), 1875–1887.
112. A. Richard, R. Judson, K. Houck, C. Grulke, P. Volarath, I. Thillainadarajah, C. Yang, J. Rathman, M. Martin, J. Wambaugh, T. Knudsen, K. Mansouri, J. Kancherla, G. Tier, A. Williams, S. Little, K. Crofton and R. Thomas, The ToxCast Chemical Landscape - Paving the Road to 21st Century Toxicology, *Chem. Res. Toxicol.*, 2016, **29**(8), 1225–1251.
113. R. Judson, A. Richard, D. J. Dix, K. Houck, M. Martin, R. Kavlock, V. Dellarco, T. Henry, T. Holderman, P. Sayre, S. Tan, T. Carpenter and E. Smith, The toxicity data landscape for environmental chemicals, *Environ. Health Perspect.*, 2009, **117**(5), 685–695.
114. International Risk Governance Council (IRGC), *Guidelines for emerging risk governance*, International Risk Governance Council (IRGC), Lausanne, 2015, available from: http://www.irgc.org.
115. R. Flage and T. Aven, Emerging risk – Conceptual definition and a relation to black swan type of events, *Reliab. Eng. Syst. Safe.*, 2015, **144**, 61–67.
116. http://www.normandata.eu.
117. https://www.modernet.info.
118. M. Nandi, J. Venton and P. J. Aston, A novel method to quantify arterial pulse waveform morphology: attractor reconstruction for physiologists and clinicians, *Physiol. Meas.*, 2018, **39**, 104008.
119. B. P. Lankadurai, E. G. Nagato and M. J. Simpson, Environmental metabolomics: an emerging approach to study organism responses to environmental stressors, *Environ. Rev.*, 2013, **21**, 180–205.
120. S. Kar and J. Leszczynski, Exploration of computational approaches to predict the toxicity of chemical mixtures, *Toxics*, 2019, **7**, 15.
121. S. Bopp, A. Richarz, A. Worth, E. Berggren and M. Whelan, Something from nothing? Ensuring the safety of chemical mixtures, Publications Office of the European Union, EUR 29258 EN, https://ec.europa.eu/jrc/en/publication/something-nothing-ensuring-safety-chemical-mixtures.
122. A. Kortenkamp and M. Faust, Regulate to reduce chemical mixture risk, *Science*, 2018, **361**(6399), 224–226.

123. H. Qian, M. Chen, K. M. Kransler and R. T. Zaleski, Assessment of chemical coexposure patterns based upon phthalate biomonitoring data within the 2007/2008 National Health and Nutrition Examination Survey, *J. Exposure Sci. Environ. Epidemiol.*, 2015, **25**, 249–255.
124. P. P. Egeghy, R. Judson, S. Gangwal, S. Mosher, D. Smith, J. Vail and E. A. Cohen Hubal, The exposure data landscape for manufactured chemicals, *Sci. Total Environ.*, 2012, **414**, 159–166.
125. E. A. Cohen Hubal, A. M. Richard, I. Shah, J. Gallagher, R. Kavlock, J. Blancato and S. W. Edwards, Exposure science and the U.S. EPA National Center for Computational Toxicology, *J. Exposure Sci. Environ. Epidemiol.*, 2010, **20**, 231–236.
126. U.S. Environmental Protection Agency, Rapid chemical exposure and dose research, https://www.epa.gov/sites/production/files/2014-12/documents/exposure_forecasting_factsheet.pdf.
127. D. Vallero, P. Egeghy, T. Buckley, K. Isaacs, M.-R. Goldsmith, H. Ozkaynak and J. Mitchell, *Mapping ExpoCast onto ToxCast*, International Society of Exposure Science, Poster, Basel, Switzerland, 2013.
128. P. P. Egeghy, L. S. Sheldon, K. K. Isaacs, H. Özkaynak, M.-R. Goldsmith, J. F. Wambaugh, R. S. Judson and T. J. Buckley, Computational exposure science: An emerging discipline to support 21st-century risk assessment, *Environ. Health Perspect.*, 2016, **124**(6), 697–702.
129. U.S. Environmental Protection Agency, Next Generation Risk Assessment: Recent advances in molecular, computational, and systems biology, EPA/600/R-14/004, 2014.
130. Government of Canada, Overview of the Chemicals Management Plan, https://www.canada.ca/content/dam/hc-sc/migration/cs-sc/alt_formats/pdf/fact-fait/overview-vue-eng.pdf.
131. European Chemicals Agency, Mapping the chemical universe to address substances of concern – Integrated Regulatory Strategy Annual Report 2019, ECHA-2019-R-10-EN, 2019.
132. S. Kim, P. A. Thiessen, E. E. Bolton, J. Chen, G. Fu, A. Gindulyte, L. Han, J. He, S. He, B. A. Shoemaker, J. Wang, B. Yu, J. Zhang and S. H. Bryant, PubChem substance and compound databases, *Nucleic Acids Res.*, 2016, **44**, D1202–D1213.
133. B. van Ravenzwaay, M. Herold, H. Kamp, M. D. Kapp, E. Fabian, R. Looser, G. Krennrich, W. Mellert, A. Prokoudine, V. Strauss, T. Walk and J. Wiemer, Metabolomics: A tool for early detection of toxicological effects and an opportunity for biology based grouping of chemicals-From QSAR to QBAR, *Mutat. Res., Genet. Toxicol. Environ. Mutagen.*, 2012, **746**, 144–150.
134. A. V. Singh, P. Laux, A. Luch, C. Sudrik, S. Wiehr, A.-M. Wild, G. Santomauro, J. Bill and M. Sitti, Review of emerging concepts in nanotoxicology: opportunities and challenges for safer nanomaterial design, *Toxicol. Mech. Methods*, 2019, **29**(5), 378–387.

135. R. L. Marchese Robinson, I. Lynch, W. Peijnenburg, J. Rumble, F. Klaessig, C. Marquardt, H. Rauscher, T. Puzyn, R. Purian, C. Åberg, S. Karcher, H. Vriens, P. Hoet, M. D. Hoover, C. O. Hendren and S. L. Harper, How should the completeness and quality of curated nanomaterial data be evaluated?, *Nanoscale*, 2016, **8**, 9919–9943.
136. A.-N. Richarz, A. Avramopoulos, E. Benfenati, A. Gajewicz, N. Golbamaki Bakhtyari, G. Leonis, R. L. Marchese Robinson, M. G. Papadopoulos, M. T. Cronin and T. Puzyn, Compilation of data and modelling of nanoparticle interactions and toxicity in the NanoPUZZLES Project, in *Modelling the Toxicity of Nanoparticles*, ed. L. Tran, M. A. Bañares and R. Rallo, Springer International Publishing, Cham, 2017, pp. 303–324.
137. R. Liu, T. Hassan, R. Rallo and Y. Cohen, HDAT: web-based high-throughput screening data analysis tools, *Comput. Sci. Discovery*, 2013, **6**, 014006.
138. N. A. Lewinsky and B. T. McInnes, Using natural language processing techniques to inform research on nanotechnology, *Beilstein J. Nanotechnol.*, 2015, **6**, 1439–1449.
139. M. Fernandez and A. S. Barnard, Identification of nanoparticle prototypes and archetypes, *ACS Nano*, 2015, **9**(12), 11980–11992.
140. A.-N. Richarz, L. Lamon, D. Asturiol and A. P. Worth, Current developments and recommendations in computational nanotoxicology in view of regulatory applications, in *Computational Nanotoxicology: Challenges, Pitfalls, and Perspectives*, ed. A. Gajewicz and T. Puzyn, Jenny Stanford Publishing Pte. Ltd., 2020, pp. 99–155.
141. M. E. Bates, S. Larkin, J. M. Keisler and I. Linkov, How decision analysis can further nanoinformatics, *Beilstein J. Nanotechnol.*, 2015, **6**, 1594–1600.

CHAPTER 2

Biological Data in the Light of Toxicological Risk Assessment

VESSELA VITCHEVA*

Medical University of Sofia, Faculty of Pharmacy, Department of Pharmacology, Pharmacotherapy and Toxicology, 2 "Dunav" str., Sofia – 1000, Bulgaria
*E-mail: vesselavitcheva@yahoo.com

2.1 Introduction

A human health risk assessment is the process to estimate the nature and probability of adverse health effects in humans who may be exposed to new and existing chemicals through different routes and environmental media, now or in the future. Risk assessment is an ever-evolving process whereby scientific information on the hazardous properties of chemicals and the extent of exposure results in a statement as to the probability that exposed populations will be harmed. The probability of harm can be expressed either qualitatively or quantitatively, depending on the nature of the scientific information available and the intent of the risk assessment.[1]

A risk assessment of the effect of chemical substances on human health normally examines the following potential toxic effects, also referred to as endpoints: acute toxicity; irritation; corrosiveness; sensitisation; repeated dose toxicity; mutagenicity; carcinogenicity; toxicity of reproduction by each of the likely routes of exposure – oral (by ingestion), dermal (by absorption

Issues in Toxicology No. 41
Big Data in Predictive Toxicology
Edited by Daniel Neagu and Andrea-Nicole Richarz
© The Royal Society of Chemistry 2020
Published by the Royal Society of Chemistry, www.rsc.org

through the skin) and inhalation. It also examines the different human populations affected (workers, consumers, human population exposed indirectly through environment).

Safety assessment and evaluation of chemicals that are part of our everyday life (industrial chemicals, pesticides, cosmetics, food additives, and pharmaceuticals) are mainly based on information of their toxicity from experimental studies, by far the greatest part of the data about the toxic and/or adverse effects of the evaluated compounds has traditionally relied on studies carried out in animals (*in vivo*). Other studies use cell lines, cell cultures *etc.* (*in vitro*). More recently, alternative methods based on computational toxicology, such as (quantitative) structure–activity relationships ((Q)SARs), predicting these endpoints based on model development from the existing toxicity data, have been included in legislation for chemical safety assessment, such as REACH.

In risk assessment, new data are collected and existing data that are available in databases are evaluated. This chapter gives an overview on the different types of studies carried out to generate biological data, as required for toxicological hazard assessment, for example as an information requirement of the European REACH legislation. The main toxicity tests providing information about the systemic effects of the compounds are reviewed, based on the exposure duration – acute, sub-chronic and chronic. Toxicity tests designed to evaluate specific endpoints and target organs, *e.g.*, reproductive and developmental effects, carcinogenicity, genotoxicity, neurotoxicity, immunotoxicity are also discussed. Additional attention is given to the alternative methods of toxicity. Traditional experimental studies are mostly generated in a rather time-consuming process; however, new high-throughput *in vitro* technologies have increased the volume and speed of data generation. The chapter continues with a brief overview of toxicokinetic and toxicodynamic studies and how these data are applied in the light of modes of action (MoA) and potential adverse outcome pathways (AOPs). Examples of sources of toxicological data, as well as the main steps in data quality assessment and data interpretation are reviewed.

2.2 Data Generated by *In Vivo* Toxicity Testing

Toxicity testing in animals is conducted to identify possible adverse effects resulting from exposure to an agent and to develop dose–response relationships that allow evaluation of responses at other exposure concentrations. *In vivo* toxicity studies are carried out following internationally approved guidelines and guidance documents (OECD, US EPA, FDA, NTP *etc.*), which aim to harmonise the main steps and variables (*e.g.*, species, sex, age, duration of exposure *etc.*) of the testing procedures for different types of tests so that the test results are comparable and provide the same level of information. OECD Test Guidelines health and environmental effects and other endpoints are available from the Website https://www.oecd.org/env/ehs/testing/oecdguidelinesforthetestingofchemicals.htm.

The main source of data for systemic effects of a certain compound are toxicity tests tracing its effects after acute and multiple (repeated dose) administration – sub-chronic and chronic tests.

2.2.1 Acute Toxicity Testing

Acute systemic toxicity testing involves an assessment of the general toxic effects of a single dose or multiple doses of a chemical or product, within 24 hours by a particular route (oral, dermal, inhalation) and that occur during a subsequent 14 day observation period. The test animals are typically rodents (rats and mice) and the endpoints they are observed for include not only mortality but clear signs of toxicity, deviant behaviour and changes in growth. Historically, the primary focus of an acute toxicity test was to determine the chemical dose that caused death in 50% of the test animals (LD_{50}). Today, acute toxicity tests are also used to determine dosing regimens for longer-term toxicity testing and to evaluate more fully the effects of acute exposure. Acute toxicity data are also required by many regulatory frameworks to enable a substance to be ranked for classification purposes and hazard assessment.

2.2.1.1 *Test Guidelines*

Currently, there are three OECD standard Test Guidelines (TGs) for acute oral toxicity testing: the Fixed Dose Procedure (OECD TG 420 [2]), Acute Toxic Class Method (OECD TG 423 [3]) and Up-and-Down Procedure (OECD TG 425 [4]), which have substituted the conventional acute toxicity test (OECD TG 401) in order to minimise the pain and discomfort of the test animals, and to obtain more information on the pathogenesis of toxicity. All three methods permit estimation of the LD_{50} with a confidence interval and the results allow a substance to be ranked and classified according to the Globally Harmonised System (GHS).[5] The preferred animal species is the rat and the test substance is administered in a single dose by gavage. Animals are observed daily for a total of 14 days. Observations include changes in skin and fur, eyes and mucous membranes, respiratory, circulatory, autonomic and central nervous systems, motor activity and behaviour pattern. Attention should be given to tremors, convulsions, salivation, diarrhea, lethargy, sleep and coma. Individual weights of animals have to be determined before and after the substance administration. All animals should be subjected to gross necropsy and pathological changes should be recorded. The three methods differ slightly by procedure.

The Fixed Dose Procedure (OECD TG 420)[2] uses rats (five per sex, females preferred) and fixed dose levels as follows: 5, 50, 300, and 2000 mg kg^{-1}. Death is not an endpoint. The initial dose level is selected as the dose expected to produce some signs of toxicity without causing severe toxic effects or mortality based on *in vivo/in vitro* data (if no information exists, the starting dose will be 300 mg kg^{-1}). Further groups of animals may be dosed at higher or lower fixed doses, depending on the presence or absence of signs of toxicity. This procedure continues until the dose causing evident toxicity is identified.

The Acute Toxic Class Method (OECD TG 423)[3] consists of a stepwise procedure with the use of three animals of a single sex (normally females) per step. The endpoint is lethality. Absence or presence of compound-related mortality of the animals dosed at one step will determine the next step (*i.e.*, no further testing is needed; dosing of three additional animals with the same dose; dosing of three additional animals at the next higher or lower dose level). The starting dose level is selected from one of four fixed levels (5, 50, 300, and 2000), as being the dose that is most likely to produce mortality in some of the dosed animals. When there is no information on the substance, the starting dose is recommended to be 300 mg kg^{-1}.

The Up-and-Down Procedure (OECD TG 425)[4] consists of a single ordered dose progression in which animals of a single sex (normally females) are dosed, one at a time. The first animal receives a dose step below the level of the best estimate of the LD_{50} (when no information is available to make a preliminary estimate of the LD_{50}, the suggested starting dose is 175 mg kg^{-1}). If the animal survives, the dose for the next animal is increased by a factor of 3.2 times the original dose; if it dies, the dose for the next animal is decreased by a similar dose progression.

2.2.1.2 *Limitations of the Data Generated from Acute Toxicity Tests*

Acute toxicity studies are based on the assumption that acute toxicity and lethality in animal models are relevant to humans. For many chemicals, however, this assumption is inadequate for several reasons. First of all, dose extrapolations from animals to humans are not simple: smaller rodents generally have a far greater rate of metabolism than humans and therefore clear a chemical more rapidly, reducing total exposure relative to dose. In order to extrapolate from the animal to humans, adjustment or uncertainty factors, taking into consideration the intra-species and inter-species differences, are applied. Second, species specificity has to be taken into account because differences in toxic responses are critical for extrapolating from animal data to human risk. This is due to the fact that metabolic and biological differences may lead to responses in animals or humans that are absent in the other species.

2.2.1.3 *Use of Acute Toxicity Data*

Acute toxicity data can be used beyond toxicity ranking. For example, the information on the onset of toxicity – sudden, delayed, time-limited or continuous – can provide insights into the time course of absorption, distribution, and clearance of a toxicant. Comparing data from different routes of exposure (*e.g.*, oral and parenteral) gives some idea of relevant bioavailability. Using the data from acute toxicity studies, scientists can develop hypotheses about pathogenesis and target organs affected by acute exposure.

In conclusion, the acute toxicity data can help to evaluate and manage the risks associated with potential exposure to noxious agents. Within the classic *in vivo* testing schemes, acute toxicity tests provide a relatively quick overview to screen large numbers of chemicals and identify those that warrant further toxicity testing.

2.2.2 Toxicity Testing after Repeated Dose Administration (Short-term, Sub-chronic and Chronic Toxicity Tests)

Repeated dose toxicity testing includes sub-chronic, chronic and combined chronic toxicity and carcinogenicity studies. The main goal is to investigate the adverse effects occurring as a result of repeated daily dosing of a chemical for a specified period of time. The repeated dose studies are an important part of the data package produced to perform quantitative risk assessment of industrial chemicals, biocides, pesticides, pharmaceuticals. The results obtained from these studies give information on general characteristics of toxicity, the target organs, the dose–response relationship for each toxicity endpoint, responses to toxic metabolites formed in the organism, delayed responses, cumulative effects, the margin between toxic/non-toxic dose, information on reversibility/irreversibility of the effect, and for determination of the No Observed Adverse Effect Level (NOAEL) or No Observed Effect Level (NOEL) of toxicity. The latter are used to establish thresholds or guidelines for human exposure.

2.2.2.1 Test Guidelines

The principles and the experimental procedure of the *in vivo* repeated dose toxicity tests are determined by several OECD Test Guidelines, summarised in Table 2.1.

Table 2.1 OECD test guidelines for sub-chronic and chronic toxicity testing.

Test method	Test guideline
28 Day oral toxicity study in rodents	OECD 407 [6]
90 Day oral toxicity study in rodents	OECD 408 [7]
90 Day oral toxicity study in non-rodents	OECD 409 [8]
Dermal toxicity: 21/28 day study (rat, rabbit or guinea pig)	OECD 410 [9]
Dermal toxicity: 90 day study (rat, rabbit or guinea pig)	OECD 411 [10]
Inhalation toxicity: 28 day study in rodents	OECD 412 [11]
Inhalation toxicity: 90 day study in rodents	OECD 413 [12]
Chronic toxicity studies in rodents	OECD 452 [13]
Combined chronic toxicity and carcinogenicity studies in rodents	OECD 453 [14]

For the *in vivo* repeated dose toxicity tests, OECD TGs require measurement and evaluation of a number of endpoints, including: body weights and food consumption, clinical signs (such as changes in skin, fur, eyes, secretions, posture, and response to handling), motor activity, grip strength, sensory reactivity to stimuli, clinical pathology (clinical chemistry and hematology), and ophthalmology. At study termination, a gross necropsy is conducted on all animals, and selected organs are weighed. A full histopathologic analysis is conducted on all animals in the control and high-dose groups, on all animals that were killed or died during the study, and on all gross lesions. If exposure-related changes are detected, the analysis is extended to all treatment groups. Target organs are examined in all animals. For the carcinogenicity bioassays, gross necropsy and histopathology are extensive because the primary focus is on detecting neoplasms. In 2018, a number of endocrine-related measurements, particularly relevant to thyroid function, were added to the 90 day oral toxicity study test guideline. The Inhalation Toxicity TGs were updated in 2017 to include the testing and characterisation of effects of nanomaterials.

2.2.2.2 *Use of Short-term and Sub-chronic Toxicity Test Data*

The exposure durations are typically 28 days or 90 days for short-term and chronic toxicity studies, respectively. Administration of the chemical (oral, inhalation, or dermal route) is usually determined by the route of potential or actual human exposure. Those studies are not designed to assess effects that have a long latency period, such as cancer, but do provide information that can be used in setting doses for chronic toxicity and carcinogenicity studies. However, they can provide initial or definitive data for risk assessment purposes. The main limiting factor of those studies is sometimes the smaller sample size, which reduces the sensitivity of the study to detect adverse effects.

2.2.2.3 *Use of Chronic Toxicity and Carcinogenicity Test Data*

The chronic toxicity study provides information on the potential health hazards likely to arise from repeated exposure over a considerable part of the entire lifespan (in rodents). The study provides data on the toxic effects of the substance and on the characterisation of the dose–response relationship that can be used to predict the chronic toxicity effects at human exposure levels. It also indicates the target organs, identifies the NOAEL or the point of departure for establishment of a Benchmark Dose (BMD) and last but not least, the data can be used to test hypotheses regarding the mode of action of a compound.

2.2.3 Toxicity Tests for Specific Endpoints

2.2.3.1 *Reproductive and Developmental Toxicity Tests*

The general purpose of reproductive and developmental toxicity assays is to investigate the adverse effects of a chemical on sexual function and fertility in adult males and females, to evaluate the competence of breeding pairs to produce phenotypically normal offspring, as well as to gain information on possible developmental toxicity of the tested compound in the offspring. These tests are conducted to predict human hazard potential, but the data from these models may or may not be relevant for predicting human risk. The predictive power of the tests is very often limited by differences in the underlying biology. A notorious example of how species differences are important is the exposure to thalidomide, to which rats are highly resistant and humans are very sensitive regarding developmental toxicity as apparent by severe limb malformations.[15]

The data derived from these tests should allow the determination of No Observed Adverse Effect Levels (NOAELs), Lowest Observed Adverse Effect Levels (LOAELs) and/or benchmark doses for the various endpoints and/or can be used to characterise effects detected in previous repeated dose studies and/or serve as a guide for subsequent testing.

Reproductive and developmental toxicity testing includes a broad category of endpoints because of the multiple stages of exposure and the variety of possible effects. The main reproductive and developmental studies are summarised in Table 2.2.

The prenatal developmental toxicity study (OECD TG 414)[16] is usually performed with female rats and rabbits and is used to examine embryonic and fetal toxicity as a consequence of exposure during pregnancy, usually beginning at implantation and continuing through the end of organogenesis or to full term. The pregnant animals are then killed and the fetuses are weighed, sexed, and examined. The studied endpoints include: growth retardation, anatomic variations, external, visceral, and skeletal alterations and lethality. Recently, a number of endocrine-related measurements in the dams and fetuses were added.

Table 2.2 OECD test guidelines for reproductive and developmental toxicity testing.

Test method	Test guideline
Prenatal developmental toxicity test	OECD 414[16]
Reproductive/developmental toxicity screening test	OECD 421[17]
Combined repeated dose toxicity study with reproductive/developmental toxicity screening test	OECD 422[18]
Two-generation reproduction toxicity study	OECD 416[19]
Extended one-generation reproductive toxicity study	OECD 443[20]
Developmental neurotoxicity study	OECD 426[21]

For reproductive toxicity, a variety of tests are available. The Reproductive/Developmental Toxicity Screening Test (OECD TG 421)[17] and the Combined Repeated Dose Toxicity Study with Reproductive/Developmental Toxicity Screening Test (OECD TG 422)[18] provide insights into gonadal function, fertility, pregnancy, parturition, and prenatal and postnatal developmental toxicity. These are screening-level assays used to make decisions about the need for further testing.

Long considered the definitive test of reproductive toxicity is the Two-Generation Reproduction Toxicity Study (OECD TG 416),[19] which involves continuous dosing of parental rats and their offspring with the test substance (usually in the diet, but other routes are possible) for two entire generations, up through weaning of the F2 pups at three weeks of age. The study is designed to provide general information concerning the effects of a test substance on the integrity and performance of the male and female reproductive systems, including gonadal function, the estrous cycle, mating behaviour, conception, gestation, parturition, lactation, and weaning, and the growth and development of the offspring. The study also provides information about the effects of the test substance on neonatal morbidity, mortality, and preliminary data on prenatal and postnatal developmental toxicity and serves as a guide for subsequent tests. In addition to studying growth and development of the F1 generation, the integrity and performance of the male and female reproductive systems, as well as growth and development of the F2 generation are investigated.

The retrospective analysis of the data obtained from the two-generation reproduction toxicity study for the second generation (F2) indicates little impact on the final outcome of the study in terms of regulation. Therefore, the Extended One-Generation Reproductive Toxicity study (OECD TG 443)[20] was developed to evaluate specific life stages not covered by other types of toxicity studies and test for effects that may occur as a result of pre- and postnatal chemical exposure. This study provides information on the pre- and postnatal effects of chemicals on development, as well as a thorough evaluation of systemic toxicity in pregnant and lactating females and young and adult offspring. The key developmental endpoints include: offspring viability, neonatal health, developmental status at birth, and physical and functional development until adulthood. The information obtained is expected to identify specific target organs in the offspring. The endpoints determined to evaluate the integrity and performance of the adult male and female reproductive systems are: gonadal function, the estrous cycle, epididymal sperm maturation, mating behavior, conception, pregnancy, parturition, and lactation.

The Developmental Neurotoxicity Study (OECD TG 426)[21] Test Guideline has been developed to address endpoints that had not been covered. Since a number of chemicals are known to produce neurotoxicity in humans and other species, the purpose of this test is to determine *in utero* and early postnatal developmental neurological effects of the test substance. A developmental neurotoxicity study can be conducted as a separate study, incorporated into a reproductive toxicity and/or adult neurotoxicity study or added onto a prenatal developmental toxicity study. Rat is the preferred species. The neurotoxicity endpoints in dams include: unusual body posture (*e.g.*, hunched back), activity level (*e.g.*, decreased or increased exploration of the standard area), changes in gait (*e.g.*, waddling, ataxia), reactivity to handling, placing or other environmental stimuli, as well as the presence of clonic or tonic movements, convulsions, tremors, stereotypies (*e.g.*, excessive grooming, unusual head movements, repetitive circling), bizarre behaviour (*e.g.*, biting or excessive licking, self-mutilation, walking backwards, vocalisation). Offspring are evaluated for neurotoxic effects including gross neurologic and behavioral abnormalities, and the evaluation of brain weights and neuropathology during postnatal development and adulthood.

2.2.3.2 *Neurotoxicity as a Target Endpoint*

Neurotoxicity of a chemical as a target endpoint can be investigated during the standard acute, sub-chronic and chronic toxicity tests, according to the respective guidelines. If the data obtained from these studies or a structure–activity relationship screening suggests potential neurotoxicity, a more thorough evaluation of this endpoint can be carried out in accordance with the Neurotoxicity Study in Rodents (OECD TG 424).[22]

The neurotoxic potential of chemicals can be assessed using a wide array of methods, including behavioural, physiologic, anatomic and neurochemical. Behavioural alterations are discerned by changes in motor activity, sensations of sight, hearing, or touch, and cognitive functions, such as learning, memory and attention. At the physiological level, a chemical may change (reduce or accelerate) the speed of neurotransmission or change the threshold of neuronal activation. Anatomic changes can include alterations of the cell body, the axon, or the myelin sheath of neurons or of the thickness of cell layers in specific brain regions. Neurochemical effects of a chemical may affect the synthesis, release, re-uptake or metabolism of certain neurotransmitters.

2.2.3.3 *Immunotoxicity as a Target Endpoint*

The current immunotoxicity testing guidance for drugs, high production–volume chemicals, and pesticides specifies the use of animal models to assess potential biomarkers of immune system effects (*e.g.*, lymphoid organ and bone marrow indices, histopathology) or actual measures of immune function (*e.g.*, responses to challenge with antigens or pathogens).[23]

Two main types of immunotoxic response may result from chemical exposure: immunosuppression and antigenicity, which is more commonly known as allergy or hypersensitivity. Immunotoxicity testing for hypersensitivity has been done for years in guinea pigs (OECD TG 406).[24] The expense and technical difficulties associated with guinea pig tests along with the better understanding of the immunobiological events that result in the acquisition of skin sensitisation led to a development and validation of an alternative mouse model – the mouse local lymph node assay (LLNA) (OECD TG 429).[25] This test has obtained regulatory acceptance for screening for skin sensitisers and predicting the human health risk.[26] The LLNA detects proliferation of lymphocytes in lymph nodes on application of potential skin sensitisers to the ears of mice. The test includes a quantitative measure of cell proliferation as a function of incorporation of a radioisotope into the DNA of dividing lymphocytes. The updated TG summarises performance standards for evaluating new or modified test methods functionally and mechanistically similar to the LLNA and includes the option of a reduced LLNA approach, which could use up to 40% fewer animals.

Potential immunosuppressive effects are often identified during subchronic toxicity testing. The toxicological endpoints that indicate immunosuppressive effects on the immune system include a decrease in whole and differential blood-cell counts, changes in lymphoid organ weights, and histopathological changes in the spleen, thymus, lymph nodes, and bone marrow.

2.2.3.4 Genotoxicity as a Target Endpoint

Genotoxicity refers to adverse effects on DNA, genes, and chromosomes and the purpose of the genotoxicity tests is to detect compounds that induce genetic damage directly or indirectly by various mechanisms.[27] Genotoxicity tests should enable a hazard identification with respect to damage to DNA and the success of repair. Permanent damage to DNA in the form of gene mutations, larger scale chromosomal damage, recombination and numerical chromosome changes is generally considered to be an essential factor for causing heritable effects and in the multi-step process of malignancy, a complex process in which genetic changes may play only a part. Compounds that are positive in tests that detect these kinds of damage have the potential to be human carcinogens and/or mutagens. The tests detect mutagenic potential, aneugenicity or clastogenicity. There are *in vitro* bacterial or mammalian cell tests (see Section 2.3.1); however, *in vivo* tests are also part of the test battery for genotoxicity, *e.g.*, the *In Vivo* Mammalian Alkaline Comet Assay (OECD TG 489),[44] Mammalian Erythrocyte Micronucleus Test (OECD TG 474),[45] Mammalian Bone Marrow Chromosomal Aberration Test (OECD TG 475)[46] or the Transgenic Rodent Somatic and Germ Cell Gene Mutation Assays (OECD TG 488).[47]

An overview of toxicity endpoints and corresponding test methods used in regulatory safety assessment can also be found on the Alttox.org website at http://alttox.org/mapp/toxicity-endpoints-tests.

2.3 Data Generated by *In Vitro* Studies

The cost of animal testing along with concerns of animal welfare led to the development and validation of various alternatives to animal testing based on the three Rs strategy – reduction, replacement, refinement – introduced in 1959 by Russel and Burch.[28] The existence of tens of thousands of chemicals for which little or no data are available makes the traditional *in vivo* testing strategy impractical. Screening with *in vitro* systems in order to prioritise chemicals for further, more detailed toxicological studies, is more practical. The US Environmental Protection Agency (EPA), for example, has made a significant investment in this approach through the ToxCast research programme,[29] screening a large set of environmental chemicals using relatively inexpensive and rapid high-throughput screening (HTS) assays. The goal is to develop predictive toxicity signatures, and to use them to prioritise chemicals for further testing.[30]

Apart from screening and prioritising purposes, the alternative, new methodology approaches are an important tool to gain a better, mechanistic understanding of chemical-induced toxicity and to enhance our understanding of the hazard potential of chemicals. In particular, tests in human cell lines aim to improve the human relevance of toxicity testing.

In vitro test methods are used in regulatory risk assessment, for example, to name two different applications, in the REACH regulation or for the assessment of food additives by the US Food and Drug Administration (FDA).[31,32] In order to be accepted by the regulatory agencies, the *in vitro* model systems have to be formally scientifically validated. Examples of *in vitro* methods adopted as OECD Test Guidelines are given in the following. A list of further validated *in vitro* tests for different endpoints is available at http://alttox.org/mapp/table-of-validated-and-accepted-alternative-methods. The European Commission Tracking System for Alternative methods towards Regulatory acceptance (TSAR) (https://tsar.jrc.ec.europa.eu) tracks the progress of non-animal methods for assessment of chemicals towards acceptance as recognised test methods. The European Union Reference Laboratory for alternatives to animal testing (EURL ECVAM) Database Service on Alternative Methods to Animal Experimentation (DB-ALM, https://ecvam-dbalm.jrc.ec.europa.eu) compiles method descriptions and protocols of alternative methods in biomedical sciences and toxicology, both in research and for regulatory purposes.

2.3.1 *In Vitro* Test Methods for Genotoxicity/Mutagenicity Testing

One of the first questions addressed when evaluating the human safety of a certain compound is whether it poses any genotoxicity and carcinogenicity risks to human population and if so, what are the possible mechanisms underlying these effects. The answer to this question is first given on the basis of the results obtained from *in vitro* tests of genotoxicity/

carcinogenicity. No single test, however, is capable of detecting all relevant genotoxic mechanisms of action, thus the usual approach is to carry out a battery of *in vitro* tests and, if needed, specific *in vivo* tests (see Section 2.2.3.4) for genotoxicity. The most commonly used *in vitro* methods for genotoxicity/mutagenicity and carcinogenicity testing are discussed in the following.

2.3.1.1 Bacterial Reverse Mutation Test (Ames Test)

The Ames test (OECD TG 471)[33] is a fast and sensitive assay of the ability of a chemical compound or mixture to induce mutations in DNA. The assay is based upon the reversion of mutations in the histidine (*his*) operon in the bacterium *Salmonella typhimurium*. The *his* operon encodes enzymes required for the biosynthesis of the amino acid histidine. Strains with mutations in the *his* operon are histidine auxotrophs – they are unable to grow without added histidine. Revertants that restore the *His*$^+$ phenotype will grow on minimal medium plates without histidine. This provides a simple, sensitive selection for revertants of *his* mutants. The bacterial reverse mutation test is commonly employed as an initial screen for genotoxic activity and, in particular, for point mutation-inducing activity. An extensive database has demonstrated that many chemicals that are positive in this test also exhibit mutagenic activity in other tests. However, there are examples of mutagenic agents that are not detected by this test. The main reasons ascribed for these shortcomings are:

- Salmonella is a prokaryote, thus is not a perfect model for humans.
- The original test does not account for metabolites that are formed by in the hepatic system.
- Chemical structure of the compounds. For example, drugs that contain the nitrate moiety sometimes come back positive for Ames when they are indeed safe. Nitroglycerin is an example that gives a positive Ames yet is still used in treatments today.
- With the Ames test, very high dose concentrations are used and nitrate compounds that can potentially generate nitric oxide (NO), an important signal molecule, will give a false positive result.

2.3.1.2 In Vitro *Mammalian Chromosomal Aberration Test*

The purpose of the *in vitro* chromosomal aberration test (OECD TG 473)[34] is to identify agents that cause structural chromosome aberrations in cultured mammalian cells. A variety of cell lines, strains or primary cell cultures, including human cells, may be used (*e.g.*, Chinese hamster fibroblasts, human or other mammalian peripheral blood lymphocytes). Cell cultures are exposed to the test substance both with and without metabolic activation. Structural aberrations may be of two types – chromosome or chromatid. The majority of chemical mutagen-induced aberrations are of the chromatid type, but chromosome-type aberrations also occur. The observed endpoint – an increase in the number of polyploid cells – may indicate that the test substance has the potential to inhibit mitotic processes and to induce numerical chromosomal aberrations. When an increase in the number of cells with endoreduplicated chromosomes is observed, the potential of the test substance to inhibit cell cycle progression can be discussed. A test substance for which the results do not meet these criteria is considered non-mutagenic in this system.

2.3.1.3 In Vitro *Mammalian Cell Gene Mutation Test*

The *in vitro* mammalian cell gene mutation test using the *Hprt* and *xprt* genes (OECD TG 476)[35] can be used to detect gene mutations induced by chemical substances. Cultures of different established cell lines or cell strains can be used. In these cell lines, the most commonly-used genetic endpoints measure mutation at hypoxanthine-guanine phosphoribosyl transferase (HPRT), a transgene of xanthine-guanine phosphoribosyl transferase (XPRT) and, according to the previous guideline, at thymidine kinase (TK). Cells deficient in thymidine kinase (TK) due to the mutation $TK^{+/-}$, $TK^{-/-}$ are resistant to the cytotoxic effects of the pyrimidine analogue trifluorothymidine (TFT). Thymidine kinase proficient cells are sensitive to TFT, which causes the inhibition of cellular metabolism and halts further cell division. Thus, mutant cells are able to proliferate in the presence of TFT, whereas normal cells, which contain thymidine kinase, are not. Data should include cytotoxicity and viability determination, colony counts and mutant frequencies for the treated and control cultures. Colonies are scored using the criteria of small and large colonies on at least one concentration of the test substance (highest positive concentration) and on the negative and positive control. Mutant cells that have suffered the most extensive genetic damage have prolonged doubling times and thus form small colonies. The induction of small colony mutants has been associated with chemicals that induce gross chromosome aberration. Mutant frequency should be expressed as the number of mutant cells per number of surviving cells.

Many compounds that are positive in this test are mammalian carcinogens; however, there is not a perfect correlation between this test and carcinogenicity. Correlation is dependent on the chemical class and there is increasing evidence that there are carcinogens that are not detected by this test because they appear to act through other, non-genotoxic mechanisms or mechanisms not readily detected in these cells.

2.3.1.4 In Vitro *Mammalian Cell Micronucleus Test*

The *in vitro* micronucleus test (OECD TG 487)[36] is a genotoxicity test for the detection of micronuclei in the cytoplasm of interphase cells. Micronuclei may originate from acentric chromosome fragments (*i.e.*, lacking a centromere), or whole chromosomes that are unable to migrate to the poles during the anaphase stage of cell division. The assay detects the activity of clastogenic and aneugenic test substances in cells that have undergone cell division during or after exposure to the test substance. Positive results from the micronucleus assay indicate that the test substance induces chromosome breakage loss, in cultured mammalian cells. Negative results indicate that, under the test conditions used, the test substance does not induce chromosome breaks and/or gain or loss in cultured mammalian cells.

2.3.2 *In Vitro* Test Methods for Carcinogenicity

Cell transformation assays (CTAs) have long been proposed as *in vitro* methods for the identification of potential chemical carcinogens. These assays measure induction of phenotypic alterations characteristic of tumorigenic cells, and cells transformed *in vitro* have been shown to induce tumors when injected into immunosuppressed experimental animals. CTAs mimic some key stages of *in vivo* multi-step carcinogenesis and have been shown to have a good concordance with rodent bioassay results, detecting both genotoxic and non-genotoxic carcinogens. CTAs are currently used by the chemical, agrochemical, cosmetic and pharmaceutical industries and academia for screening purposes and to investigate basic mechanisms of carcinogenicity, but they are not widely accepted for regulatory purposes due to a number of reservations.

2.3.3 *In Vitro* and *In Chemico* Test Methods for Skin Sensitisation

For the endpoint of skin sensitisation, non-animal methods were developed and validated and adopted in an amendment of the REACH legislation Annexes in 2016 as a standard information requirement, replacing the previous *in vivo* standard Test Guidelines.

The *in vitro* and *in chemico* assays are based on mechanistic knowledge on the effect of skin sensitisation in the organism, and cover different key events in the adverse outcome pathway (AOP, see Section 2.4) of skin sensitisation from the molecular initiating event (MIE) to the adverse effect in the whole organism.[48] Therefore, these test methods should be organised in combination or together with other information.

The *in chemico* Direct Peptide Reactivity Assay (DPRA) addresses the first key event or MIE of skin sensitisation, *i.e.*, peptide reactivity (OECD TG 442C).[49] The second key event, *i.e.*, the keratinocyte induction of cytoprotective gene pathways linked to skin sensitisation, is addressed by the *in vitro* ARE-Nrf2 luciferase test methods. Currently, two methods are included in OECD TG 442D[50] – the KeratinoSens™ and the LuSens methods. Performance standards are available for validation of similar test methods.[51]

The activation of dendritic cells (DC), *i.e.*, the third key event of the skin sensitisation AOP, is covered by three *in vitro* assays: the human Cell Line Activation Test or h-CLAT method, the U937 Cell Line Activation Test or U-SENS and the Interleukin-8 Reporter Gene Assay or IL-8 Luc assay (OECD TG 442E).[52]

2.4 Mechanistic Understanding – Adverse Outcome Pathways

Effects of chemicals or their metabolites in biological systems are a result of interactions of the biologically effective dose of the toxicant with a molecular target. Chemicals can cause many types of toxic effects by a variety of mechanisms. At a molecular level, for example, a toxicant can interact with DNA or RNA. At a cellular level, it can interfere with receptor–ligand binding or bind to biomolecules, such as proteins or lipids.

A major area of a recent advance in hazard assessment and risk characterisation has been an increasingly common understanding of the concepts of "toxicity pathways", "adverse outcome pathways" (AOPs) and "modes of action" (MoAs).[39,59,60] They all capture and formally describe a mechanistic understanding of biological systems and the various ways that toxicants can interfere with them to cause adverse effects. Toxicity pathways are defined as "a sequence of intracellular events which regulate normal biology which, when sufficiently perturbed by a xenobiotic, leads to an adverse outcome at the level of the cell, and possibly the whole organism".[40] Multiple biological processes are involved, including absorption, molecular events on a cellular level, events at target tissues, inhibition or activation of toxicity pathways, adaptation to stressors and finally, at high degrees of perturbation, adverse consequences.

Improved understanding of toxicological processes, together with advances in toxicogenomics, bioinformatics, systems biology and computational toxicology, facilitate the change in the approach to regulatory toxicity testing and risk assessment. In this context, the AOP framework supports the collection, organisation and evaluation of relevant chemical, biological and toxicological data for use in human health risk assessment (see also Chapter 9). It is a framework for a knowledge-based safety assessment that relies on understanding toxicity, rather than simply observing its effects. This concept lies at the heart of collaborations and research initiatives, such as Tox21 (http://www.epa.gov/ncct/Tox21) or SEURAT-1 (http://www.seurat-1.eu).

The AOP concept is based on the mechanistic understanding of toxicological processes on different biological levels, defined as a series of key events, starting from an initial interaction of a chemical with a biological target (molecular initiating event). For example, the binding of a chemical compound to a protein in the cell causes changes at the cellular level, which lead to effects at the tissue and organ levels, and finally culminate in an adverse outcome for the individual (human health effects) or population. These key events and processes at the different levels of biological organisation are described by the mode-of-action (MoA).[41]

It has to be borne in mind that there are numerous sequential steps in producing a toxic response, and biomarkers of early changes may not reflect the critical toxicodynamic process. Some effects measured may be related to adaptive changes (*e.g.*, induction of liver enzymes after multiple administration), which are reversible after treatment cessation. In order to serve as a surrogate marker of toxic effect, the effect measured should be representative, both qualitatively and quantitatively, of the critical toxic endpoint.

Overall, the AOP concept supports the paradigm change in toxicology towards 21st Century toxicology based on a mechanistic understanding of toxic effects and perturbations of pathways[61] and moving towards targeted *in vitro* assays to support chemical hazard and risk assessment.

2.5 Data from New Approach Methodologies

Data generated with new methodologies, contributing towards increasing the speed and variety of data generated for toxicological assessments of chemicals, include high-throughput *in vitro* data.

An example is ToxCast (Toxicity Forecaster, http://www.epa.gov/ncct/toxcast), a multi-year effort launched by the US EPA in 2007 and part of the Tox21 initiative. Within the frame of ToxCast, over 4500 chemicals from a broad range of sources, including industrial and consumer products, food additives, and potentially "green" chemicals that could be safer alternatives to existing chemicals, have been evaluated. Chemicals were evaluated in over 700 high-throughput assay endpoints that cover a range of high-level cell responses and approximately 300 signaling pathways. ToxCast results are contributing to the federal agency collaboration of Toxicity Testing in the 21st Century (Tox21, http://www.epa.gov/ncct/tox21). Tox21 pools chemical research, data and screening tools from multiple federal agencies including the US EPA, NIH, National Toxicology Program/National Institute of Environmental Health Science, National Center for Advancing Translational Sciences and the US FDA. Tox21 is compiling high-throughput screening data on ten thousand chemicals, called the Tox21 10K library, about their potential to disrupt biological pathways that may result in toxicity. So far, about 70 quantitative high-throughput screening assays have been run.

Furthermore, omics technologies, such as transcriptomics, proteomics, metabolomics, *etc.*, contribute to new information on toxicological effects, in particular relating directly to humans, although they are currently rarely used in a regulatory context yet. Frameworks for their regulatory application, including Good Laboratory Practice requirements, are being discussed.[56–58] These technologies generate large volumes of data, contributing to the new big data in toxicology. For an overview of the technologies, related data resources and application examples, please refer to Chapter 7.

2.6 Toxicokinetic Data

Toxicokinetic studies are undertaken as a part of toxicological evaluation of compounds to aid better understanding of the relationship between the external exposure and the tissue level of the toxic agent at the target site, and to define the absorption, distribution, metabolism, and elimination (ADME) of the tested compound that can be used in human risk assessment. These studies also assess the effects of sex, species, and age on ADME. Metabolism studies, which are part of the toxicokinetic studies, are used to identify biotransformation products and to examine the potential for accumulation of compounds or metabolites in tissues with repeated dosing. Thus, the data collected during a toxicokinetic study focus on time profiles of parent chemical and metabolite concentrations in plasma and other tissues and can include rates of chemical absorption, metabolism, and excretion, chemical related changes in blood chemistry, bioavailability, protein binding, and depletion of cofactors.[37]

Appropriate toxicokinetic data can be derived from:

- *in vivo* studies that define the kinetics of the circulating chemical or active metabolite, under the experimental conditions in animals
- *in vivo* experimentation in human volunteers given very low doses of the chemical under evaluation
- *in vitro* measurements of critical processes, *e.g.*, enzyme activity
- using computational prediction models, developed based on available data, *e.g.*, for metabolite formation, based on chemical structures.

These data, if obtained from well-conducted and described toxicokinetic studies, provide a strong scientific foundation for estimating equivalent human dose in various risk assessment models. Furthermore, toxicokinetic data are used to understand mechanisms of toxicity and to quantify interspecies differences (*i.e.*, those between test species and humans) and variability among humans (*i.e.*, those in different subsets and individuals within the population) in the internal or target organ dose.

Another important application of the toxicokinetic data is the development of physiologically-based pharmacokinetic (PBPK) models. PBPK models can be used to predict the target organ dose, with models based on (animal) data with appropriate interspecies scaling. In cases where measured kinetic parameters are not sufficiently predictive of the anticipated exposure scenario, PBPK models are often helpful in extrapolations across routes, over large dose ranges or where tissue uptake is non-linear.[38]

Toxicokinetics data are also generated in high-throughput settings (see Chapter 11).

2.7 Sources of Toxicological Data

The practice of modern toxicology has generated and continues to generate an enormous amount of data with a variety of biological tests, with increasing speed and volume by recent high-throughput and high-content methodologies. The information for the toxicological profile of a chemical is stored in public or proprietary databases and repositories.

Proprietary data are generated by industry and are provided to regulatory bodies for the purpose of evaluating the safety of compounds for marketing. In most cases, these data are strictly confidential, unless the owner has given permission for their publication or has published them. The internal data are generally perceived to be more homogeneous than literature data. This is due to the fact that the companies are operating under standard operating procedures (*e.g.*, US FDA red book, National Toxicology Program protocols *etc.*) and results are meant to be compared to results from other compounds with a similar structure.

Examples of publicly available data resources of toxicological and bioactivity data are given in the following (see also Chapter 4 for toxicological databases and Chapter 7 for omics data resources).

2.7.1 TOXicology Data NETwork (TOXNET)

TOXNET (http://toxnet.nlm.nih.gov) is a toxicology data network maintained by the U.S. National Library of Medicine and providing databases on toxicology, hazardous chemicals, environmental health, information on special toxic effects of chemicals in humans and/or animals, and citations from the scientific literature. It includes for example the following databases:

TOXLINE (Toxicology Literature Online, http://toxnet.nlm.nih.gov/newtoxnet/toxline.htm) is a bibliographic database containing a range of citations from specialised journals and other sources. It provides references on the biochemical, pharmacological, physiological, and toxicological effects of drugs and other chemicals. Most of the bibliographic citations include abstracts and/or indexing terms and Chemical Abstract Service (CAS) Registry Numbers.

CCRIS (Chemical Carcinogenesis Research Information System, http://toxnet.nlm.nih.gov/cgi-bin/sis/htmlgen?CCRIS) is developed and maintained by the National Cancer Institute (NCI). It contains over 9000 chemical records with carcinogenicity, mutagenicity, tumour promotion, and tumour inhibition test results. Data are derived from studies cited in primary journals, current awareness tools, NCI reports, and other special sources, with test results reviewed by experts in carcinogenesis and mutagenesis. The database provides historical information from the years 1985 to 2011 and is no longer updated.

CPDB (Carcinogenic Potency Database, http://toxnet.nlm.nih.gov/cgi-bin/sis/htmlgen?CPDB.htm) provides standardised analyses of the results from 6540 chronic, long-term animal cancer tests conducted since the 1950s and reported in the general published literature or by the National Cancer Institute and the National Toxicology Program. The database covers the years 1980 to 2011 and is no longer updated.

CTD (Comparative Toxicogenomics Database, http://toxnet.nlm.nih.gov/cgi-bin/sis/htmlgen?CTD) contains manually curated data describing cross-species chemical–gene/protein interactions and chemical– and gene–disease relationships. They provide insights into the molecular mechanisms underlying variable susceptibility and environmentally influenced diseases, as well as into complex chemical–gene and protein interaction networks. CTD is updated several times a year.

GENE-TOX (Genetic Toxicology Data Bank, http://toxnet.nlm.nih.gov/cgi-bin/sis/htmlgen?GENETOX) includes genetic toxicology test results on over 3200 chemicals from the US EPA. The database covers the years 1991 to 1998 and is no longer updated.

HSDB (Hazardous Substances Data Bank) (http://toxnet.nlm.nih.gov/cgi-bin/sis/htmlgen?HSDB) provides toxicity data for over 5700 potentially hazardous chemicals. It has information on emergency handling procedures, industrial hygiene, environmental fate, human exposure, detection methods, and regulatory requirements. The data are fully referenced and reviewed by a scientific review panel.

2.7.2 Registry of Toxic Effects of Chemical Substances (RTECS)

RTECS (http://ccinfoweb.ccohs.ca/rtecs/search.html) is a database that contains critical toxicity data on more than 174 000 chemical substances, maintained by the Canadian Centre for Occupational Health and Safety (CCOHS).

2.7.3 Kyoto Encyclopedia of Genes and Genomes (KEGG)

KEGG (http://www.genome.jp/kegg/) is a large database providing genomic and molecular-level information in view of understanding high-level functions of biological systems, such as cells, the whole organism and ecosystems. KEGG is a computer representation of biological systems, built of molecular building blocks of genes and proteins (genomic information) as well as chemical substances (chemical information) that are integrated with knowledge on interactions, reactions and relation networks (systems information). The database also contains disease and drug information (health information) as perturbations to the biological system.

2.7.4 Aggregated Computational Toxicology Resource (ACToR)

ACToR[53] (https://actor.epa.gov/actor/home.xhtml) is the US EPA's online warehouse of publicly available chemical toxicity data about potential chemical risks to human health and the environment. ACToR aggregates data from thousands of public sources, including the US EPA, National Institutes of Health, US FDA, corresponding agencies in Canada, Europe, and Japan, and academic sources, on over 500 000 environmental chemicals. ACToR contains chemical structure information, physical–chemical properties, *in vitro* assay data, tabular *in vivo* data, summary toxicology calls (*e.g.*, statements that a chemical is considered to be a human carcinogen) and links to online toxicology summaries. The data from ACToR are used by the US EPA to assess the toxicity data landscape for environmental chemicals.[30] It includes data from ToxCast. ACToR has been integrated in the iCSS ToxCast Dashboard (https://actor.epa.gov/dashboard).

2.7.5 European Chemicals Agency (ECHA) Data

The European Chemicals Agency (ECHA) makes available information on the chemicals manufactured and imported in Europe, covering their hazardous properties, classification and labelling, and information on their safe use (https://echa.europa.eu/information-on-chemicals). The data include (non-confidential) information from registration dossiers submitted for the Regulations of Registration, Evaluation, Authorisation and Restriction of Chemicals (REACH), Classification, Labelling and Packaging (CLP), Biocidal Products (BPR) and Prior Informed Consent (PIC). The data are searchable by chemical name, EC or CAS number.

2.7.6 COSMOS Next Generation

The COSMOS Database (http://www.cosmostox.eu) has been created by the European Union COSMOS research project as a publicly available source of high-quality chemistry and toxicity data and continues to be a public resource as COSMOS Next Generation (https://ng.cosmosdb.eu/https://cosmosdb.eu). In 2019, it has access to over 44 000 chemical structures and includes an inventory of cosmetics ingredients comprising more than 19 000 substances and 5500 individual chemical structures. Data are available for over 12 500 toxicological studies for 1660 compounds across 27 endpoints relating to human health.[54] It includes amongst others the US

FDA PAFA dataset, donated by the Office for Food Additives Safety (OFAS) at the US FDA Center for Food Safety and Applied Nutrition (CFSAN) and the oRepeatToxDB dataset compiled from different sources and containing a total amount of 340 *in vivo* oral repeated dose toxicity studies for 228 chemicals with the full plethora of observed toxicological effects together with the corresponding sites at which the effects occurred. COSMOS NG also includes a downloadable, updated dataset for the development of Threshold of Toxicological Concern (TTC), which has been enriched with data for cosmetics ingredients.[55]

2.7.7 Data from Assessment Reports

Most of the publicly available *in vivo* toxicology data are in the form of narrative reports from which detailed tabular data may or may not have been extracted. Examples are the original NTP, IRIS, ECHA, OECD Screening Information Data Sets (SIDS) reports, or SCCS/SCCP reports.

The *National Toxicology Program (NTP)* (http://ntp.niehs.nih.gov) evaluates substances for a variety of health-related effects from short-term and/or longer-term studies, generally two-year toxicology and carcinogenesis studies, mainly in rodents. The results are published in the NTP Toxicity Report series or in peer-reviewed scientific journals, both publicly available.

The *Integrated Risk Information System (IRIS)* (http://www.epa.gov/iris/intro.htm) from the US EPA is a human health assessment programme that evaluates information on health effects that may result from exposure to various substances in the environment. Through the IRIS Programme, the US EPA provides the highest quality science-based human health assessments to support the Agency's regulatory activities. The IRIS database is prepared and maintained by the US EPA's National Center for Environmental Assessment (NCEA) within the Office of Research and Development (ORD). It contains information on more than 550 chemical substances with information on human health effects. The heart of the IRIS database is its collection of searchable documents that describe the health effects of individual substances, containing descriptive and quantitative information on the following categories:

- *Non-cancer effects*: oral reference doses and inhalation reference concentrations (RfDs and RfCs, respectively) for effects known or assumed to be produced through a non-linear (possibly threshold) mode of action. In most instances, RfDs and RfCs are developed for the non-carcinogenic effects of substances.
- *Cancer effects*: descriptors that characterise the weight of evidence for human carcinogenicity, oral slope factors, and oral and inhalation unit risks for carcinogenic effects. Where a non-linear mode of action is established, RfD and RfC values may be used.

The *European Chemicals Agency (ECHA)* makes available (non-confidential) information and toxicity data from the submitted registration dossiers (https://echa.europa.eu/information-on-chemicals).

The *European Food Safety Authority (EFSA)* makes available a summary of human, animal health and environmental hazard assessments for a number of assessed substances in EFSA's chemical hazards database OpenFoodTox (https://www.efsa.europa.eu/en/microstrategy/openfoodtox). It provides information on substance characterisation and a summary of toxicological endpoints and reference values.

When a chemical or chemical category assessment has been agreed by *OECD* member countries, several final documents are available to the public: the OECD-generated *profile* (called Screening Information Dataset (SIDS)) contains brief summaries of SIDS endpoints as well as the major conclusions of the hazard assessment. The *assessment report* (SIDS Initial Assessment Report (SIAR) or Initial Targeted Assessment Report (ITAR)) contains longer discussions of the evaluated endpoints as well as conclusions. The *dossier* contains separate summaries for each study available on the chemical or category. The dossier is usually included with the profile and assessment report, although for some assessments (*e.g.*, some that rely on national or regional programme documents), dossiers are not available.

The *Scientific Committee on Consumer Safety (SCCS)*, formerly the Scientific Committee of Consumer Products (SCCP), provides opinions on health and safety risks (chemical, biological, mechanical and other physical risks) of non-food consumer products such as cosmetic products and their ingredients, toys, textiles, clothing, personal care and household products, and services, *e.g.*, tattooing, artificial sun tanning. The reports are publicly available at (https://ec.europa.eu/health/scientific_committees/consumer_safety_en).

2.8 Quality of Data

The quality of the available data is fundamental to successful risk assessment because poor data will result in an inadequate assessment or make the most conservative assessment necessary. Furthermore, the quality and accuracy of the models developed to predict toxicity depend on the quality of the underlying data. The quality of the study, the method, the reporting of the results, and the conclusions that are drawn, must be carefully evaluated. The evaluation of data quality includes the assessment of relevance, reliability and adequacy, terms that were defined by Klimisch *et al.*[42] who proposed a systematic approach for evaluating the quality and reliability of toxicological data.

- *Relevance* covers the extent to which data and tests are appropriate for a particular hazard identification or risk characterisation.

- *Reliability* deals with evaluating the inherent quality of a test report or publication relating to preferably standardised methodology and the way the experimental procedure and results are described to give evidence of the clarity and plausibility of the findings. Reliability of data is closely linked to the reliability of the test method used to generate the data.
- *Adequacy* defines the usefulness of data for hazard/risk assessment purposes.

On the basis of these criteria, Klimisch *et al.*[42] developed a scoring system to assess the reliability of data, subdividing them into four categories:

1 *Reliable without restrictions*: "studies or data [...] generated according to generally valid and/or internationally accepted testing guidelines (preferably performed according to Good Laboratory Practice (GLP)) or in which the test parameters documented are based on a specific (national) testing guideline [...] or in which all parameters described are closely related/comparable to a guideline method."
2 *Reliable with restrictions*: "studies or data [...] (mostly not performed according to GLP), in which the test parameters documented do not totally comply with the specific testing guideline, but are sufficient to accept the data or in which investigations are described which cannot be subsumed under a testing guideline, but which are nevertheless well documented and scientifically acceptable."
3 *Not reliable*: "studies or data [...] in which there were interferences between the measuring system and the test substance or in which organisms/test systems were used which are not relevant in relation to the exposure (*e.g.*, unphysiological pathways of application) or which were carried out or generated according to a method which is not acceptable, the documentation of which is not sufficient for assessment and which is not convincing for an expert judgment."
4 *Not assignable*: "studies or data [...] which do not give sufficient experimental details and which are only listed in short abstracts or secondary literature (books, reviews, *etc.*)."

Where there is more than one study for each endpoint, the greatest weight is attached to the studies that are the most relevant and reliable (those with Klimisch scores 1 and 2) and those studies are usually used to obtain the point of departure (POD) value (NOAEL/LOAEL/BMD) used to estimate the risk that the compound in question may pose to humans. Concerning the data reliability, it should also be noted that some types of data classified as "not reliable" or those for which it is not possible to assign reliability may be used as supporting data when needed, for example in a Weight of Evidence approach.

As already mentioned, most of the data used in classic human toxicological risk assessment are from laboratory animals, and an evaluation of the relevance of these animal data to humans has to be performed also. When data are available, dose–response relationships in the animal studies (or the severity of the effect, when only a single dose has been tested) are also assessed as part of the risk assessment process. This evaluation may be aided by the use of information (when available) on the toxicokinetics and toxicodynamics of the substance in both humans and the animal species used. When the data suggest that the endpoint observed is associated with species-specific mechanisms or modes of action (*e.g.*, α2u-globuline nephropathy in male rats)[43] and if this is the only adverse effect observed, then the study has to be classified as "irrelevant" to humans. However, this conclusion should be justified using well-documented evidence.

2.9 Data Interpretation

Evaluating biological data from different sources is a constant challenge when conducting data searches and comparisons. It is generally accepted that *in vitro* data are somehow easier to compare; however, when they are obtained from more than one source, this can be an issue for comparability due to the different experimental conditions that can give rise to different responses. Regarding the interpretation of the data from *in vivo* experimental studies, a whole variety of issues exist, considering the high number of the variables that have to be taken into account, for example the species and strain used, the study design, including the route and frequency of treatment, the dose range, the duration of the study *etc.* All these variables can influence the type and the nature of the response observed in the experiments. The study design should be consistent with GLP or other guidelines, ensuring the quality and adequacy of the study. In addition, biological variability and interindividual differences have to be taken into account. Moreover, when it comes to the interpretation of the toxicological effects and phenotypes, further complications can be related to the fact that most reports from *in vivo* experiments list only those parameters that have changed under treatment when compared to the control. This creates a problem when trying to identify compounds that do not cause a specific effect so that they can be compared to those that do. The absence of evidence does not necessarily mean the absence of the effect. However, when the data are clear in showing an effect of exposure to a chemical, the results are easily accepted, and appropriate supporting studies can also confirm the possible underlying mechanism(s).

There are several steps that have to be considered for data interpretation of experimental studies:

- Comparison of the adverse changes observed with the normal control group. The controls in any experiment give a baseline of experimental normality against which all the tested groups are assessed. When treatment-related differences from controls have been detected, the assessment should seek to indicate whether these differences are due to the direct influence of the tested compound

- Determination of the presence of dose–response relationships, which is a particularly important criterion in assigning an observed difference to the treatment. The presence of the effect in similar studies or mechanistic evidence from related data would also support a relationship with the tested chemical
- Determination of the level of significance of the effects observed. Data should be analysed taking into account the three types of levels: statistical, biological and toxicological or clinical significance.

Statistical significance means that the test group is different from the controls in a numerical size and that the difference in means is large enough for the effects of variance to be overcome. A statistically significant difference, however, is not always of *biological significance*. The latter relates to a change that may be important for the animal but that is not necessarily adverse and that is probably reversible. For example, normal hepatic adaptation to treatment is often manifested with a minimal hepatocellular hypertrophy. Cessation of the treatment or exposure usually leads to disappearance of those changes at the end of the recovery period. Therefore, the biological significance does not equate to toxicological significance. *Toxicological significance* stands for changes which, if persisting, may impact the well-being or the survival of the exposed animals. Although reversibility may mean that an adverse change is not of toxicological significance, the degree of change is important. For example, the increase in both alanine and aspartate aminotransferases (ALT and AST) is a good indicator for hepatic toxicity. However, if their change is not accompanied with morphological changes, *e.g.*, necrosis or centrilobular hypertrophy, the effect is of biological but not of toxicological significance.

This interpretation of experimental observations, to produce the final outcome as toxicological data to be recorded, is an expert judgement process that is difficult to automate, as compared to the generation of large-scale *in vitro* or omics data, which relies on automatic data analysis approaches (see Chapters 7 and 8).

2.10 Use of Biological Data for Predictive Toxicology

Overall, the types of biological data generated and recorded in view of toxicological assessment of chemicals are very diverse, including, amongst others:

- quantitative dose descriptors: $LC5_0/LD_{50}$; NO(A)EL/LO(A)EL; RfDs/RfCs
- qualitative description of the severity and potency of the effect (*e.g.*, skin/eye irritation, skin sensitisation)
- narrative description (*e.g.*, of histopathological changes)
- binary yes/no statements
- hazard classifications

- dose–response relationships
- assay results: *in chemico*, *in vitro*, cell assays, 3D organoids *etc*
- omics and gene array read-outs
- statistical *vs.* time-resolved recordings (kinetics).

Classic toxicological data are generated "slowly", but have moved into the big data era with high-throughput and high-content data, the new technologies accelerating the volume, speed of generation and, even more, the variety of toxicological data. Especially, the variety is one of the big data characteristics that are a challenge for toxicological database storage and data integration (see Chapters 4 and 5), in particular when linked to the chemical structure of the compounds to be assessed toxicologically, processing chemical structures being a chemo-informatics challenge as such (see Chapter 3).

Generally, these biological data can be used in the context of

- classic toxicological hazard and risk assessment
- *e.g.*, fulfilling regulatory information requirements, which can be different for different legislations
- elucidation of toxicological mechanisms; once these are known, targeted assays can be used to replace classic tests for specific endpoints
- use the data generated for one substance to extrapolate to similar substances without available data (read-across)
- development of computational models to predict the outcome of the tests and the hazard of chemicals, for example (quantitative) structure–activity relationships ((Q)SARs).

Examples of tests and assays for the different endpoints of genetic toxicity, the respective availability of data and related database sources in view of use for model development, as well as already existing predictive models for some of the endpoints, have been reviewed in detail and are summarised in Hasselgren *et al.*[62]

The predictive models are also stored in repositories, as well as the actual prediction results, which are compiled in databases for a number of chemical compounds, for reference and retrieval (see Chapter 6).

References

1. R. C. James, D. A. Warren, N. C. Halmes and S. M. Roberts, in *Principles of Toxicology: Environmental and Industrial Applications*, ed. P. L. Williams, R. C. James and S. M. Roberts, John Wiley & Sons INC, New York, 2nd edn, 2000, pp. 437–477.
2. OECD (Organisation for Economic Cooperation and Development), *Acute Oral Toxicity – Fixed Dose Procedure*, Chemicals Testing Guidelines No. 420, Paris, December 17, 2001.

3. OECD (Organisation for Economic Cooperation and Development), *Acute Oral Toxicity – Acute Toxic Class Method*, Chemicals Testing Guidelines No. 423, Paris, December 17, 2001.
4. OECD (Organisation for Economic Cooperation and Development), *Acute Oral Toxicity: Up-and-Down Procedure*, Chemicals Testing Guidelines No. 425, Paris, Organisation for Economic Cooperation and Development, October 3, 2008.
5. United Nations - Economic Commission for Europe (UN/ECE), *Globally Harmonised System of Classification and Labelling of Chemicals (GHS)*, UN, New York and Geneva, 2013.
6. OECD (Organisation for Economic Cooperation and Development), *Repeated Dose 28-day Oral Toxicity Study in Rodents*, Chemicals Testing Guidelines No. 407, Paris, October 3, 2008.
7. OECD (Organisation for Economic Cooperation and Development), *Repeated Dose 90-Day Oral Toxicity Study in Rodents*, Chemicals Testing Guidelines No. 408, Paris, June 25, 2018.
8. OECD (Organisation for Economic Cooperation and Development), *Repeated Dose 90-Day Oral Toxicity Study in Non-rodents*, Chemicals Testing Guidelines No. 409, Paris, September 21, 1998.
9. OECD (Organisation for Economic Cooperation and Development), *Repeated Dose Dermal Toxicity: 21/28-day Study*, Chemicals Testing Guidelines No. 410, Paris, May 12, 1981.
10. OECD (Organisation for Economic Cooperation and Development), *Subchronic Dermal Toxicity: 90-day Study*, Chemicals Testing Guidelines No. 411, Paris, May 12, 1981.
11. OECD (Organisation for Economic Cooperation and Development), *Subacute Inhalation Toxicity: 28-Day Study*, Chemicals Testing Guidelines No. 412, Paris, June 25, 2018.
12. OECD (Organisation for Economic Cooperation and Development), *Subchronic Inhalation Toxicity: 90-day Study*, Chemicals Testing Guidelines No. 413, Paris, June 25, 2018.
13. OECD (Organisation for Economic Cooperation and Development), *Chronic Toxicity Studies*, Chemicals Testing Guidelines No. 452, Paris, June 25, 2018.
14. OECD (Organisation for Economic Cooperation and Development), *Combined Chronic Toxicity/Carcinogenicity Studies*, Chemicals Testing Guidelines No. 453, Organisation for Economic Cooperation and Development, Paris, June 25, 2018.
15. C. Therapontos, L. Erskine, E. R. Gardner, W. D Figg and N. Vargesson, Thalidomide induces limb defects by preventing angiogenic outgrowth during early limb formation, *Proc. Natl. Acad. Sci. U. S. A.*, 2009, **106**(21), 8573–8578.
16. OECD (Organisation for Economic Cooperation and Development), *Prenatal Development Toxicity Study*, Chemicals Testing Guidelines No. 414, Paris, June 25, 2018.

17. OECD (Organisation for Economic Cooperation and Development), *Reproduction/Developmental Toxicity Screening Test*, Chemicals Testing Guidelines No. 421, Paris, July 29, 2016.
18. OECD (Organisation for Economic Cooperation and Development), *Combined Repeated Dose Toxicity Study with the Reproduction/Developmental Toxicity Screening Test*, Chemicals Testing Guidelines No. 422, Paris, July 29, 2016.
19. OECD (Organisation for Economic Cooperation and Development), *Two-Generation Reproduction Toxicity*, Chemicals Testing Guidelines No. 416, Paris, January 22, 2001.
20. OECD (Organisation for Economic Cooperation and Development), *Extended One-Generation Reproductive Toxicity Study*, Chemicals Testing Guidelines No. 443, Paris, Organisation for Economic Cooperation and Development, June 25, 2018.
21. OECD (Organisation for Economic Cooperation and Development), *Developmental Neurotoxicity Study*, Chemicals Testing Guidelines No. 426, Paris, Organisation for Economic Cooperation and Development, October 16, 2007.
22. OECD (Organisation for Economic Cooperation and Development), *Neurotoxicity Study in Rodents*, Chemicals Testing Guidelines No. 424, Paris, July 21, 1997.
23. R. Luebke, Immunotoxicant Screening and Prioritization in the Twenty-first Century, *Toxicol. Pathol.*, 2012, **40**, 294–299.
24. OECD (Organisation for Economic Cooperation and Development), *Skin Sensitisation*, Chemicals Testing Guidelines No. 406, Paris, July 17, 1992.
25. OECD (Organisation for Economic Cooperation and Development), *Skin Sensitisation : Local Lymph Node Assay*, Chemicals Testing Guidelines No. 429, Paris, July 22, 2010.
26. G. F Gerberick, C. A. Ryan, R. J. Dearman and I. Kimber, Local lymph node assay (LLNA) for detection of sensitization capacity of chemicals, *Methods*, 2007, **41**, 54–60.
27. European Medicines Agency (EMA), ICH Topic S 2B. Genotoxicity: a standard battery for genotoxicity testing of pharmaceuticals. March 1998; CPMP/ICH/174/95.
28. W. M. S. Russell and R. L Burch, *The Principles of Humane Experimental Technique*, Methuen, London, 1959. Reprinted by UFAW, 1992: England. ISBN 0 900767 78 2.
29. D. J. Dix, K. A. Houck, M. T. Martin, A. M. Richard, R. W. Setzer and R. J. Kavlock, The ToxCast program for prioritizing toxicity testing of environmental chemicals, *Toxicol. Sci.*, 2007, **95**, 5–12.
30. R. Judson, A. Richard, D. J. Dix, K. Houck, M. Martin, R. Kavlock, V. Dellarco, T. Henry, T. Holderman, P. Sayre, S. Tan, T. Carpenter and E. Smith, The Toxicity Data Landscape for Environmental Chemicals, *Environ. Health Perspect.*, 2009, **117**, 685–695.

31. Food and Drug Administration (FDA). 2006, Select Committee on GRAS Substances (SCOGS) Database Overview, Available http://www.cfsan.fda.gov/~dms/opascogs.html, accessed 8 August 2008.
32. Food and Drug Administration (FDA), 2007, Cumulative Estimated Daily Intake/Acceptable Daily Intake Database, Available http://www.cfsan.fda.gov/~dms/opa-edi.html, accessed 8 August 2008.
33. OECD (Organisation for Economic Cooperation and Development), Chemicals Testing Guidelines No. 471, Paris, July 21, 1997.
34. OECD (Organisation for Economic Cooperation and Development), Chemicals Testing Guidelines No. 473, Paris, July 29, 2016.
35. OECD (Organisation for Economic Cooperation and Development), *In Vitro Mammalian Cell Gene Mutation Tests Using the Hprt and Xprt Genes*, Chemicals Testing Guidelines No. 476, Paris, July 29, 2016.
36. OECD (Organisation for Economic Cooperation and Development), *In Vitro Mammalian Cell Micronucleus Test*, Chemicals Testing Guidelines No. 487, Paris, July 29, 2016.
37. J. R. Buchanan, L. T. Burka and R. L. Melnick, Purpose and guidelines for toxicokinetic studies within the National Toxicology Program, *Environ. Health Perspect.*, 1997, **105**, 468–471.
38. WHO/PCS Guidance Document for the Use of Data in Development of Chemical-Specific Adjustment Factors (CSAFs) for Interspecies Differences and Human Variability in Dose/Concentration–Response Assessment, July 2001, Available at http://www.who.int/ipcs/publications/methods/harmonization/en/csafs_guidance_doc.pdf.
39. M. T. D. Cronin and A. N. Richarz, Mode of Action Working Group: Use of Mode of Action Related to Repeated Dose Systemic Toxicity – A Framework for Capturing Information, in *Towards the Replacement of in Vivo Repeated Dose Systemic Toxicity Testing*, ed. T. Gocht and M. Schwarz, Imprimerie Mouzet, France, 2012, vol. 2, pp. 284–289.
40. M. Whelan and M. Andersen, *Toxicity Pathways from Concepts to Application in Chemical Safety Assessment*, JRC Scientific and policy reports EUR 26389 EN, 2013, , DOI: 10.2788/49626.
41. OECD (Organisation for Economic Co-operation and Development), *Revised Guidance Document on Developing and Assessing Adverse Outcome Pathways*, Series on Testing and Assessment No. 184, OECD, Paris, France, 2017.
42. H. J. Klimisch, M. U. Andreae and U. Tillmann, A systematic approach for evaluating the quality of experimental toxicological and ecotoxicological data, *Regul. Toxicol. Pharmacol.*, 1997, **25**, 1–5.
43. A. M. Doi, G. Hill, J. Seely, J. R. Hailey, G. Kissling and J. R. Bucher, alpha 2u-globulin nephropathy and renal tumors in national toxicology program studies, *Toxicol. Pathol.*, 2007, **35**, 533–540.
44. OECD (Organisation for Economic Co-operation and Development), *Test No. 489: In Vivo Mammalian Alkaline Comet Assay*, OECD Guidelines for the Testing of Chemicals, Section 4, OECD Publishing, Paris, 2016.

45. OECD (Organisation for Economic Co-operation and Development), *Test No. 474: Mammalian Erythrocyte Micronucleus Test*, OECD Guidelines for the Testing of Chemicals, Section 4, OECD Publishing, Paris, 2016.
46. OECD (Organisation for Economic Co-operation and Development), *Test No. 475: Mammalian Bone Marrow Chromosomal Aberration Test*, OECD Guidelines for the Testing of Chemicals, Section 4, OECD Publishing, Paris, 2016.
47. OECD (Organisation for Economic Co-operation and Development), *Test No. 488: Transgenic Rodent Somatic and Germ Cell Gene Mutation Assays*, OECD Guidelines for the Testing of Chemicals, Section 4, OECD Publishing, Paris, 2013.
48. OECD (Organisation for Economic Co-operation and Development), *The Adverse Outcome Pathway for Skin Sensitisation Initiated by Covalent Binding to Proteins. Part 1: Scientific Evidence*, Series on Testing and Assessment No. 16, OECD: Paris, France, 2012.
49. OECD (Organisation for Economic Co-operation and Development), *Test No. 442C: In Chemico Skin Sensitisation: Assays Addressing the Adverse Outcome Pathway Key Event on Covalent Binding to Proteins*, OECD Guidelines for the Testing of Chemicals, Section 4, OECD Publishing, Paris, 2019.
50. OECD (Organisation for Economic Co-operation and Development), *Test No. 442D: In Vitro Skin Sensitisation: ARE-Nrf2 Luciferase Test Method*, OECD Guidelines for the Testing of Chemicals, Section 4, OECD Publishing, Paris, 2018.
51. OECD (Organisation for Economic Co-operation and Development), *Guidance Document No 213. Performance Standards for Assessment of Proposed Similar or Modified in Vitro Skin Sensitisation ARE-NrF2 Luciferase Test Methods*, Series on Testing and Assessment, Paris, 2015.
52. OECD (Organisation for Economic Co-operation and Development), *Test No. 442E: In Vitro Skin Sensitisation: In Vitro Skin Sensitisation Assays Addressing the Key Event on Activation of Dendritic Cells on the Adverse Outcome Pathway for Skin Sensitisation*, OECD Guidelines for the Testing of Chemicals, Section 4, OECD Publishing, Paris, 2018.
53. R. S. Judson, *et al.*, Aggregating Data for Computational Toxicology Applications: The U.S., "Environmental Protection Agency (EPA) Aggregated Computational Toxicology Resource (ACToR) System", *Int. J. Mol. Sci.*, 2012, **13**(2), 1805–1831.
54. C. Yang, D. Hristozov, A. Tarkhov, T. Kleinöder, I. Boyer, M. Cronin, E. Fioravanzo, H. Kim, B. Heldreth, A. Mostrag-Szylchtying, J. Rathman, A. Richarz, C. Schwab, V. Vitcheva and A. Worth, COSMOS DB as an international share point for exchanging regulatory and toxicity data of cosmetics ingredients and related substances, *Toxicol. Lett.*, 2015, **238**(Suppl.), S382.
55. C. Yang, S. M. Barlow, K. L. Muldoon Jacobs, V. Vitcheva, A. R. Boobis, S. P. Felter, K. B. Arvidson, D. Keller, M. T. D. Cronin, S. Enoch, A. Worth

and H. M. Hollnagel, Thresholds of Toxicological Concern for cosmetics-related substances: New database, thresholds, and enrichment of chemical space, *Food Chem. Toxicol.*, 2017, **109**, 170–193.
56. R. Buesen, *et al.*, Applying 'omics technologies in chemicals risk assessment: Report of an ECETOC workshop, *Regul. Toxicol. Pharmacol.*, 2017, **91**(Suppl. 1), S3–S13.
57. J. Bridges, U. G. Sauer, R. Buesen, L. Deferme, K. E. Tollefsen, T. Tralau, B. van Ravenzwaay, A. Poole and M. Pemberton, Framework for the quantitative weight-of-evidence analysis of 'omics data for regulatory purposes, *Regul. Toxicol. Pharmacol.*, 2017, **91**(Suppl. 1), S46–S60.
58. H.-M. Kauffmann, *et al.*, Framework for the quality assurance of 'omics technologies considering GLP requirements, *Regul. Toxicol. Pharmacol.*, 2017, **91**(Suppl. 1), S27–S35.
59. G. T. Ankley, R. S. Bennett, R. J. Erickson, D. J. Hoff, M. W. Hornung, R. D. Johnson, D. R. Mount, J. W. Nichols, C. L. Russom, P. K. Schmieder, J. A. Serrrano, J. E. Tietge and D. L. Villeneuve, Adverse outcome pathways: a conceptual framework to support ecotoxicology research and risk assessment, *Environ. Toxicol. Chem.*, 2010, **29**(3), 730–741.
60. G. T. Ankley and S. W. Edwards, The Adverse Outcome Pathway: A Multifaceted Framework Supporting 21st Century Toxicology, *Curr. Opin. Toxicol.*, 2018, **9**, 1–7.
61. D. Krewski, *et al.*, Toxicity Testing in the 21st Century: A Vision and a Strategy, *J. Toxicol. Environ. Health, Part B*, 2010, **13**(2–4), 51–138.
62. C. Hasselgren, *et al.*, Genetic toxicology *in silico* protocol, *Regul. Toxicol. Pharmacol.*, 2019, **107**, 104403, 1–21.

CHAPTER 3

Chemoinformatics Representation of Chemical Structures – A Milestone for Successful Big Data Modelling in Predictive Toxicology

NIKOLAY KOCHEV*[a], NINA JELIAZKOVA[b] AND IVANKA TSAKOVSKA[c]

[a]Department of Analytical Chemistry and Computer Chemistry, University of Plovdiv, 24 Tzar Assen Str., 4000 Plovdiv, Bulgaria; [b]Ideaconsult Ltd, 4 A. Kanchev Str., Sofia 1000, Bulgaria; [c]Institute of Biophysics and Biomedical Engineering – BAS, Acad. G. Bonchev Str., Bl.105, 1113 Sofia, Bulgaria
*E-mail: nick@uni-plovdiv.net

3.1 Introduction to Chemoinformatics Approaches for Chemical Structure Representation

Chemoinformatics[1,2] is a well-established interdisciplinary field that supplies tools for solving problems connected to studied chemical objects by means of informatics approaches. The term "chemoinformatics" emerged at the end of the 90s within the context of pharmaceutical industry activities.

Issues in Toxicology No. 41
Big Data in Predictive Toxicology
Edited by Daniel Neagu and Andrea-Nicole Richarz
© The Royal Society of Chemistry 2020
Published by the Royal Society of Chemistry, www.rsc.org

It is a successor of the term "mathematical chemistry", which has a strong focus on graph theory applications on the molecule structure, and developed rapidly in the 80s and 90s. Chemoinformatics also includes the rich inheritance of the chemometrics[3] field and a variety of data mining approaches with a focus on the chemical structure as well as all methods for handling 3D chemical structures and large databases. In this regard, the chemoinformatics tools are essential for the efficient handling of large data collections used in computational toxicology and medicinal chemistry. Chemoinformatics approaches obey the logic of the basic information chain: data → information → knowledge. The data associated with the studied objects is mined to gain chemical information and, consequently, the chemical information is used to generate chemical knowledge formalised as models, relations, classifiers, *etc.* The chemical knowledge is used to make efficient decisions with regards to the toxicity of the studied chemical objects by academic and industry researchers as well as regulatory agency officers. The chemical structures play a central role in these methodologies.

A rational drug design approach is deployed on top of the chemical information chain where molecule properties such as physicochemical, biological activities and toxicity are modelled as functions of the structure. Rational predictions based on (quantitative) structure–activity relationship ((Q)SAR) modelling[4] have expanded to big data processing, which can be illustrated as searching a needle in a haystack.[5] The molecules with desired properties are searched for in a vast chemical space. A portion of it, called "drug-like molecules subspace", is estimated to be of a size of 10^{60}–10^{200} molecules. Currently, this trend has developed towards big information integration of large chemical, biological and genomics databases, knowledge bases and information resources that require application of chemoinformatics and chemogenomics approaches.

An adequate representation/formalisation of the structures is required for each level of the information chain, in order to handle the chemical information with computerised tools. There is a powerful principle in chemoinformatics: the adequateness of structure information treatment determines the effectiveness of chemical modelling.[1] Typically, the physicochemical or biological properties are considered as functions (models) of the structure of the molecule. On the other hand, the structure representation itself can also be considered as a model by which aspects of the reality (theoretical and experimental) are represented. The structure formalisation is the starting point (*i.e.*, the basic model) for any other activities and is reductionistic by its nature because only particular aspects of the chemical reality are formalised. The structure representation methods reflect certain chemistry models or theories and are not necessarily compatible with each other. This incompatibility poses challenges for converting chemical objects from one format to another, since the background chemistry paradigm may differ to some extent. The major challenges concerning structure representations are discussed in Section 3.8.

Figure 3.1 shows the relationships between several types of widely used chemical objects.

The connection table[2] (CT) plays a central role in all chemoinformatics applications. This is the basic topological representation of the chemical structure and is described in detail in Section 3.4. In many chemoinformatics systems, the connection table is considered as the basic part of the molecule structure (sometimes considered as the "full" structure when only topology is taken into account) and, typically, together with 2D coordinates and 3D coordinates is applied to formalise the chemical structure. The information of the molecule (chemical compound) includes the chemical structure as well as properties and descriptors. The sets of molecules, or ensembles, are used to represent combinatorial libraries, structure data sets such as tautomers and conformers, chemical reactions as well as the substances that include an ensemble of chemical objects to form their composition. The chemical substance representation is described in detail in Section 3.9.

The Markush structure[6] is a representation introduced in order to search chemical structure patents and simultaneously describes several chemical structures (see Figure 3.2). Typically, these special types of chemical objects are not valid structures as they may represent a fragment, search pattern defined by logical expressions, a construction of macro "atoms" where an atom represents a particular fragment, or Markush structures with

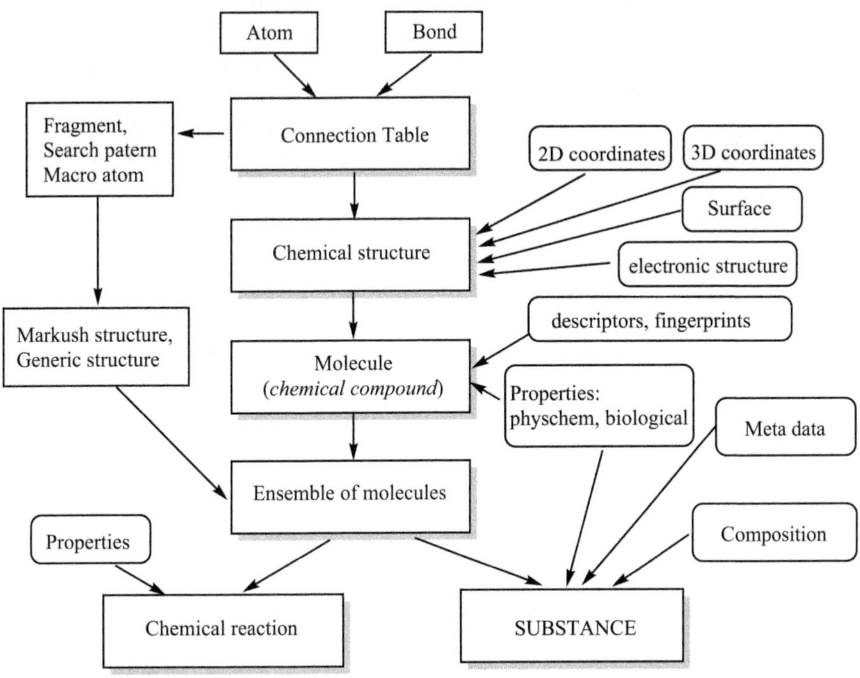

Figure 3.1 Relationships between chemical objects.

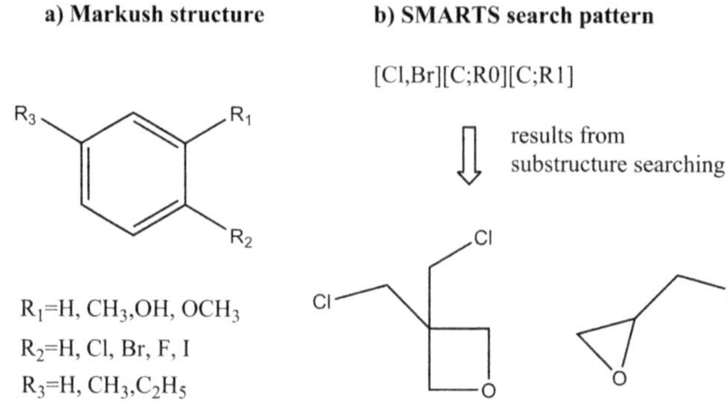

Figure 3.2 Special types of chemical objects: (a) Markush structure; (b) substructure search pattern defined by SMARTS and example of the result from a search in the Ambit-LRI data base (https://ambitlri.ideaconsult.net/tool/ui/_search).

variable parts in them. The search patterns and the fragments generally are not saturated and contain free valences. In some cases, the latter may look like valid structures but they are, rather, valid portions (fragments) of a structure and therefore must not be mistaken for the chemical objects that represent real molecules. The chemoinformatics solutions for handling these special chemical objects are sophisticated.[1]

The Markush structure consists of a constant part and a variable part defined by means of different types of variable components: R-groups, variable positions, variable group frequency and homology variation. The first type of variation (the R-groups) is the most often used and is easy to implement in various file formats and linear notations. An example of a Markush structure with 3 R-groups is shown in Figure 3.2a. Group R_1 includes four possible substituents, group R_2 defines five options and group R_3 defines four options, respectively. The number of all possible structures defined by this Markush structure is the number of all combinations of the substituent options for groups R_1, R_2 and R_3, *i.e.*, $4*5*4 = 80$ structures. Among these structures are the molecule of benzene (R_1 = H, R_2 = H, R_3 = H), the molecule of toluene (R_1 = CH3, R_2 = H, R_3 = H), the molecule of phenol (R_1 = OH, R_2 = H, R_3 = H), *etc.* The search pattern defined by a SMARTS[7] (SMiles ARbitrary Target Specification) string is given in Figure 3.2b. This special chemical object represents a fragment of three atoms: a halogen atom (Cl or Br) and two carbons atoms, one carbon being part of a cycle and the other carbon must not be in a cycle. Figure 3.2b also shows two molecules from the Ambit-LRI database[8] that comply with the search pattern conditions. Markush structures and search patterns can be used to define a domain of the chemical space where a particular QSAR method is applied for toxicity prediction as well as to define a set of similar compounds where the structural similarity principle is used for toxicity prediction.

Considering the hierarchy from Figure 3.1, the objects with significant practical importance are the chemical structure and the molecule. The classical model of the molecule can be defined as a triad consisting of a chemical structure, descriptors and properties (see Figure 3.3). Typically, a QSAR/QSPR (quantitative structure–activity relationship/quantitative structure–property relationship) model for toxicity or other endpoints can be described in terms of this classical triad, *i.e.*, the toxicity (which is a property of the molecule) is considered as a function of the structure: $P = f(S)$. In most cases, modelling is performed using molecular descriptors that are considered as a function of the structure *i.e.*, $P = f_1(D)$ and $D = f_2(S)$.

The majority of public chemical database implementations is based on the classical structure paradigm – the molecule triad. This paradigm is sufficient generally for scientific research purposes, providing a platform for virtual screening and modelling of the properties of the studied molecules. This classical structure paradigm has been used for several decades and it is still nowadays the predominant base layer for the public chemical databases. However, this approach is too restrictive for many challenging cases within the scope of chemical information management by academia, industry and regulatory bodies. The substance paradigm (Figure 3.3) is a more sophisticated data model used to represent adequately the studied chemical objects. The substance includes not a single structure but a composition that can be defined by many structures playing various roles: major constituent, additives, impurities, *etc.* Further, the substance object includes metadata concerning the performed experiments and measurements. A more accurate composition representation and rich metadata support is needed for the efficient treatment of the data coming from many different biological assays including toxicity experiments. The substance presentation paradigm is discussed in Section 3.9.

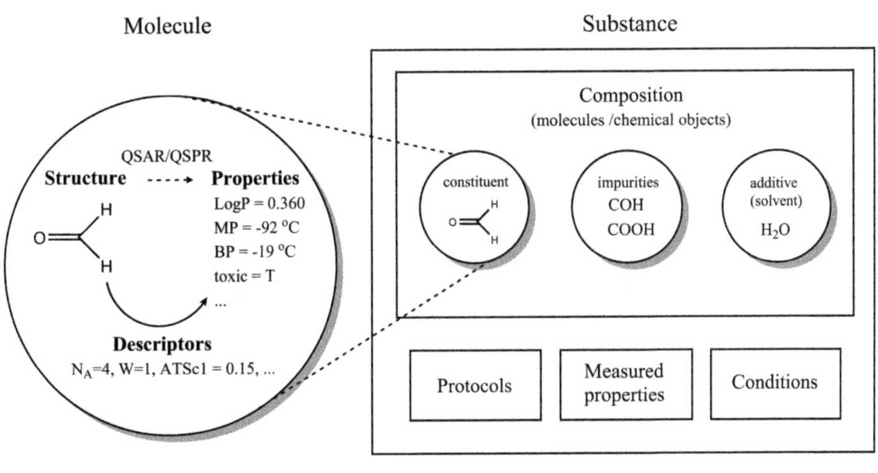

Figure 3.3 Classical structure paradigm *vs.* substance paradigm for the molecule of formaldehyde.

3.2 Structure Representation Characteristics

The structure representation is a key element for the manipulation of big chemical data and the implementation of efficient chemoinformatics software tools. In order to systemise the various methods for structure representation, each structural representation is considered and characterised in several main aspects:[1]

(i) technique for representation (chemical information formalisation method),
(ii) level (dimension) of structure representation,
(iii) area of application, and
(iv) H atom management.

Generally, the third aspect (the area of application) is determined on the basis of characteristics (i) and (ii). On the other hand, the structural representation technique and level can be adjusted specifically for a particular task according to the domain of application.

In Figure 3.4, the three basic characteristics (technique, level and application) are extended with one additional characteristic – the treatment of H atoms, depicted as a dashed arrow. The first three characteristics will be

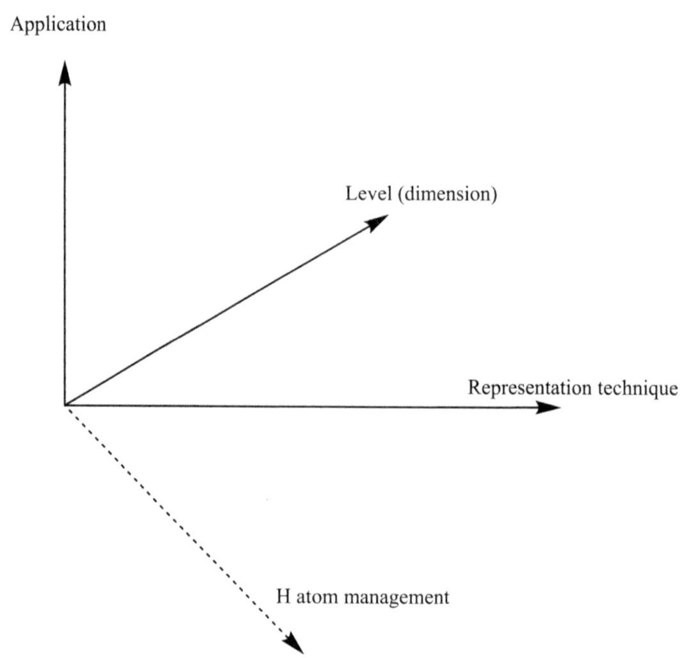

Figure 3.4 Structure representation basic characteristics.

described in detail in later sections. The treatment of the H atoms deserves special attention. The large number of structures in the "haystack" of the chemical space is due to the combinatorial explosion of the organic structure isomers; thus, the organic compounds are the major part of the modern big databases. About 60–70% percent of the atoms of the organic compounds are hydrogen atoms. That is why it is important to have an efficient way for handling the H atoms' information. There are two basic approaches for handling H atoms in a structure:

(i) *Explicit H atoms* – the H atoms are treated as the other atoms and are numbered explicitly. An example for explicit H atoms is given in the 3D structure of Figure 3.5 and in the connection table of Figure 3.8a.
(ii) *Implicit H atoms* – the H atoms are described as attributes to the other atoms. H atoms are treated by "default" and are not numbered. An example for implicit H atoms is given in the connection table of Figure 3.5 as well as the redundant connection table of Figure 3.8b.

Typically, the approach based on implicit H atoms reduces the size of the representation data object by about 60%, which is why it is a widely adopted approach for topological representations, especially in large chemical databases. Also, many of the topological descriptors are defined on H depleted chemical graphs, *i.e.*, they are based on structure representations with implicit H atoms.

Figure 3.5 shows various structure representations for the molecule of proline (pyrrolidine-2-carboxylic acid) grouped on several levels: 0D/1D (constitution), 2D (topology), 3D geometry, 4D conformations and some specialised techniques. All these representation levels are discussed in detail in the following sections.

A structure representation is unambiguous if the chemical object can be reconstructed exactly from it. Typically, most techniques for structure representations on topological (2D) and geometrical (3D) levels are unambiguous. The molecular formula is an example of ambiguous structure representation on a constitutional level. For example, the formula C_5H_{12} represents the molecule of pentane; however, the three possible isomers (normal pentane, isopentane and neopentane) are not defined. A given structure representation is unique if the algorithm technique guarantees the same result each time. Typically, most techniques are based on atom numbering (*e.g.*, connection table, see Figure 3.5). Unique representations are obtained by means of canonicalisation rules that guarantee the unique ordering of the atoms (*e.g.*, Morgan algorithm[9]). Unique representations are used as a lookup key for fast chemical object identification in large databases.

0D/1D

IUPAC nomenclature

pyrrolidine-2-carboxylic acid
proline

Constitutional descriptors 0D

$N_{Atoms}=8$ $N_{O}=2$
$N_{Bonds}=8$ $N_{N}=1$
$N_{C}=5$
$N_{O,Rel}=0.25$

Constitutional descriptors 1D

$N_{OH}=1$
$N_{C=O}=1$
$N_{CC(C)C}=1$
$N_{Rings}=1$

2D

Connection table

Atoms		Bonds		
1	O	1	2	2
2	C	2	3	1
3	O	2	4	1
4	C	4	5	1
5	C	4	8	1
6	C	5	6	1
7	C	6	7	1
8	N	7	8	1

2D diagram

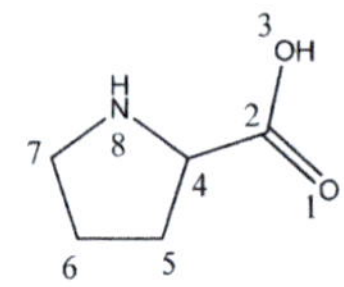

Linear notations

SMILES:
OC(=O)C1CCCN1

InChI: 1S/C5H9NO2/c7-5(8)4-2-1-3-6-4/h4,6H,1-3H2,(H,7,8)

Topological descriptors
W = 62
Zagreb = 36

3D

3D structure geometry

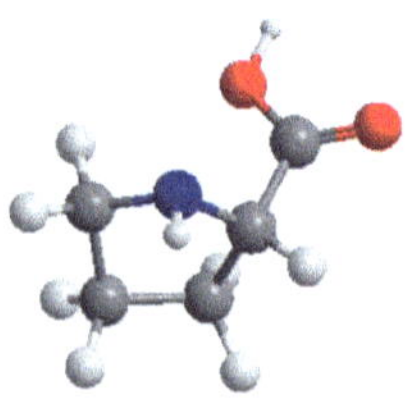

3D surface

Geometric descriptors

PPSA = 273.60
PNSA = 124.39

4D

Conformations

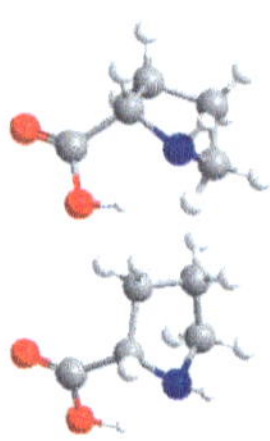

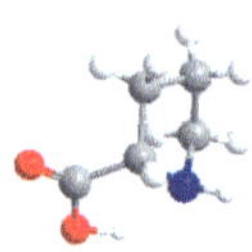

...

Geometric descriptor mean over all conformers

$\mathbf{PPSA}_{mean}$ = 234.67
$\mathbf{PNSA}_{mean}$ = 117.06

Registry numbers
CAS# 609-36-9
ChEMBL72275
ChemSpider 594
DrugBank DB00172

Hash-codes
InChI key:
ONIBWKKTOPOVIA-UHFFFAOYSA-N

Fingerprints
MACCS:
0000000000000000000001000
0000000000000000000000000
00001000000000000000000
000000001100000011000110
0010001000000100000001001
0100000101000000101000000
010001011011110100110

Figure 3.5 Structure representation of the molecule of proline by means of various techniques on levels 0D–4D.

3.3 Constitutional Representations

The representations on a constitutional level describe what components a given structure consists of and how many of them are present. Constitutional levels are formally labeled as 0D and 1D (the distinction between 0D and 1D is not always done). The topological representation at 2D level, described in detail in the next section, is the basic level for structure formalisation and can be used as a comparison point. Level 0D structure representation means that no topology is described. Accordingly, 1D level describes partial topology whereas the full topological information is described at 2D level. From the information point of view, 0D and 1D levels are below the topological information content (2D) whereas 3D and 4D levels are beyond the topology – they describe much richer information content of the molecule geometry and conformations.

The constitutional representations are typically descriptors[10] of the type N_X, *i.e.*, this is the number of occurrences of a substructure, a fragment or feature of type X. Depending on the complexity of X, the descriptor may by 0D or 1D. For example, the molecular formula is composed of several constitutional descriptors, N_C, N_H, N_O, *etc.*, which are typical 0D descriptors. Group counts for a predefined list of structure fragments and functional groups are widely used 1D descriptors and they are the basis for what is known as additive schemas[4] (or group contribution methods for modelling). The groups/fragments can also be defined in a more generic way as topological patterns where all possible patterns are searched exhaustively. Popular patterns are B-groups of type A ~ B, G-groups of type A ~ B ~ C, topological torsions of type A ~ B ~ C ~ D, D-groups of type A ~ B(~C) ~ D. The group counts for the structure of proline are given in Figure 3.6.

The constitutional descriptors may describe more generic counts, *e.g.*, N_{Atoms}, N_{Bonds} and N_{Rings}. Also, any arithmetic combination of constitutional descriptors defines another constitutional descriptor, *e.g.*, the relative number of oxygen atoms: $N_{O,rel} = N_O/N_{Atoms}$. In this sense, the molecular weight (MW) is also a constitutional descriptor. The constitutional

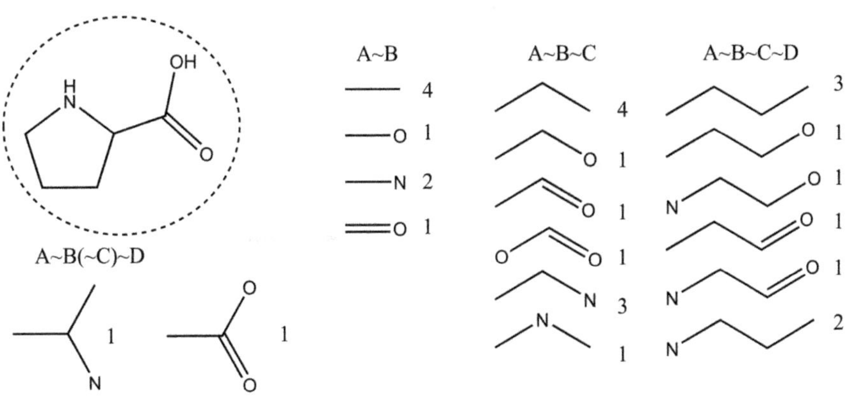

Figure 3.6 Various group patterns for the structure of proline with their counts.

descriptors are generally easy to apply for software implementation and for interpretation and though they are not so informationally rich, they are quite often used for QSAR modelling and virtual screening. The constitutional descriptors are calculated very promptly, require minimal storage memory space and are efficient enough for similarity principle application and to define simple rules for screening (fast rejection of undesired chemical objects).

3.4 Topological Representations

The topological representations are handled by the methods of graph theory,[11] which is a well-established mathematical field with various applications in informatics, chemistry, biology, *etc.* In fact, the major catalyst for the development of chemoinformatics in the 80s and 90s was the active involvement of graph theory for structure representation, structure manipulation, molecular descriptors, structure searching and modelling. A graph[11] G[E,V] is defined as a set of vertices (or nodes) and a set of edges (connections) that describe the interrelations of the graph vertices. Generally, any system or object that is constructed by components (sub-objects) having definite relations with each other can be represented by a graph. Graph-theoretical approaches are used for various chemoinformatics tasks where the logic behind many algorithms is to search a solution in a space of possible solutions represented as nodes of a graph; the solution searching is done by traversing the graph. One of the fundamental applications of graph theory is topological representation of chemical structures. An example of a graph representation for the molecule of proline is given in Figure 3.7.

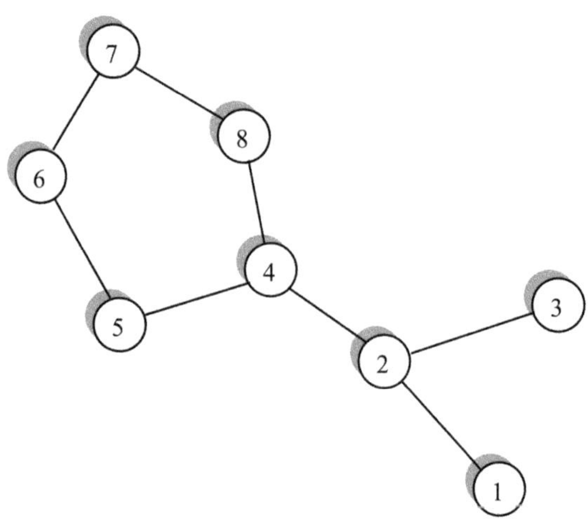

Figure 3.7 Labelled (numbered) graph with eight vertices that represents the structure of proline.

The analogy of atoms to graph vertices and, accordingly, bonds to the graph edges is quite straightforward and intuitive. The latter is the foundation for the topological representation of a chemical structure. Topologically, a structure can be represented by a molecular graph of the type S[A, B], having two sets: a set of atoms, A, and a set of bonds, B. For the example in Figure 3.7, the topology (connectivity) of the molecule of proline is described by the set of atoms (graph vertices), $A = \{1,2,3,4,5,6,7,8\}$ and the set of bonds (graph edges), $B = \{(1,2), (2,3), (2,4), (4,5), (4,8), (5,6), (6,7)\}$. This representation is called labelled or numbered graph since each vertex has its own formal number. The concrete numbering is not important but what counts is the connectivity description. For example, the graph from Figure 3.7 can be numbered by means of 5040 ($5040 = 7! = 1.2.3.4.5.6.7$) different combinations but each one will represent the same graph (the same topological information). All these representations are isomorphic to each other and represent the same chemical object.

The representation of structural information as numbered graphs is sufficient for representing alkanes and cycloalkanes. Furthermore, the graph theory introduces different types of the vertices (also called vertex colours in original mathematical terminology) and types of the edges. Such a "labelled coloured graph" is sufficient to describe the topology of any organic molecule.

In the last four decades, various techniques for structure representation have been established. A practical chemoinformatics technique for storing/representing chemical graphs is the *connection table* (CT) (see Figure 3.8). The connection table is used by almost every chemical software task: structure representation, structure storage, transformation, visualisation, calculation of molecular descriptors and properties. The first (or main) form of a connection table corresponds directly to the chemical graph (Figure 3.8a). It is composed of a list of atoms with atom types and a list of bonds represented as triples (atom1, atom2, bond-type). In the second form of the connection table, the first topological layer[10] (environment) of each atom is included along with the atoms types (see Figure 3.8b). The second connection table form does not have a separate bond list because the first topological atom layers describe the neighbour atoms for each atom and the connection types to the neighbours, thus connectivity is fully represented. For example, the first topological layer of atom 2 is represented as <1,2>, <3,1>, <4,1>, *i.e.*, atom 2 is connected to atom 1 with a double bond (<1,2>), connected to atom 3 with a single bond (<3,1>) and connected to atom 4 with a single bond (<4,1>). The second connection table form is also called redundant connection table since the connectivity is described twice – each bond is described respectively in the row of each atom that participates in the bond. The second form of connection table is quite useful for most of the chemoinformatics topological algorithms; therefore, it is most commonly implemented as a standard internal structure representation or as a software functionality that renders the information on run-time. On the other hand, most file formats usually use the standard first connection table form or its logic directly to represent the chemical structures.

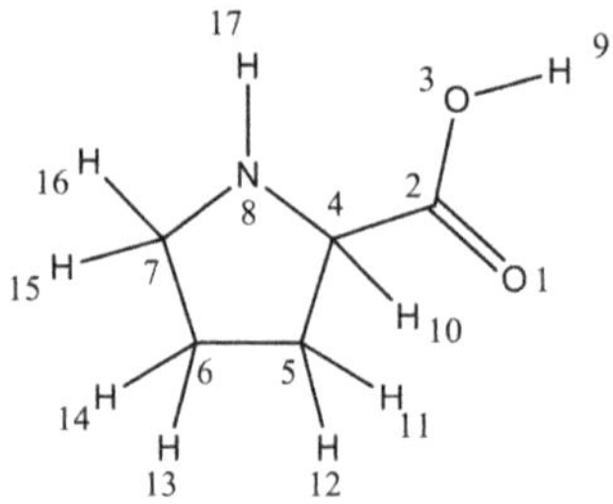

(a)
Connection table with explicit hydrogens

Atoms		Bonds		
1	O	1	2	2
2	C	2	3	1
3	O	2	4	1
4	C	4	5	1
5	C	4	8	1
6	C	5	6	1
7	C	6	7	1
8	N	7	8	1
9	*H*	*3*	*9*	*1*
10	*H*	*4*	*10*	*1*
11	*H*	*5*	*11*	*1*
12	*H*	*5*	*12*	*1*
13	*H*	*6*	*13*	*1*
14	*H*	*6*	*14*	*1*
15	*H*	*7*	*15*	*1*
16	*H*	*7*	*16*	*1*
17	*H*	*8*	*17*	*1*

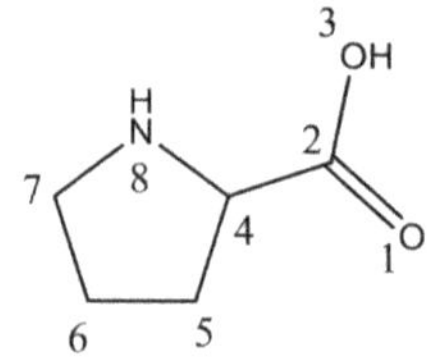

(b)
Redundant connection table
(implicit H atoms)

1	O	<2,2>		
2	C	<1,2>	<3,1>	<4,1>
3	O	<2,1>		
4	C	<2,1>	<5,1>	<8,1>
5	C	<4,1>	<6,1>	
6	C	<5,1>	<7,1>	
7	C	<6,1>	<8,1>	
8	N	<4,1>	<7,1>	

Figure 3.8 Connection table variants for the structure of proline: (a) connection table with explicit H atoms, (b) redundant connection table with implicit H atoms.

Apart from 2D level designation, the term "2D" is used in the sense of a structure representation technique – so-called 2D structure diagrams. In this context, 2D representation designates a "beautifully" drawn chemical structure. 2D atom coordinates are obtained observing a set of rules (*e.g.*, hexagonal grids, regular polygons, *etc.*) in order to obtain a visually clear structure. 2D atom coordinates do not represent crucial chemical information but are used only for structure drawing as a vector graphic. The 2D diagram practically visualises the structure topology and represents the graph geometrically. Therefore, the term "2D" should not be associated with any geometrical information but with topological data. However, stereochemistry information may be derived from the 2D diagram, *e.g.*, the double bond *cis*/*trans* configuration and the chirality of an atom. In addition, some stereo templates

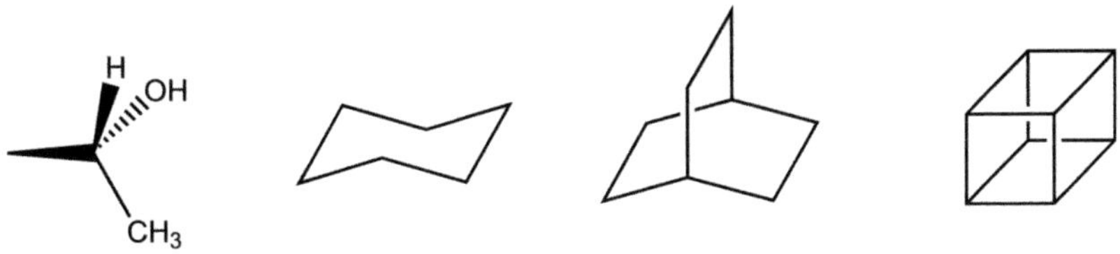

Figure 3.9 Stereochemistry representation and some partial 3D templates by means of 2D diagrams.

can be used for popular conformations such as the cyclohexane chair and boat conformations (see Figure 3.9) *etc.* Currently, many efficient algorithms for automatic generation of 2D coordinates are available in various chemoinformatics software packages such as CACTVS,[12] ChemAxon Marvin,[13] ACDLabs ChemSketch,[14] CDK[15] library, *etc.*

The connection table is a fundamental feature for the internal representation of structural data in the majority of chemical software systems. There are limitations of the conventional connection table representation for some classes of chemical objects, *e.g.*, aromatic systems, tautomers, coordination systems, biomacromolecules, *etc.* Usually, the connection table is enriched with additional lists of atom and bond attributes as well as sets of special rules, which are applied when the problematic cases mentioned are dealt with. A quite efficient structure representation on top of the connection table has been developed by the group of Gasteiger *et al.* – the RAMSES[1] representation. Here, the information for the chemical bonds is replaced with a more complicated list of electron systems.

3.5 Linear Notations

The linear (or line) notation is a sequence of alphabetic and numeric characters and additional allowed specialised symbols ordered according to a set of syntactical and logical (semantical) rules. As a result, a string representation of the chemical object is obtained. The systematic and trivial chemical nomenclatures in a more general sense can be considered as a particular case of linear notations. However, they are not efficient for computer treatment and they do not comply with the basic requirements listed next. Although over the last decades, software packages have been developed for automatic conversion from the International Union of Pure and Applied Chemistry (IUPAC) nomenclature[16] to connection tables and *vice versa*, nomenclature is used as additional information with a secondary goal in chemical databases. Typically, linear notations are techniques on a 2D topological level (see Figure 3.5) and they comply with basic requirements such as:

- unambiguity,
- compact representation,
- full topology description (representation of the connectivity),

- easy development of computer handling algorithms,
- easy manual treatment (directly by the user) if possible,
- unique representation (if possible) or canonicalisation rules for unique representation.

Line notations are useful because they can be dealt with efficiently directly by the human user as well by software algorithms. The growth of the field of chemoinformatics is characterised by the development of many line notations. Presently, two of the most popular line notations are SMILES (*S*implified *M*olecular *I*nput *L*ine *E*ntry *S*ystem)[17] and InChI (IUPAC *In*ternational *Ch*emical *I*dentifier).[18] There is a special extension of SMILES to represent substructures: SMARTS (SMiles ARbitrary Target Specification).[7] Line notations represent essentially the same information as connection tables, but it is stored in a different way – as a string. The use of line notations requires software parsers that practically generate a connection table for a given line notation string.

Line notations are useful for storage in databases as well as for a direct input of chemical data by the user. Because there is a large diversity of line notations, one should pay attention to what purpose a given notation is used for. Different notations are used to represent different types of chemical objects (see object hierarchy in Figure 3.1): compounds, search queries (a query is totally different from the compound!), reactions, sets of compounds, combinatorial libraries (*e.g.*, Markush structures), *etc.* SMILES and InChI notations are used primarily for representing the structures of chemical compounds. InChI and SMILES are widely used and both notations are academic, scientific and industrial standards for structure representation.

3.5.1 SMILES Linear Notation

The SMILES[17] (*S*implified *M*olecular *I*nput *L*ine *E*ntry *S*ystem) linear notation is intuitive and easy to learn. It is based on the valence model of the chemical compounds. There are four basic syntactical/semantical rules for the description of: (i) atoms and their basic properties, (ii) bonds, (iii) branching and substituents, and (iv) ring closures. Once these rules are mastered, they can be used for representing virtually any organic compound. There are additional features in SMILES for the representation of stereochemistry information. The basic SMILES syntax is supported at http://daylight.com. There is also an OpenSMILES initiative,[19] which is compatible with the original SMILES standard. Examples for SMILES notations are given in Table 3.1.

The SMILES syntax supports basic stereochemistry: double bond *cis-/trans* configuration, several types of chirality – tetrahedral chirality, allene atom chirality, octahedral chirality, *etc.* It is important to note that SMILES defines chirality *relatively* while the chemical chirality traditionally is defined

Table 3.1 Representation of chemical structures by means of SMILES line notation.

SMILES	Compound
C	Methane
CC=C	Propene
CCO	Ethanol
CC(C)C	2-Methylpropane
C1CCCCC1	Cyclohexane

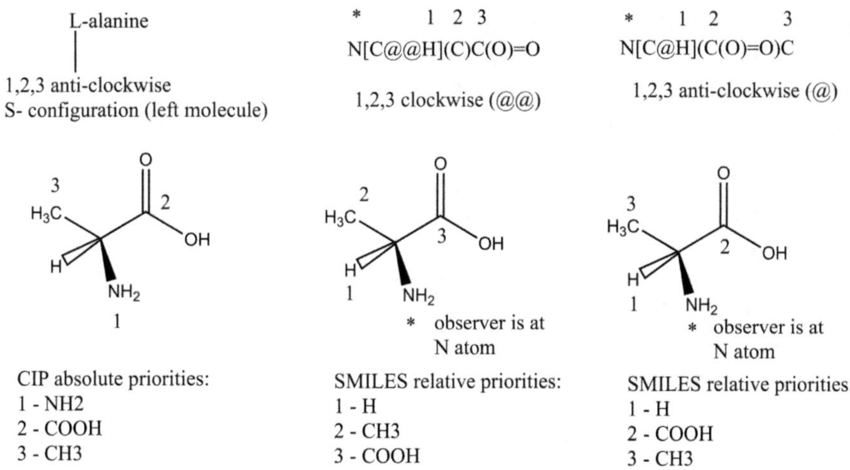

Figure 3.10 Stereochemistry information representation for the molecule of L-alanine by means of absolute CIP rules and the SMILES line notation relative approach.

absolutely, *e.g.*, Cahn–Ingold–Prelog (CIP) priority rules.[20] In SMILES, the ligand atoms are ordered relatively, *i.e.*, the order is determined by the atom appearance in the line notation, while with CIP rules, the ligands are prioritised absolutely (see Figure 3.10). The relative ordering-based stereochemistry representation in SMILES describes the same stereochemistry information (*e.g.*, R or S configuration) but in a different manner. There is no direct correspondence between designations @/@@ (SMILES) and R/S (CIP rules) because they are based on a different logic. As can be seen in Figure 3.10, several relative ways for describing the chirality of L-alanine are possible. Usually, a software tool is required (although it can be done manually by the user) to convert from relative to absolute stereochemistry representation and *vice versa*.

The SMILES line notation is not unique by default and there are many implementations of algorithms for unique SMILES in various software systems. Therefore, when working with unique SMILES representations, it is important to use the same software tool for all operations within off-line database filling and on-line utilisation and services.

3.5.2 InChI Linear Notation

The InChI[18] (IUPAC *In*ternational *Ch*emical *I*dentifier) linear notation is a non-proprietary identifier for chemical substances. InChI was developed by IUPAC in order to represent the chemical objects in a more suitable way for computer processing of chemical structural information. The InChI linear notation is a unique representation (*i.e.*, it can be used for identification), it can encode absolute stereochemistry, delocalisation, aromaticity, and zwitterions. In addition, basic tautomerism can be encoded on request. The InChI generation algorithm includes three major steps: input structure normalisation, canonicalisation and serialisation of the data into a series of bytes. InChI is divided into several layers: (i) formula, (ii) connectivity information, (iii) isotope information, (iv) stereochemistry, and (v) tautomer information (optional). The InChI line notation for the molecule of L-alanine is:

1/C3H7NO2/c1-2(4)3(5)6/h2H,4H2,1H3,(H,5,6)/t2-/m0/s1.

The InChI linear notation is predominantly used by means of software tools and is not convenient to be manually created.

On top of the InChI notation, IUPAC developed the InChIKey, which is a hash code based on the InChI information (more details on hash codes are given in Section 3.7). The InChIKey is a simple representation and facilitates the searching of molecules *via* internet search engines as well as fast identification in big chemical databases. The InChIKey of L-alanine is:

QNAYBMKLOCPYGJ-UWTATZPHSA-N.

3.5.3 SMARTS Linear Notation

The SMARTS[7] (*SM*iles *AR*bitrary *T*arget *S*pecification) linear notation is an extension of the SMILES line notation and is used for the description of queries for substructure searching. Although both may look identical in many cases, SMILES describes valid molecule(s) while SMARTS defines structure fragments that can be searched against other target molecules. The basic topology information (*i.e.*, a graph with vertices and edges) within SMARTS is defined in the same way as SMILES defines it, but instead of describing atoms and bonds, the SMARTS syntax allows the definition of atom expressions and bond expressions. An atom or bond expression is defined by means of atomic/bond primitives and logical operations. In this way, SMARTS can be used for very specific definitions of structural fragments. A fragment consisting of an aliphatic carbon or aliphatic nitrogen atom connected with a halogen atom is defined as [C,N]-[Cl,Br,F,I]. The SMARTS string [#6,#7]-[Cl,Br,F,I] defines a query for any carbon or nitrogen atoms connected with a halogen. Different matches for the SMARTS queries are shown in Figure 3.11.

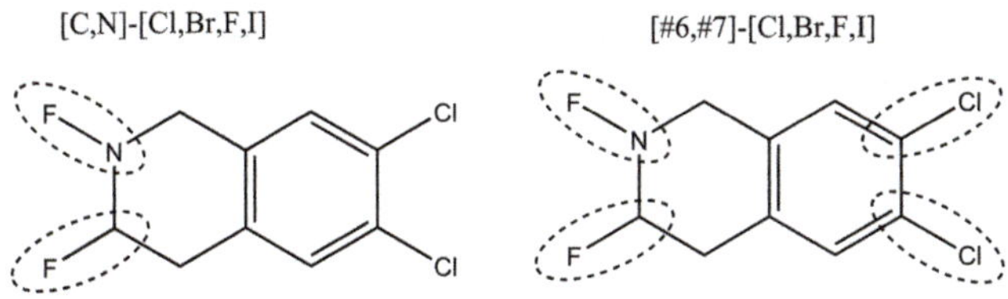

Figure 3.11 SMARTS query matches against 6,7-dichloro-2,3-difluoro-1,2,3,4-tetra hydroisoquinoline.

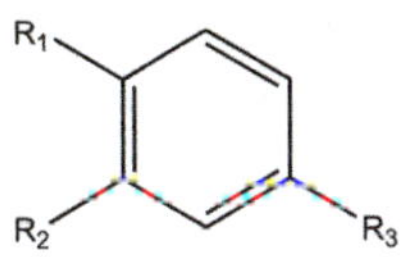

R1=H, OH, CH3, Cl, Br, F	6 substituents
R2=H, F, Cl, CH3	4 substituents
R3=H, Cl, Br	3 substituents

number of combinations 72

c1(-[H, $([OH]),$([CH3]),Cl,Br,F])c(-[H,F,Cl,$([CH3])])cc(-[H,Cl,Br])cc1

Figure 3.12 Representation of a small combinatorial set of molecules by means of a Markush structure defined with SMARTS line notation.

The SMARTS notation supports recursive atom expressions, *i.e.*, another SMARTS string can be included within a given atom expression (example $([CH3]) shown in Figure 3.12). The SMARTS linear notation can be used to define basic variation points (the R groups) of Markush structures[6] and can be used for representing combinatorial libraries. For example, the linear notation c1(-[H,$([OH]),$([CH3]),Cl,Br,F])c(-[H,F,Cl,$([CH3])])cc(-[H,Cl,Br])cc1 defines a Markush structure (see Figure 3.12) with 3 R-groups and encodes totally 6*4*3 = 72 chemical structures.

Examples of software tools that efficiently perform SMARTS-based substructure searching are CACTVS[12] and Ambit-SMARTS.[21]

In addition to the linear notations discussed, there are others that are used successfully in the chemoinformatics community. For instance, the linear notation SMIRKS[22] is an extension of SMARTS and SMILES used for the definition of generic chemical reactions and transformations. SLN[23] (Sybil Line Notation) is another example of efficient linear notation. SLN is not as popular as SMILES and InChI notations but it is a quite universal notation. SLN has all the capabilities of SMILES, SMARTS and SMIRKS plus support for many other useful chemoinformatic features.

3.6 3D and 4D Structure Representations

The 3D level of structure representation describes the molecule geometry and the associated atomic and molecular properties as well as electron densities if considered on a quantum-mechanical level. 3D geometrical information is crucial for handling important physical, chemical and biological properties.[4] For example, biological activities that result from ligand–receptor interactions are determined by the receptor 3D configuration and the 3D shape of the ligand. Thus, the ability to store and handle this information is of utmost importance.

On top of the connectivity (2D), the full geometrical configuration of a molecule is described by adding 3D Cartesian atom coordinates to the connection table, *i.e.*, a triple, (x_i,y_i,z_i), is added for each atom (see Figure 3.13).

Cartesian coordinates are needed for molecular visualisation as well as for many computational procedures. However, 3D Cartesian coordinates do not have direct chemical meaning (for example, if a molecule is shifted in the space or rotated, the coordinates are changed, but not the chemical information itself). Another important system of coordinates used for a 3D geometry description are the internal coordinates: bonds lengths (r_i), valence angles (θ_i) and dihedral angles, also called torsional angles (ω_i). Each atom in the molecule is described in terms of its atomic number, bond length, bond angle, and dihedral angle. The most popular representation of the molecule by means of internal coordinates is the Z-matrix, which is typically used as input for many computational software packages.

Usually, at the 3D level, H atoms are treated explicitly, unlike at the 2D level as already discussed. This is due to the fact that, when looking at the molecule geometry, exact 3D coordinates of all atoms (hydrogen and non-hydrogen) are needed. Different types of raster graphics play an important role for the visualisation of 3D information. The 3D coordinates describe the basic molecular skeleton. On top of the molecular skeleton, other 3D approaches are used, such as molecular surfaces that represent basic molecular shapes and various properties as functions of 3D coordinates. Molecular fields

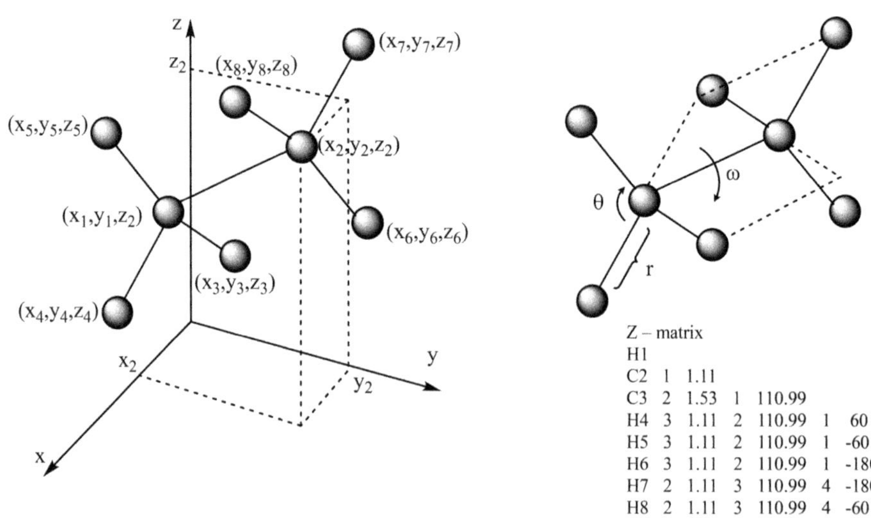

Figure 3.13 3D structure representation for the molecule of ethane by means of Cartesian coordinates and internal coordinates.

represent another approach for 3D information, which basically describes the distribution of a particular property value in the 3D space around the molecule.

The term "4D" formally designates ensembles of conformations for a given molecule. The fourth dimension is treated as a "parameter" that formalises the conformational flexibility of the molecule. In the chemoinformatics literature, the term "5D" is used as well. However, it is not a designation of a structure representation level, but is a class of QSAR methods (called 5D-QSAR)[4] that takes into account the conformational flexibility both of the ligand structure and the active receptor site.

3.6.1 3D Structure Optimisation

The experimental determination of 3D structures is performed mainly by means of X-ray crystallography[24] as well as other techniques such as nuclear magnetic resonance (NMR) spectroscopy[25] and electron microscopy,[26] often used in a combination. Currently, there are several hundred million compounds stored in various public and proprietary databases. Less than 1% of these compounds have experimentally determined 3D geometries. The largest database with experimental 3D structures is the Cambridge Structural Database (CSD)[27] with about 1 million structures. From these numbers, it is clear that the basic approach for 3D structure determination is computer-based (*in silico*) theoretical computation. In addition, the experimental procedures are not applicable at all for virtual compounds that have not been synthesised yet. Since the presumption is that some conformations of the ligands are bioactive, methods for 3D structure optimisation are an important part of the 3D structure manipulations. Energy minimisation is commonly referred to as geometry optimisation.

The procedures for geometry optimisation on a large scale are based on the method of molecular mechanics.[4] The molecular mechanics algorithms are used for fast generation of 3D structures and are implemented in tools for handling very large chemical databases. Quantum mechanics methods are more precise than molecular mechanics but are extremely slow for large-scale calculations spanning over millions of structures. That is why molecular mechanics (and related 3D generators) can be considered as the "workhorse" for 3D structure optimisation. The molecular mechanics computational approach is inspired by the ideas of vibrational spectroscopy and utilises empirically parametrised force fields. A force field, in this context, is a set of energy potential functions that represent the molecule energy as a function of its 3D geometry. Molecular mechanics constitutes a reductionist approach for the molecule energy calculation and is based on the Born–Oppenheimer approximation. The electron effects in the structure are considered indirectly by means of extended sets of atom classes used to define empirically parametrised energy potential functions associated with the geometry elements. Molecular mechanics is based on an artificially defined potential energy function called steric energy. The steric energy function, $E = f(x_1, x_2, ..., x_n)$, is expressed as a function of the molecule geometry coordinates $X = (x_1, x_2, ..., x_n)$. In analogy to vibrational spectroscopy, each internal coordinate x_i, whether bond length, valence angle or dihedral angle, has an optimal value x_{i0} and when x_i deviates from the 'ideal' value x_{i0}, the energy increases:

$$\Delta E_i = f(\Delta x_i) \tag{3.1}$$

The forces within a molecule work in such a way that the energy is minimised. The force is calculated as an anti-gradient of the potential energy:

$$F = -\nabla E(X) \tag{3.2}$$

Steric energy is obtained as a sum of several energy terms (see Figure 3.14) that take into account valent and non-valent interactions within the molecule.

The basic potential energy term is the harmonic potential, which is a quadratic function of the parameter deviation from the ideal value x_{i0}:

$$E_i = k_i(x_i - x_{i0})^2 \tag{3.3}$$

k_i is a force constant and is empirically determined together with the ideal parameter value x_{i0} as part of the force field parametrisation. The force field parameters are determined on the base of experimental data and quantum mechanical calculations. Harmonic potential approximation is used for bond stretching, angle bending and out-of-plane bending (sp^2 hybridised atoms). In addition, the cross terms $k_{ij}(x_i - x_{i0})$ $(x_j - x_{j0})$, are a generalisation of the harmonic potential. Dihedral angle deformations (or torsions) are approximated as sums of sin/cos functions:

$$E_{\text{tor}}(\omega) = \sum_{i=1}^{n} \frac{V_n}{2}\left(1 - \cos(i\omega + \gamma)\right) \tag{3.4}$$

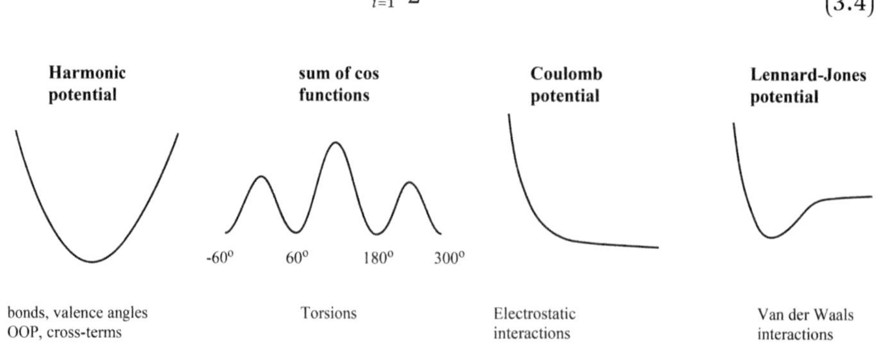

Figure 3.14 Molecular mechanics potential functions used for the definition of the steric energy.

Non-valent interactions in the molecule take into account intramolecular forces working at larger distances between atoms that are not covalently bonded. There are two basic potentials used for non-valent interactions: Coulomb potential for electrostatic forces (eqn (3.5)) and Lennard-Jones potential (designated as 6–12 potential) for van der Waals forces (eqn (3.6)).

$$E_{ij} = \frac{1}{4\pi\varepsilon_0\varepsilon}\frac{q_i q_j}{r_{ij}} \tag{3.5}$$

$$E_{vdW}(r) = \varepsilon\left[\left(\frac{r_0}{r}\right)^{12} - 2\left(\frac{r_0}{r}\right)^{6}\right] \tag{3.6}$$

Geometry optimisation and energy minimisation are closely related. The optimal geometry, X_{opt}, is obtained by minimising the energy $E_{min} = f(X_{opt})$. Minimisation procedures are iterative gradient methods based on the principle from eqn (3.2). For an iteration step, $X_n \rightarrow X_{n+1}$, the molecule geometry improvement towards the minima is expressed *via* the gradient ∇E.

$$X_{n+1} = X_n - c\nabla E(X_n) \tag{3.7}$$

Minimisation procedures with first-derivative methods such as the gradient's steepest descent (eqn (3.7)) and conjugate gradient minimisation algorithms are commonly used as well as second-derivative-based methods.

3.6.2 Conformational Search

Handling molecular conformations is important since their properties may vary significantly. Of special interest are the bioactive conformations that interact with receptors. The molecule conformations are obtained by conformational analysis, which relies on several approaches such as systematic conformational search, rule-based approach, random search, molecular dynamics simulation, *etc.* The conformational flexibility of the molecules is due to rotations around single bonds. Principal geometric parameters to be varied in the conformational analysis are dihedral angles. The other two types of internal coordinates, bond lengths and valence angles, are considered rigid and are not systematically changed – their values are "adjusted" as a result of the minimisation procedure applied after dihedral parameter systematic variation. Each rotatable single bond defines one dihedral driver, *i.e.*, only one of the several possible dihedral angles is considered as a parameter for variation. The other dihedral angles associated with the dihedral driver will also automatically change due to the rotations, since the valence angles and bond lengths are rigid. Typically, three stable minima exist around each single bond for values of the dihedral angle around 60°, 180° and 300° (or −60°) (see Figure 3.14). The conformational flexibility of cyclic systems is handled by simultaneous change of two dihedral drivers yielding "corner flapping" within a cycle polygon (Figure 3.15).

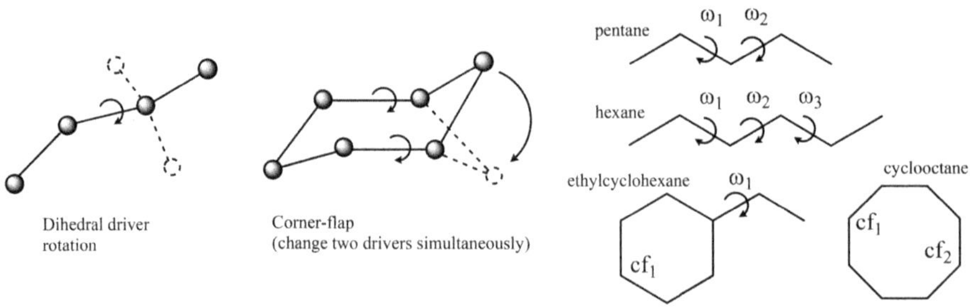

Figure 3.15 Variable parameters for conformational analysis.

Each dihedral driver has three possible positions and each "corner flapping" gives two possible positions. Therefore, the total number of parameter combinations is

$$N_{conf} = 3^{N_\omega}.2^{N_{cf}} \tag{3.8}$$

Each combination gives a starting geometry that is optimised by the standard molecular mechanics gradient minimisation procedures. For example, a conformation analysis for the molecule of pentane could be performed by means of two dihedral drivers (see Figure 3.15). The outer single bonds to the terminal methyl groups are not considered in the analysis since their rotation does not yield a new geometry (all three H atoms in the methyl group are equivalent). Thus, the pentane conformational analysis is described by 3 × 3 = 9 parameter combinations. The actual number of conformations is less than nine because of symmetry considerations. Similarly, the conformational analysis for the molecule of hexane gives 3 × 3 × 3 = 27 parameter combinations (see Figure 3.15) used as starting points for geometry optimisation. Conformational analysis for the molecule of ethylcyclohexane is performed by means of one dihedral driver and one "corner-flapping", *i.e.*, 3 × 2 = 6 combinations are considered. The cyclooctane conformers are characterised by means of two "corner-flappings", *i.e.*, 2 × 2 = 4 possible combinations.

The systematic conformational analysis can be improved and speeded up by means of rule-based approaches. A good example of an efficient conformer generator based on several hundred rules derived from the CSD[27] database is the software ROTATE.[28]

Molecular dynamics[4] constitute another approach for conformational analysis, which can be considered as an extension of the molecular mechanics method. Molecular dynamics is based on eqn (3.2), which expresses the intramolecular forces from the steric energy function $E(X)$ and the second principle of Newton's mechanics (eqn (3.9)).

$$F = ma \tag{3.9}$$

Molecular dynamics computational iteration includes consequent application of eqn (3.2) to calculate forces, then calculation of the acceleration $a = F/m$ from eqn (3.9). Finally, atom velocities and atom positions are updated on the basis of the acceleration a. Molecular dynamics simulation starts with initial atom positions and initial atoms velocities, usually obtained from a Maxwell–Boltzmann distribution. The analysis of the molecule trajectory for a particular time interval simulation gives information about the possible conformations. Usually, a molecular dynamics simulation is performed at extremely high temperatures (*i.e.*, the atom velocities are scaled up accordingly), *e.g.*, 900 K, which helps the virtual molecule to "jump over" higher rotational energy barriers and to identify more conformers in this way.

The molecular mechanics and molecular dynamics principles for non-valent energy potentials of Coloumb (electrostatic forces) and Lennard-Jones (van der Waals forces) are used in two other important chemoinformatics techniques: docking[29] for modelling ligand–receptor interactions when the structure of the receptor is known, and Comparative Molecular Field Analysis[30] (CoMFA) for 3D-QSAR modelling. CoMFA is an alternative to the docking method when the target protein is not known, and is complementary to docking when the receptor structure is known.

3.7 Molecular Descriptors, Fingerprints and Hash Codes

A molecular descriptor[10] is a number or a set of numbers that represent useful information about the molecule. The molecular descriptor's value may be experimentally derived or the result of a computational procedure. From a technical point of view, descriptors may be classified as scalar (single value) and vectors (multiple values). For simplicity, with the term "vector", we designate the case of "multiple values", where more complicated data organisation is possible such as matrix, scalar field, vector field, *etc.* As shown in Figure 3.3, molecular descriptors play an important role in the process of property modelling and are typically used as an interface (intermedium layer) between structure information and property data. Depending on the context and specific applications, descriptors may even be considered as structural representations (*e.g.*, constitutional descriptors[10] of the type N_X); others are actually measured experimental properties or calculated descriptors that may be considered as models of the structure. This manifold appearance of molecular descriptors is due to their wide range of applications, but typically they are treated as a separate layer in the molecule representation. One of the most important categories of molecular descriptors is the category of structural descriptors, which are calculated as a function of the structure $d = f(S)$. By means of a vector of several structural descriptors, the molecule, which is a complicated chemical object, is represented as a point (chemical pattern) in the n-dimensional space (see Figure 3.16). In this way, many important chemoinformatics tasks can be performed in terms of chemical patterns (points in the chemical space), for example imaging of the chemical space, similarity principle applications, chemical object distance determination, clustering, classification, virtual screening, *etc.* The descriptors are very simplified structural representations, but they are quite useful since they allow the application of mathematical modelling and informatics algorithms for the chemical patterns (as points of vector spaces) instead for the original chemical objects.

Structural descriptors can be classified (in exactly the same way as structure representations are classified) into the levels 0D, 1D, 2D, 3D and 4D. One of the main streams of chemoinformatics development was the creation of new molecular descriptors. Currently, there are more than 5000 descriptors[10] as systematised in handbooks, encyclopedias, software manuals, *etc.* Popular software applications for molecular descriptor calculation with large sets of descriptors are Dragon[31] and PaDEL[32] – based mainly on the descriptors implemented in the Chemistry Development Kit (CDK).[15]

To a great extent, 0D and 1D structure representations themselves are structural descriptors as well (discussed in Section 3.3). Topological (2D) descriptors form an important subclass of descriptors and they can be technically categorised as:

- *topological indices* – invariants of the chemical graph that do not depend on structure numbering,
- *matrices* that describe rich topological information,
- *eigenvalues, eigenvectors and characteristic polynomials* of topological matrices.

Topological indices are calculated on the basis of diverse chemical logic, but there are some major trends dominating. One big group of topological descriptors is obtained on the principle of additive schemas. The structure is

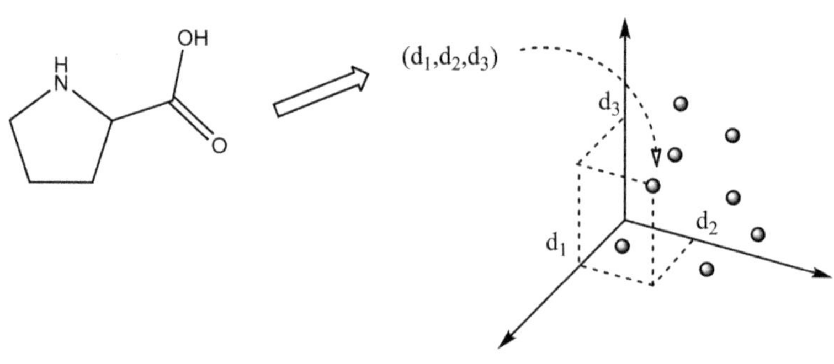

Figure 3.16 Descriptor usage: the chemical structure is represented as a vector of three descriptors, *i.e.*, a chemical pattern that is a point in the 3D vector space.

considered a set of components such as atoms, bonds, fragments, D-groups, G-groups, topological torsions, paths, *etc.* A weight is given to each component and all component weights are summarised to obtain the topological index. For example, the Zagreb[10] index is calculated as the sum of the squares of the topological degrees of all atoms (see Figure 3.17), while the Wiener[10] index is defined as the sum of all topological distances (sum over all pairs of atoms). By varying the way the molecule is "decomposed" and by using different group weights, a wide range of topological descriptors can be defined. The weighting can include atomic and physicochemical properties, thus enriching the descriptor versatility.

Another principle for descriptor calculation is based on the informational content. The molecule is "decomposed" into components and the components (atoms, bonds, paths, *etc.*) are classified into categories according to a given chemical logic. The Shannon entropy[10] (also called informational content) is calculated on the basis of component categories. The Shannon entropy is used directly as a descriptor or is normalised by using other topological invariants.

3D geometrical descriptors constitute another large descriptor group. Many of the 2D topological descriptors have 3D analogues. For example, the 3D Wiener index is defined in the same manner as the classic Wiener index, *i.e.*, it is a sum of all geometrical distances calculated over all atom pairs. In addition to the 2D analogues, geometrical descriptors include invariants specific to the 3D molecule, such as gravitational descriptors, shape descriptors, volume descriptors, quantum chemical descriptors, surface descriptors, interaction energy descriptors, field-based descriptors, electrostatic descriptors as well as descriptors that combine 2D and 3D information. Autocorrelation approaches are used for both 2D and 3D descriptors. The ATS (Autocorrelation of Topological Structure)[10] descriptor with a weighting scheme based on atomic numbers is shown as an example in Figure 3.17.

Bit-vectors, also called fingerprints,[2] are specialised techniques for coding chemical information. They rely on the utilisation of bitwise operations, which are the fastest type of computer calculations. Therefore, fingerprints are very useful for applications in large and very large chemical databases.

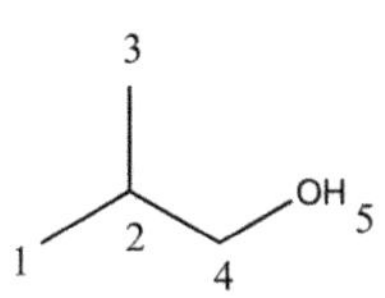

$Z = \delta_1^2+\delta_2^2+\delta_3^2+\delta_4^2+\delta_5^2 = 1^2+3^2+1^2+2^2+1^2 = 16$

$W = d_{12}+d_{13}+d_{14}+d_{15}+d_{23}+d_{23}+d_{24}+d_{25}+d_{34}+d_{35}+d_{45}$
$= 1+2+2+3+1+1+2+2+3+1 = 18$

$ATS_0 = Z_1^2+Z_2^2+Z_3^2+Z_4^2+Z_5^2 = 6^2+6^2+6^2+6^2+8^2 = 208$

$ATS_1 = Z_1Z_2+Z_2Z_3+Z_2Z_4+Z_4Z_5 = 6.6+6.6+6.6+6.8 = 156$

$ATS_2 = Z_1Z_3+Z_1Z_4+Z_3Z_4+Z_2Z_5 = 6.6+6.6+6.6+6.8 = 156$

$ATS_3 = Z_1Z_5+Z_3Z_5 = 6.6+6.8 = 84$

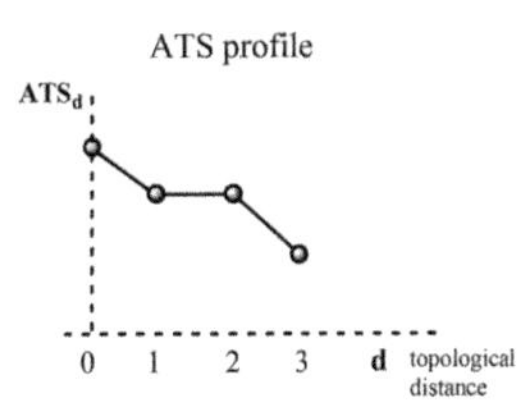

Figure 3.17 Descriptor calculation for the molecule 2-methylpropan-1-ol: Zagreb topological index (Z), Wiener index and Autocorrelation of Topological Structure (full ATS profile).

Fingerprints are used for fast database screening (a preliminary phase to substructure searching), similarity searching, virtual screening, database clustering, activity cliff detection, *etc.* Bit-vectors are calculated by means of three computational approaches (Figure 3.18):

- Each bit is associated with the presence/absence of a particular structural fragment from a preliminarily chosen list of fragments;
- The bits in the fingerprint correspond to dynamically generated fragments by an exhaustive search within the structure for all paths within a given size range. The bit positions are determined by hashing where bit collisions are allowed, *i.e.*, some bits may correspond to more than one fragment;
- The bits correspond to chosen descriptor values (binned descriptors). Usually, a group of bits codes several possible values of a particular descriptor. The latter approach can be considered as a hybrid between descriptor vector and bit-vector, having advantages of both approaches: richer chemical information and faster bit operations.

Descriptor and fingerprint representations are the basis for QSAR modelling. Descriptors provide a practical means to represent and store structure information related to the molecules and experimental data. In this context, chemical structures are keys to retrieve the toxicity information for a substance from the "big data information pool" on the basis of various modelling techniques such as linear regression, similarity approaches (*e.g.*, k-nearest neighbour algorithm kNN[3]), artificial neural network non-linear modelling, clustering, classification algorithms, rule-based approaches, *etc.*

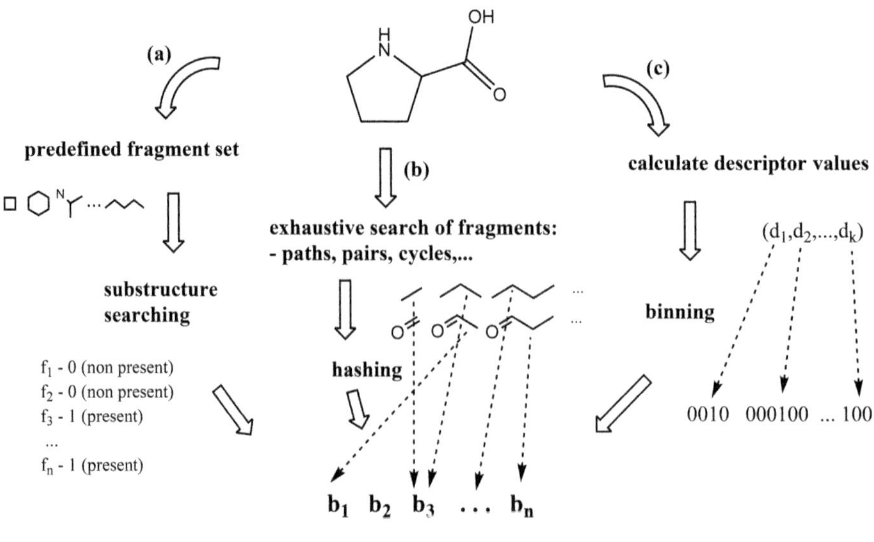

Figure 3.18 Fingerprint determination approaches: (a) based on a predefined fragment set, (b) hashed fingerprints, and (c) binned descriptor values.

Hash codes[33] are widely used informatics techniques for extremely compact coding of objects in a given field of interest. A hash code is a number or string with a predefined bit size. The hashing of a chemical structure is a particularly important technique in the context of big data management. Structure hash codes as well as hash codes for any other types of chemical objects (*e.g.*, mass spectra hash code[34]) are key techniques for fast identification of a searched object in a large database. Hash codes are also used in the process of structure registration for duplication checks as well as within software systems that generate many structures in run-time and where duplication checks are needed (*e.g.*, tautomer generators, isomer generators and virtual library generators).

Two widely used hash code calculation approaches are shown in Figure 3.19. The classic hash code approach uses the principles of pseudo-random generators, *i.e.*, the chemoinformatics hash codes are computed by determinist decomposition of the molecule to a set of components (atoms, bonds, topological spheres, *etc.*) and then the information from these fragments is used to manipulate the bits. Typically, hash codes obtained in this manner are irreversible, *i.e.*, the chemical objects cannot be decoded from the hash code, and they do not reflect the similarity principle, *i.e.*, for two similar chemical objects, the hash codes usually differ significantly. The major required characteristic is even pseudo-random distribution of the bits. This is often achieved by means of primary numbers "seeding" operations; therefore, hash code collisions are expected to be very rare in a large database. The second approach for a hash code calculation (see Figure 3.19) is to take some unique string representation of a molecule and then to use a standard (non-chemical) hashing algorithm, *e.g.*, the Secure Hash Algorithm (SHA).[35]

InChIKey is one of the most widely used publicly available structure hash codes. It is calculated on the basis of the InChI linear notation and is intended for use in chemical databases, chemoinformatics software as well as for direct web searching for structure identification. The hash code[36] implemented in the CACTVS[12] software is a quite efficient chemoinformatics hash code that is used within the large structural database of PubChem,[37] which is handling around 100 million chemical compounds.

There is another category of structural hash codes called locality sensitive hash codes[38] (LSH). Contrary to the classic hashing based on pseudo-random numbers, LSH takes into account the similarity principle. LSH is the technique of choice for big data sets and it is used for generating similarity matrices of a million compounds, for an accelerated application of the nearest neighbour search, the application of all-pair query approaches for data visualisation, clustering and data mining.

3.8 Challenging Cases for Structure Representation

So far we have discussed several techniques for structure representation on different levels: constitutional (0D and 1D), topological (2D) and geometrical (3D, 4D). The structure representation is a kind of scientific model of

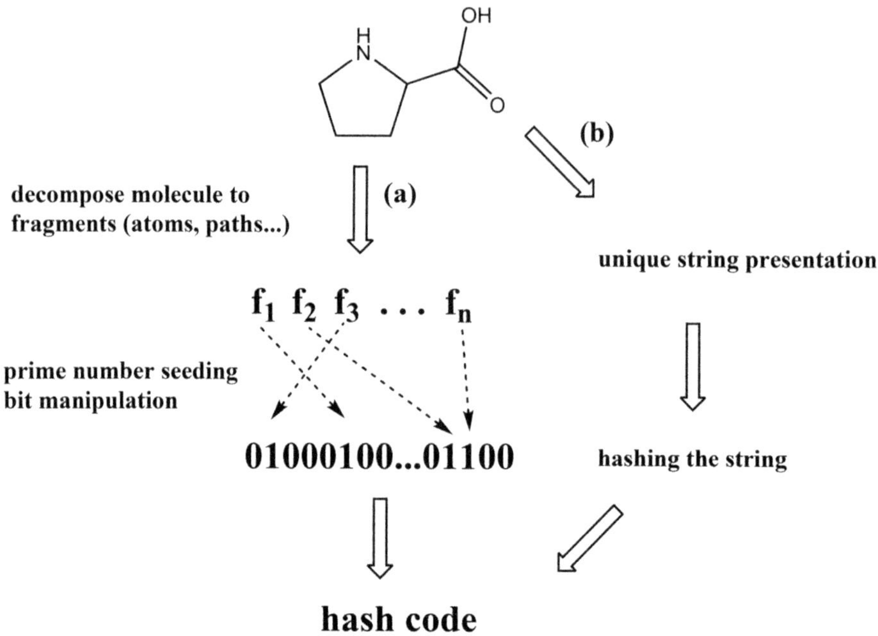

Figure 3.19 Hash code determination workflow.

the studied chemical object. Every representation is reductionist to a certain level. The complexity of the real chemical objects challenges the application of the structure representation methods. In addition, a practical (and often subjective) reason for having some difficult cases is the lack of total consensus among chemists about the nature and properties of the represented chemical objects.

These challenges have been an inseparable part of the chemoinformatics field development. For most of the challenges, sufficiently efficient solutions have been found. However, when considering the problems, one must take into account the chemical space domain the solution is intended for. There is a practical connection between the applicability domain and the quality of the developed models. The narrower the applicability domain (*e.g.*, small classes of chemical objects are included), the more precise the model and *vice versa* – the more universal the model (for example, the applicability domain may include theoretically even the entire "haystack" of 10^{100} drug-like molecules), the less precise it is. A similar trend can be observed regarding the solutions of structure representation problems. A team working on a particular implementation of a more universal representation technique (*i.e.*, intended for a big data application) invests a reasonable amount of time in order to ensure that its chemoinformatics platform works fine for typically 95% of the structures from the entire haystack. The remaining 5% are so-called difficult cases. In order to raise the percentage of "good cases" to

99%, in most of the scenarios an investment of at least twice or even five times more effort is needed. This development depends on the life span of the chemoinformatics project and the available resources.

In this section, some of the major structure representation challenges are briefly discussed. The structure representation peculiarities are important in order to achieve high quality and efficient use of big chemical data.

3.8.1 Handling Aromaticity

Aromaticity is represented within chemoinformatics software by two main approaches:

- Kekule structures with alternating single and double bonds;
- Conjugated "aromatic" systems. The atoms within the aromatic system are marked (flagged as aromatic). The bonds may also be marked as aromatic (usually set with an order of 1.5) or the bond orders may be set to describe one particular resonance Kekule structure.

Both approaches have pros and cons when treating structures. Figure 3.20 shows different aromaticity representations for the molecule *o*-cresol.

Kekule structures are problematic because they have different single and double bond distributions and their connection tables are different. Based on the comparison of the connection tables of the Kekule structure 1 and the Kekule structure 2 (see Figure 3.20), these two structures would be treated as different molecules. For example, one may search the first Kekule structure in a database and may not find it because *o*-cresol is represented as the second resonance form. A possible solution for this problem is to work with conjugated aromatic representations. For this purpose, all database structures must be converted into this form and any database query must be converted from the Kekule to the "aromatic" form. In this case, however, there are problems related to the aromaticity detection algorithms that differ from one software platform to another. More complicated polycyclic molecules (or parts of them) may be detected as aromatic in one software system and as non-aromatic in another. This is especially problematic when many different structural sets from various

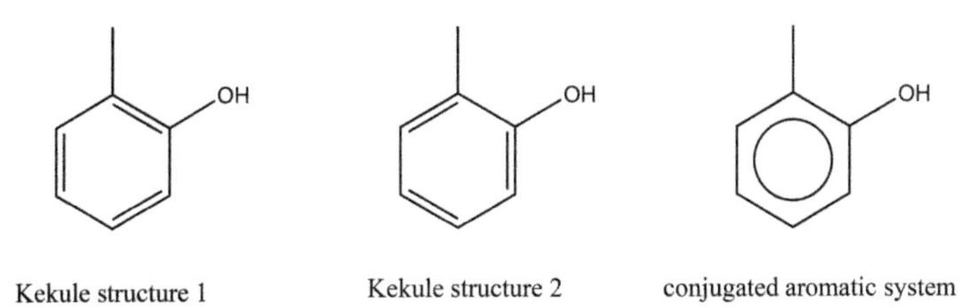

Figure 3.20 Aromaticity representation for the molecule *o*-cresol.

sources are integrated into one big database. Another issue with the aromatic representations is that the application of chemical transformations (*e.g.*, SMIRKS reactions) may be problematic because some bond orders are not exactly specified.

3.8.2 Unique Structure Representations

As discussed previously, unique representations are needed for important and routine chemoinformatics tasks related to handling chemical data such as molecule identification, molecule registration in databases, molecule duplication check and structure hash code calculation.

Structure representation variations basically stem from arbitrary atom numbering in the connection table – bearing in mind that the connection table is the foundational representation for the majority of chemoinformatics programmes. Unique representations can be obtained by using a deterministic algorithm for unique atom numbering. The application of specific rules is called canonicalisation of the structure representation. The Morgan algorithm[9] is one of the widely used algorithms for canonic numbering of atoms of chemical structures. Figure 3.21 illustrates its application for obtaining unique (or canonic) representations for the molecule proline. On the left, an arbitrarily numbered connection table is shown together with arbitrary SMILES based on this numbering. On the right, the canonic connection table and unique SMILES are shown. Canonicalisation algorithms and respectively Morgan algorithm implementations may vary from one software to another; therefore, typically, the unique representations are not so often interchanged. On the other hand, the InChI linear notation is an example of a unique representation that is widely used in many chemoinformatics systems and is an efficient means for interchange of unique structure representations.

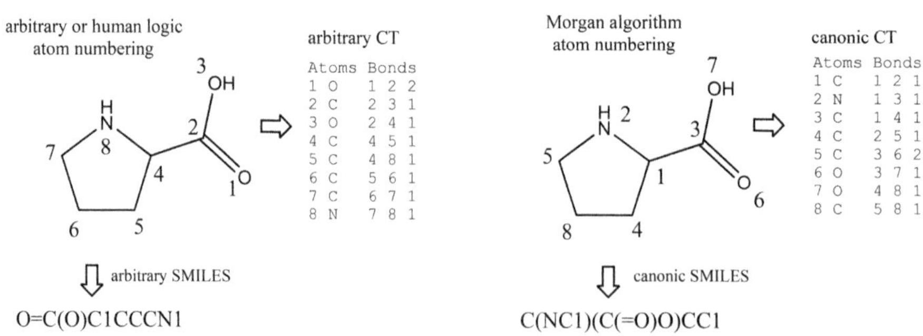

Figure 3.21 Arbitrary and canonic (unique) structure representations for the molecule proline.

3.8.3 Tautomerism

The problem of tautomerism[39,40] is one of the greatest challenges in the structure information processing chain. Chemoinformatics programmes most often handle tautomers as different structures because of their different connection tables (Figure 3.22).

This affects the core structure representation and, subsequently, the chemoinformatics processing chain. The majority of chemical database systems register tautomers as distinct structures; for example, in Chemical Abstracts Service (CAS) databases, the tautomers have distinct CAS registry numbers. It can also be observed that in some catalogues the tautomeric pairs of a given compound are unintentionally presented as different compounds at different prices. Chemical database operations, such as duplication checking during registration, storage of structures, exact, similarity and substructure searching, are highly influenced by tautomerism. The same holds true for QSAR modelling and virtual screening.[41]

To address this problem, various programmes for tautomer generation have been developed. Several efficient systems are currently available for exhaustive tautomer generation including CACTVS,[12] ChemAxon Marvin[13] and AMBIT-Tautomer.[42]

The problem of tautomerism in database management can be handled in two ways:

- creation of intra database links between all tautomers registered as different records in the database;
- use of canonic tautomers for structure registration.

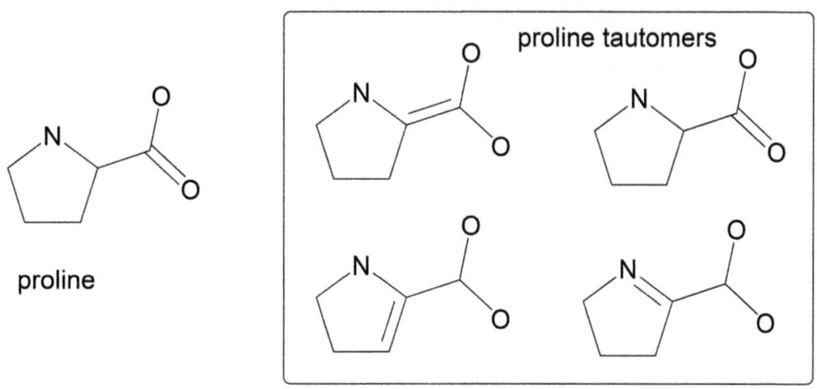

Figure 3.22 Tautomeric forms of the molecule proline generated by the Ambit-Tautomer software.

3.8.4 Miscellaneous Challenges

The connection table plays a fundamental role in the field of chemoinformatics. In this sense, the connection table is also a bottleneck and a limitation for handling some types of chemical objects. Apart from the cases of aromaticity, tautomerism and uniqueness of structure representation, there are other challenges arising from connection table limitations. Some examples are discussed below.

The representation of organometallic compounds is challenging within the connection table frame. For example, the molecule cobaltocene is difficult to represent *via* a standard connection-table-based approach. One possible solution is to represent it as a fragmented chemical object. The SMILES line notation is:

C1=CC=CC1.[Co].C2=CC=CC2.

This formal representation is a fragmented chemical object. The representation by SMILES in a non-fragmented manner would be:

C13=C4C5=C6C17[Co]3456789%10%11%12C28=C9C%10=C%11C2%12.

The latter representation uses 10 single bonds to the Co atom, which would thus be formally 10-valent. Both solutions are not satisfactory. The second solution will not be supported in chemoinformatics platforms that do not allow valences higher than 8.

Other possible approaches to represent cobaltocene are:

- to introduce coordination bond types;
- to designate the rings as macro atoms and connect the Co atom with these macro atoms;
- to represent the electron system encompassing cobaltocene.

Similar problems arise with other metallocenes and related cyclopentadienyl derivatives as well as boron-organic compounds, *etc.* The efficient treatment of these cases requires the description of the electron systems (*e.g.*, RAMSES representation of electron systems[1]) rather than the formal designations of valence bonds with formal orders.

The representation of biomacromolecules *via* standard connection tables is challenging because of their large size (hundreds and most often thousands of atoms). The main approach applied for an efficient representation is the use of macro atoms. A protein sequence can be represented using preliminary defined Sybil Line Notation (SLN) macro atoms such as:

{Ala:NHCH(CH3)C(=O)<*v* = 1,9>}
{Gly:NHCH2C(C=O)<*v* = 1,6>}.

An example sequence represented as SLN notation is:

HOGlyAlaAlaGlyAlaH.

A more compact bimolecular representation is the FASTA format based on the International Union of Biochemistry (IUB)/IUPAC[43] standard for amino acid and nucleic acid codes: A – adenosine, C – cytidine, G – guanine, *etc.* The sequence of *phycoerythrobilin:ferredoxin oxidoreductase* represented in FASTA is:

MNSERSDVTLYQPFLDYAIAYMRSRLDLEPYPIPTGFESNSAVVGKGKN-
QEEVVTTSYAFQTAKLRQIRAAHVQGGNSLQVLNFVIFPHLNYDLPFF-
GADLVTLPGGHLIALDMQPLFRDDSAYQAKYTEPILPIFHAHQQHLSWG-
GDFPEEAQPFFSPAFLWTRPQETAVVETQVFAAFKDYLKAYLDFVEQA-
EAVTDSQNLVAIKQAQLRYLRYRAEKDPARGMFKRFYGAEWTEEYIHGFLF-
DLERKLTVVK

Support of the full Markush[6] structures standard is quite challenging. Most chemoinformatics software system implementations support only the R-groups, which is the first type of Markush variation, that is substituent variation (an example of this type of variations represented *via* SMARTS line notation is given in Figure 3.12). The other three types of Markush variations – position variation, group frequency variation and homology variation – present a challenge regarding both the formal representation and the usage of these representations with other chemoinformatics algorithms.

3.9 Chemical Substances

The triad model for molecule representation, *i.e.*, structure, descriptors and properties (see Figure 3.3), is the basis for many public chemical databases and the majority of chemoinformatics activities including most QSAR modelling applications. However, regulatory bodies typically require information on chemical substances as produced by industry and not as single chemical structures. Furthermore, nanomaterials[44,45] are important research targets, as the extensive research activities in the last decade have shown. Nanotechnology is an increasingly dynamic area in materials science research and development, introducing novel materials with unique properties related to their size on the nanometer scale. Nanomaterial representation in chemical databases is a primary subject of the new and rapidly evolving field of nanoinformatics. A new paradigm is needed to address these challenges, namely representing information on chemical substances and materials.

According to the IUPAC definition, a *substance is a matter of constant composition best characterised by the entities (molecules, formula units, atoms) it is composed of. Physical properties such as density, refractive index, electric conductivity, melting point etc. characterise the chemical substance.*

A substance is defined in the European Union regulation REACH[46] by Article 3 and in the Classification, Labelling and Packaging (CLP) Regulation by Article 2 as: *a chemical element and its compounds in the natural state or obtained by any manufacturing process, including any additive necessary to preserve its stability and any impurity deriving from the process used, but excluding any solvent which may be separated without affecting the stability of the substance or changing its composition.*

According to REACH and CLP, a substance may contain: one or more *main constituents* that make up a significant part of that substance. The main constituents are clearly distinguished from other types of constituents such as:

- *impurities*, *i.e.*, all unintentional constituents originating from the manufacturing process or from the starting material(s). These can be the result of secondary or incomplete reactions occurring during the production and are present in the final substance even if not sought by the manufacturer.
- *additives*, *i.e.*, all constituents that are intentionally added to stabilise the substance and only for this purpose.

In this context, there is a difference between a substance and a mixture. A mixture consists of several substances. The REACH regulation categorises well-defined substances (monoconstituent and multiconstituent) and UVCBs (*U*nknown or *V*ariable composition, *C*omplex reaction products or *B*iological materials). The REACH definition of a substance encompasses all forms of substances and materials on the market, including nanomaterials.

A nanomaterial composition may require a definition of many components and complex relations between these components. For example, a typical nanomaterial contains a core and several shells (coatings) around the core, and modern nano-medicinal materials may have ten or more shells.

Figure 3.23 shows the data model of the Ambit chemoinformatics platform used for the representation of substances.

Substances are characterised by their "composition" and are identified by their names and IDs. Each constituent may be represented *via* the standard triad approach (see also Figure 3.3). The substance composition is not just a simple list of constituents but complex relations between the constituents may exist. Each constituent may have its own properties (typically, nominal physicochemical properties) that are distinguished from the substance properties. The substance's physicochemical and biological

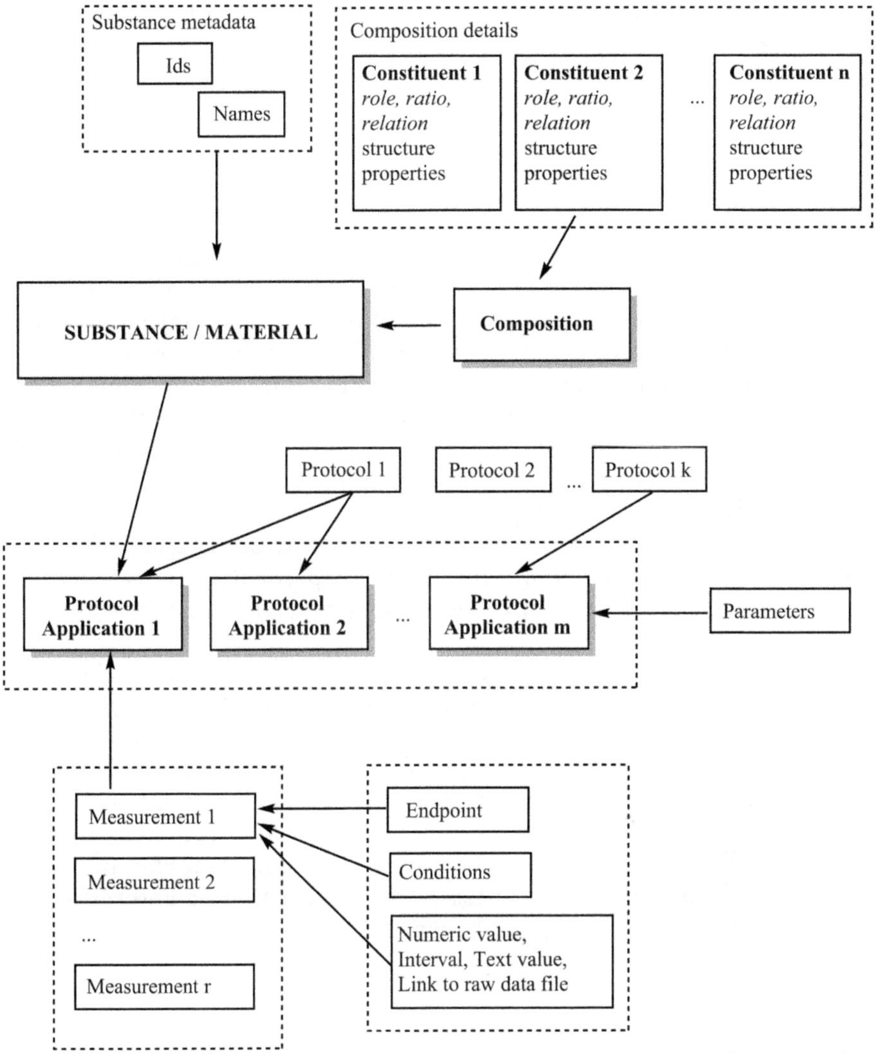

Figure 3.23 Ambit data model implemented on the basis of the substance paradigm.

properties are described by means of rich experimental data and supporting metadata (used for the interpretation of the experimental values), usually represented in a multi-layer fashion, *e.g.*, the ISA[47] (Investigation-Study-Assay) data model. A test protocol for a substance/material is described by the "protocol application" entity. Each protocol application consists of a set of "measurements" for a defined "endpoint" under given "conditions". The measurement result can be a numeric value with or without specified uncertainty, an interval, a string value, or a link to a raw data file (for example, a microscopy image). This data model is successfully used to represent chemical substances and safety data from European Chemicals Agency (ECHA) dossiers[48,49] and nanomaterials.[50]

3.10 Conclusions

In summary, the chemical structure is key for storing, retrieving or using chemical information. Moreover, structural features are essential for predictive modelling. The unambiguous identification of chemicals is a prerequisite, as is the speed of searching, retrieving and processing information (*e.g.*, database screening, substructure searching, similarity searching, database clustering, classification). This is even more important in the big data era with an increasing amount of information on chemical compounds available, which need to be linked to and between chemicals, as well as across different large databases and big data sets.

Therefore, the representation of a chemical structure is crucial for handling chemical data chemoinformatically. In particular, 3D geometrical information is also essential for considering biological activities such as ligand–receptor interactions.

This chapter gives an overview of the different aspects of chemical structure representation, from level 0D to 4D, and different techniques for chemical information formalisation. Structural descriptors have been introduced as well as fingerprints and hash codes. We have seen the specific challenges in unique structure representation, particularly issues such as aromaticity and tautomerism, as well as the chemical substance concept. These all have to be taken into account for chemical software applications and in the context of storing, searching and retrieving chemical-related data in big databases and their use for predictive modelling purposes.

References

1. J. Gasteiger and T. Engel, *Chemoinformatics: A Textbook*, WILEY-VCH Verlag, 2003, DOI: 10.1002/3527601643.
2. A. R. Leach and V. J. Gillet, *An Introduction to Chemoinformatics*, Springer, 2007, DOI: 10.1007/978-1-4020-6291-9.
3. D. L. Massart, B. G. M. Vandenginste, S. N. Deming, Y. Michotte and L. Kaufman, *Chemometrics: A Textbook*, Elsevier, 1988.
4. J. Gasteiger, *Handbook of Chemoinformatics: From Data to Knowledge in 4 Volumes*, 2003, DOI: 10.1002/9783527618279.
5. P. W. Kenny and J. Sadowski, Structure modification in chemical databases, in *Chemoinformatics in Drug Discovery*, ed. T. I. Oprea, Wiley, 2005, pp. 271–285, DOI: 10.1002/3527603743.

6. J. M. Barnard, A comparison of different approaches to Markush structure handling, *J. Chem. Inf. Comput. Sci.*, 1991, **31**(1), 64–68.
7. SMARTS theory manual, Available at http://daylight.com/dayhtml/doc/theory/theory.smarts.html.
8. Ambit-LRI data base, Available at https://ambitlri.ideaconsult.net/tool/ui.
9. J. Figueras, Morgan revisited, *J. Chem. Inf. Comput. Sci.*, 1993, **33**(5), 717–718.
10. R. Todeschini and V. Consonni, *Handbook of Molecular Descriptors*, WILEY-VCH Verlag, 2000.
11. A. T. Balaban, Applications of graph theory in chemistry, *J. Chem. Inf. Comput. Sci.*, 1985, **25**(3), 334–343.
12. W. D. Ihlenfeldt, Y. Takahashi, H. Abe and S. Sasaki, Computation and Management of Chemical Properties in CACTVS: An Extensible Networked Approach toward Modularity and Compatibility, *J. Chem. Inf. Comput. Sci.*, 1994, **34**, 109–116.
13. ChemAxon Marvin Available at https://chemaxon.com/products/marvin.
14. ACDLabs ChemSketch, Available at http://www.acdlabs.com/resources/freeware/chemsketch/.
15. C. Steinbeck, C. Hoppe, S. Kuhn, R. Guha and E. Willighagen, Recent Development of the Chemistry Development Kit (CDK) - An Open-Source Java Library Library for Chemo- and Bioinformatics, *Curr. Pharm. Des.*, 2006, **12**(17), 2111–2120.
16. IUPAC nomenclature, Available at https://iupac.org/what-we-do/nomenclature/.
17. D. Weininger, SMILES, a Chemical Language and Information System. 1. Introduction to Methodology and Encoding Rules, *J. Chem. Inf. Comput. Sci.*, 1988, **28**, 31–36.
18. InChI, Available at https://iupac.org/who-we-are/divisions/division-details/inchi.
19. OpenSMILES, Available at http://opensmiles.org/.
20. L. C. Cross and W. Klyne, Rules for the Nomenclature of Organic Chemistry: Section E: Stereochemistry, *Pure Appl. Chem.*, 1976, **45**, 11–30.
21. N. Jeliazkova and N. Kochev, AMBIT-SMARTS: Efficient Searching of Chemical Structures and Fragments, *Mol. Inf.*, 2011, **30**(8), 707–720.
22. SMIRKS – A Reaction Transform Language, Available at http://daylight.com/dayhtml/doc/theory/theory.smirks.html.
23. R. W. Homer, J. Swanson, R. J. Jilek, T. Hurst and R. D. Clark, SYBYL Line Notation (SLN): A Single Notation To Represent Chemical Structures, Queries, Reactions, and Virtual Libraries, *J. Chem. Inf. Model.*, 2008, **48**(12), 2294–2307.
24. W. Massa, *Crystal Structure Determination*, Springer Berlin/Heidelberg, 2004.
25. B. Fürtig, C. Richter, J. Wöhnert and H. Schwalbe, NMR spectroscopy of RNA, *ChemBioChem*, 2003, **4**(10), 936–962.

26. R. Erni, M. D. Rossell, C. Kisielowski and U. Dahmen, Atomic-resolution imaging with a sub-50-pm electron probe, *Phys. Rev. Lett.*, 2009, **102**(9), 096101.
27. The Cambridge Crystallographic Data Centre, Available at http://www.ccdc.cam.ac.uk/.
28. ROTATE Classic – Generation of Sets of Diverse Conformations, Available at https://www.mn-am.com/products/rotate.
29. K. M. Elokely and R. J. Doerksen, Docking Challenge: Protein Sampling and Molecular Docking Performance, *J. Chem. Inf. Model.*, 2013, **53**(8), 1934–1945.
30. J. L. Melville and D. H. Jonathan, On the Stability of CoMFA Models, *J. Chem. Inf. Comput. Sci.*, 2004, **44**(4), 1294–1300.
31. Talete srl, DRAGON for Windows, Software for Molecular Descriptor Calculations, Version 5.4, 2006, http://www.talete.mi.it//.
32. C. W. Yap, Software News and Update PaDEL-Descriptor: An Open Source Software to Calculate Molecular Descriptors and Fingerprints, *J. Comput. Chem.*, 2011, **32**(7), 1466–1474.
33. W. T. Wipke, S. Krishnan and G. I. Ouchi, Hash Functions for Rapid Storage and Retrieval of Chemical Structures, *J. Chem. Inf. Comput. Sci.*, 1978, **18**(1), 32–37.
34. G. Wohlgemuth, S. S. Mehta and R. F. Mejia, *et al.*, SPLASH, a hashed identifier for mass spectra, *Nat. Biotechnol.*, 2016, **34**(11), 1099–1101.
35. Q. H. Dang, Secure Hash Standard, National Institute of Standards and Technology, Federal Inf. Process. Stds, 2015, Available at https://doi.org/10.6028/NIST.FIPS.180-4.
36. W. D. Ihlenfeldt and J. Gasteiger, Hash codes for the identification and classification of molecular structure elements, *J. Comput. Chem.*, 1994, **15**(8), 793–813.
37. S. Kim, P. A. Thiessen and E. E. Bolton, *et al.*, PubChem Substance and Compound databases, *Nucleic Acids Res.*, 2016, D1202–D1213.
38. Y. Cao, T. Jiang and T. Girke, Accelerated similarity searching and clustering of large compound sets by geometric embedding and locality sensitive hashing, *Bioinformatics*, 2010, **26**(7), 953–959.
39. R. A. Sayle, So you think you understand tautomerism?, *J. Comput.-Aided Mol. Des.*, 2010, **24**(6–7), 485–496.
40. A. Gobbi and M.-L. Lee, Handling of tautomerism and stereochemistry in compound registration, *J. Chem. Inf. Model.*, 2012, **52**(2), 285–292.
41. V. H. Masand, D. T. Mahajan, T. Ben Hadda, R. D. Jawarkar, A. M. Alafeefy, V. Rastija and M. A. Ali, Does tautomerism influence the outcome of QSAR modeling? *Med. Chem. Res.*, 2014, **23**(4), 1742–1757, DOI: 10.1007/s00044-013-0776-0.
42. N. T. Kochev, V. H. Paskaleva and N. Jeliazkova, Ambit-Tautomer: An Open Source Tool for Tautomer Generation, *Mol. Inf.*, 2013, **32**(5–6), 481–504.

43. IUPAC-IUB Joint Commission on Biochemical Nomenclature (JCBN), Nomenclature and symbolism for amino acids and peptides. Recommendations 1983, *Biochem. J.*, 1984, **219**(2), 345–373.
44. European Commission, Commission Staff Working Paper, Types and uses of nanomaterials, including safety aspects, Brussels, 2012, Available at https://ec.europa.eu/health/sites/health/files/nanotechnology/docs/swd_2012_288_en.pdf.
45. The Opensource Handbook of Nanoscience and Nanotechnology, Available at http://en.wikibooks.org/wiki/Nanotechnology.
46. European Chemicals Agency, Guidance in a Nutshell Identification and naming of substances under REACH and CLP, version 2.0, 2017, https://echa.europa.eu/documents/10162/23036412/nutshell_guidance_substance_en.pdf/cca556cd-4f68-4b38-a29b-58cb6da31a93.
47. S.-A. Sansone, P. Rocca-Serra, A. Gonzalez-Beltran and D. Johnson, *ISA Model and Serialization Specifications 1.0*, 2016, DOI: 10.5281/zenodo.163640.
48. Q. Li, V. Koch and U. Jensch, *et al.*, Development of the Clariant CompTox Suite: A chemoinformatic system to support read-across and category formation within the scope of safety assessments of substances, *Toxicol. Lett.*, 2015, **238**(2), S172–S173.
49. N. Jeliazkova, V. Koch and Q. Li, *et al.*, Linking LRI AMBIT chemoinformatic system with the IUCLID substance database to support read-across of substance endpoint data and category formation, *Toxicol. Lett.*, 2016, **258**, S114–S115.
50. N. Jeliazkova, C. Chomenidis and P. Doganis, *et al.*, The eNanoMapper database for nanomaterial safety information, *Beilstein J. Nanotechnol.*, 2015, **6**, 1609–1634.

CHAPTER 4

Organisation of Toxicological Data in Databases

DAVID BOWER, KEVIN CROSS AND GLENN MYATT*

Leadscope, Inc., 1393 Dublin Road, Columbus, Ohio 43215, USA
*E-mail: gmyatt@leadscope.com

4.1 Introduction

Databases containing historical toxicity studies are used extensively to assess the safety of chemicals. These databases are being used directly to answer questions such as "*Has this compound been tested before?*" or "*What were the experimental conditions where a specific finding was previously seen?*" These databases can also be used to support prediction of different types of potential hazards in the absence of actual laboratory data when all that is known is the chemical structure. Questions such as "*Is this compound potentially mutagenic?*" or "*What is the estimated NOAEL for this compound?*" are being asked of these databases or information derived from these databases. To answer these questions, toxicology databases are searched to identify similar chemicals that have historical toxicity studies. The results from these studies are then used to generate a prediction for a specific type of hazard or safe dose level. This approach to gap filling is usually referred to as *read-across*.[1–4] These databases are also analysed and computational models are developed from them, which are used to make predictions. These models include Quantitative Structure–Activity Relationship (*QSAR*) models and expert rule-based models (*alerts*).[5] Prior to model building, it is usual to export and process the

Issues in Toxicology No. 41
Big Data in Predictive Toxicology
Edited by Daniel Neagu and Andrea-Nicole Richarz
© The Royal Society of Chemistry 2020
Published by the Royal Society of Chemistry, www.rsc.org

data into a form that is more amenable to computational modeling. Information from any look-up, read-across, QSAR models or alerts is used as part of a weight-of-evidence assessment and used to generate an opinion on the safety of chemicals.

Toxicity databases are an important resource to support the development of new types of toxicity studies, especially alternatives to the current battery of *in vivo* tests. This analysis has led to understanding a number of important findings including details of the threshold of toxicological concern (TTC).[6] These databases also provide support for validation of new methods, including new *in vitro* tests, high content screening, bioreactors, and different omics technologies.[7] It is important to integrate this information with historical toxicity study data for benchmarking and validation of new approaches.

There are a wide variety of established toxicity studies that have been used and are currently being routinely used to support hazard and risk assessment (see Chapter 2). These studies are generated during chemical product development, to support regulatory approval, for general assessment of chemicals exposed to humans or in the environment as well as for general research. These studies include acute and chronic toxicity studies, developmental and reproductive studies, carcinogenicity and genotoxicity studies, skin sensitisation and irritation studies, as well as environmental exposure-related studies. These toxicology studies are performed by a variety of organisations, from national laboratories and commercial companies involved in chemical R&D, to contract research organisations and academic laboratories. The results are often reported either in a publication or in study reports covering the chemical tested, the experimental protocol, the study findings and overall outcome. As part of an organisation's operation, they may use an electronic database to capture some of this information to support workflow management as well as Good Laboratory Practice (GLP) information and other regulatory compliance issues. The data may also have been collated into a local database or entered into an Excel™ spreadsheet or other document as part of a research project.

The data generated across diverse organisations is in many different non-standardised forms (mostly paper reports with some electronic databases or other electronic documents) covering different types of studies. Information may have been captured at different levels, for example only high-level endpoints may have been collected (such as Lowest Observed Adverse Effect Level or LOAEL values or positive/negative/equivocal calls) or details on individual animals treated with histopathology, clinical chemistry and hematology details recorded. The experiment design and report may have been recorded inconsistently, using different terms to mean the same experimental design concept or finding. This makes the task of developing a unified method for organising this data challenging since disparate data generated by potentially different organisations will need to be mapped into a harmonised form. This type of common organisation relies on ontologies and controlled vocabularies to ensure consistency across all studies. Since many of the historical studies are not in an electronic form, a mechanism for harvesting paper records to create consistently recorded studies is a requirement for creating such a database. In addition, any electronic databases (public or proprietary) will need to be mapped efficiently onto this database organisation. A common data exchange format is required to bring all this information together from different study types, multiple sources and organisations.

This standardised information will need to be included into some type of database management system that stores and efficiency indexes the data (such as a MySQL database).[8] Consistent with a common exchange format, any toxicity database organisation ideally handles information on the diverse experimental protocols and study designs, the test results, and the overall study results. The database should include details on compounds that were tested and link records tested across common chemicals uniquely identified through a chemical registration system. It should be possible to integrate the database with new and emerging methods including omics and high-content screening. The organisation of the data should also support its intended uses such as search and retrieval of relevant results and experimental conditions, generation of reports, extraction of data for computational analysis and modelling, methods to support read-across and export capabilities. It is becoming increasingly common to provide well-organised electronic data on toxicity studies to regulatory agencies as part of the regulatory approval process. These requirements for organising such a toxicity database to both manage the data as well as to support a variety of applications are summarised in Figure 4.1.

Organising the chemical and toxicity data in a harmonised manner is the most effective approach to supporting hazard and risk assessment of new or existing chemicals in a timely manner. A database system that contains chemical structures, toxicity study and endpoint data, as well as related

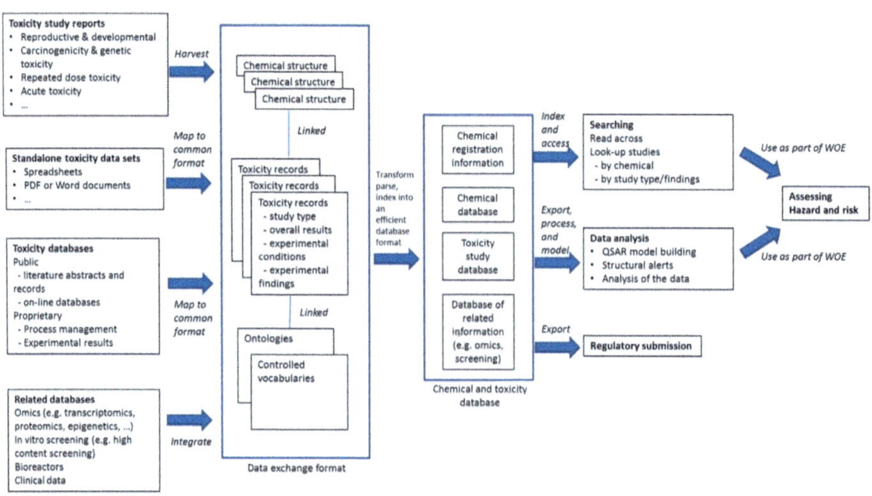

Figure 4.1 Competing needs for organising a toxicological database.

data will enable searching across chemicals, study types and experimental results. This type of search and retrieval will enable QSAR model building, identification of new structural alerts, and read-across approaches that can provide valuable input into an evidence-based assessment to hazard and risk. A well-structured harmonised database will also support the electronic submission of toxicity study results now being required by many regulatory authorities.

The following chapter covers a variety of approaches to data exchange and database design for organising toxicity and related data: SD files, ToxML, SEND, IUCLID, and ISA-tab. The chapter highlights a number of standalone databases covering reproductive and developmental studies, acute and chronic studies, as well as carcinogenicity and genotoxicity studies. A number of databases are described for project management and regulatory compliance as well as for housing information on alternative approaches. The chapter is concluded with a description of how these systems can integrate diverse toxicology-related data in a database, taking into consideration the further rapidly increasing volume and variety of the data in the future.

4.2 Standards for Exchange and Organising Toxicology Data

4.2.1 SD Files

The Structure Data (SD) file format has been the *de facto* standard for communication of chemical structure information since its development by Molecular Design Limited (MDL) (now a part of Biovia) in the 1990s.[9] It was originally designed to describe the connectivity of a single chemical structure in a text-readable format (called a Connection Table File), known as a MOL file. The SD file format extended this concept allowing for representation of multiple structures in a single file and data associated with each structure.[10] All data are stored as textual strings consisting of name-value pairs allowing for a representation of endpoint data where the endpoints are not predefined. Data values may or may not be present in each structure record. Although the format intended that only a single value be represented for an endpoint, users have included repeated data values per endpoint and made other user modifications without MDL endorsing them as extensions to the format.

Figure 4.2 illustrates the use of an SD file to capture information on the chemical structure, the chemical identifiers, the toxicity study type and endpoints and mode of action information.[11] In this example modified from a DSSTox[12] record, the first part of the SD file contains information on the atoms and bonds of the chemical structure along with their connectivity and coordinates (to support search and display). Following the definition of the chemical structure, a series of fields are shown containing the data on this chemical structure, including the chemical identifiers, calculated or measured properties of the chemical, the type of toxicity study and overall call

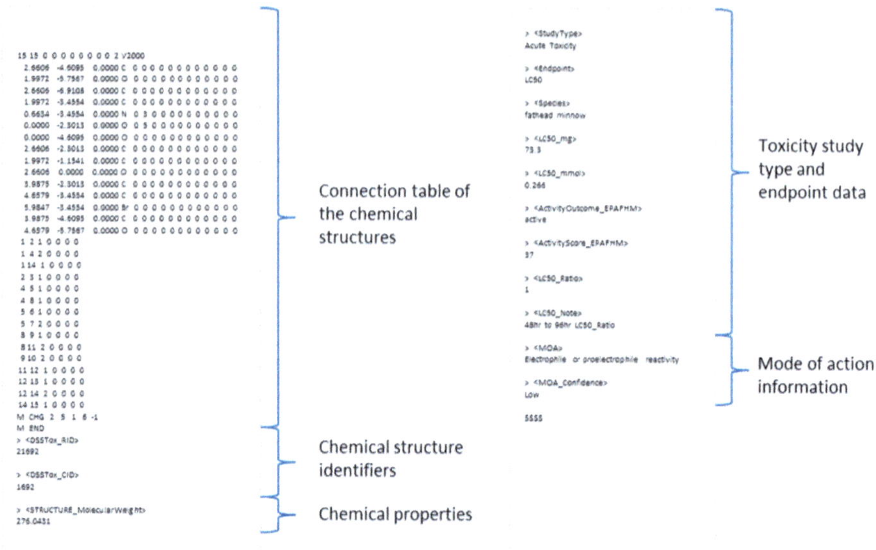

Figure 4.2 An example of an SD file containing information on the chemical structure and data.

as well as other relevant information, such as mode of action data (shown in this example). The "$$$$" signifies a delimiter separating this chemical record from the next, which would contain similar data fields for the next chemical.

In subsequent years, MDL extended the format to include structure query representation, variable structure representation (Markush structures), polymers, reactions, and 3-D information, resulting in a series of formats (RXN files, RG files, RD files) supported within the SD file. They also developed an XML-based format (XD files) for transferring structures with associated hierarchical data. This format includes metadata and source information to aid in the correct interpretation of the enclosed data, supports structure record formats other than the MOL format, and does not require the presence of structure fields per record.

Despite these improvements, the popularity of XML has spawned many other customer-driven formats that include MOL records yet do not conform to the Document Type Definition (DTD) developed by MDL for XD files. The lack of adoption of the XD file format for representing toxicology data was most likely due to the absence of a rich data definition representation specific for representing toxicology data (or data for any other scientific domain). Toxicity experiments and study information are well-defined in terms of their compound–study–test–treatment level relationships. A rich data representation needs to include the entire study design and parameters, compound details, results from several studies and tests as well as experimental results such as compound and study level calls, observations, and other result findings.

4.2.2 ToxML (DB)

ToxML (http://www.leadscope.com/toxml.php) is an open standard designed to facilitate the exchange, integration and storage of complex hierarchical toxicological study data using a well-structured electronic format.[13] The initial and early versions were developed cooperatively in agreement with diverse representatives and input from regulatory, commercial and academic entities.

ToxML is defined in Extensible Markup Language (XML) format in conjunction with a controlled vocabulary to ensure the consistency and normalisation of data. The XML format provides a useful and logical approach to modelling the data contained in toxicological studies as it is a hierarchical data representation. In addition to the hierarchical structure of XML, the format is an accepted industry standard in data modelling that supports both textual and binary data. ToxML currently covers a broad range of toxicology studies including, but not limited to, genotoxicity studies (bacterial mutagenesis, *in vitro* and *in vivo* chromosome aberration, micronucleus and mammalian mutagenesis), single-(acute) and repeat dose studies (carcinogenicity, chronic and subchronic), reproductive and developmental studies, and skin irritation, skin penetration and skin sensitisation studies.

The chemical substance of interest serves as the fundamental key (or root node) in ToxML making it the pivotal entity to which all chemical property and toxicological study data is associated within a given ToxML record. The ToxML hierarchy is further organised by study, test, and treatment level information progressing from a higher level of study information to more detailed results information. The conceptual depiction of the data organisation in ToxML can be seen in Figure 4.3.

The compound section of ToxML contains fields meant to unambiguously identify and characterise the chemical substance itself as well as sections to capture any summarised or derived data corresponding to toxicological studies associated with the chemical substance. Highlighted in Figure 4.3, moving from left to right along the top, are three of the most important fields associated with the chemical compound itself, namely the Structure, IDs and Names.

The Structure field is used for storing a machine-readable structural representation of the substance. The primary format used for the structural representation is a MOL file record described previously. In ToxML, the complete MOL file is separated into the Connection Table (CTAB) portion and the Header Block. Simplified molecular-input line-entry system (SMILES) format is another commonly used chemical structure file format that is supported. Instances occur where there is either no known or pragmatic structural representation for a chemical substance or mixture that has been studied, such as plant extracts and animal venoms. Due to these situations, the structure field cannot be made a required field. A ToxML Structure example for the substance formaldehyde is shown below:

```
<Structure>
  <Molfile>
    <ConnectionTable><![CDATA[  2  1  0  0  0  0  0  0  0  0
    V2000
  −0.7500    0.0000   0.0000 C  0  0  0  0  0  0  0  0  0  0  0  0
  0.7500   0.0000   0.0000   O  0  0  0  0  0  0  0  0  0  0  0  0
  1  2  2  0  0  0  0
M  END
]]></ConnectionTable>
  <Header>
    <Line1>LS-134</Line1>
  </Header>
  </Molfile>
  <Smiles>
  <Smiles>
    <Value>C=O</Value>
  </Smiles>
  </Smiles>
</Structure>
```

The IDs field is designed to capture typed identification data that unambiguously identifies the chemical substance, such as a CAS Registry Number (http://www.cas.org/content/chemical-substances/faqs). There has been a

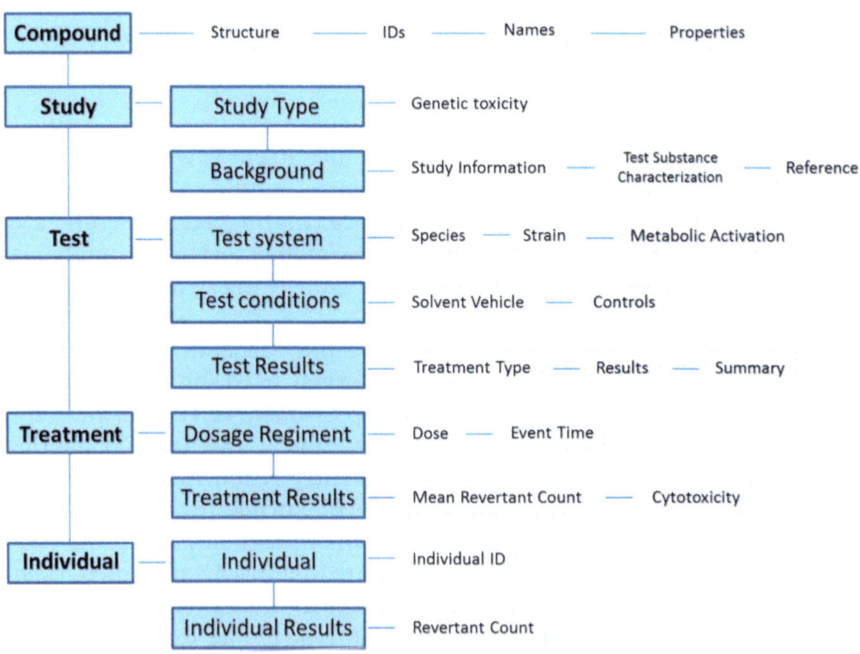

Figure 4.3 ToxML conceptual organisation.

dramatically increase in the availability of both public and private chemical substance databases, many of which have their own substance identifiers. This increased availability and use of different types of substance identifiers is taken into account by ToxML by incorporating a list of typed identifier values allowing a compound to have multiple identifiers of differing types. A sample ID section of ToxML can be seen next:

```
<Ids>
  <Id type="reg">LS-134</Id>
  <Id sources="ntp,isscan,tox21,rtecs" type="cas">50-00-0</Id>
  <Id sources="rtecs" type="rtecs">LP8925000</Id>
</Ids>
```

Names are arguably just as important as identifiers in the ToxML hierarchy. Like the IDs section of ToxML, and for the same reasons, the names section of ToxML is comprised of a list of typed values as well. In the chemical industry, the types of names often encountered are either systematic or trade names. Systematic names contain rule-based descriptions that unambiguously define a substance in terms that a chemist understands and can reproduce the exact picture of the chemical substance. Trade names, on the other hand, are simply names used in marketing, sales and as common references for substances. A very good example of systematic *versus* trade names can be seen with Tylenol:

```
<Names>
  <Name type="tradename">Tylenol</Id>
  <Name type="tradename">Paracetamol</Id>
  <Name type="iupac">N-(4-hydroxyphenyl)acetamide</Id>
</Names>
```

In this example, the entry listed with type="iupac" is indicating that the value is a name derived for chemical structure using nomenclature rules established by the International Union of Pure and Applied Chemistry (IUPAC). The other two entries in the list, Tylenol and Paracetamol, are common names under which this over-the-counter pain relieving and antipyretic medication is marketed and sold but do not contain any fundamental information about the physical structure of the substance.

The toxicological study portion of ToxML contains numerous levels of hierarchical data and is much more complex than the areas dedicated to identifying chemical substances. In general, the hierarchy progresses in a manner that naturally follows the scientific method moving from high-level study design information to distinct experimental testing data and finally experimental results and overall conclusions.

The study level contains information fundamental to the type, background, compliance and overall results of the particular toxicological study data being captured. The information at this level should be thought of as information that is common across virtually all toxicological studies and scientific studies in general. A few key sections of ToxML at this level include the Study Type, Background, Study Calls, and Study Call Criteria. The Study Type is a required field that defines the specific type of toxicological study (or assay) for which data is being captured including values such as "Bacterial Mutagenesis Study" or "Carcinogenicity Study" and will have implications on the sections that will appear within the lower, more detailed levels.

Background information can provide researchers with a valuable insight as to how reliable a particular study record might be considered. The Background section within a given study provides an area to record information such as a study report number, title, protocol, start date, end date, compliance and references. Researchers may be interested to know if a particular protocol was followed or whether or not the study was conducted under a Good Laboratory Practice directive. The time periods during which toxicological studies are conducted can also be a vital piece of information due to the fact that testing methods are refined, improved and evolve over time as new insights are gained. A sample of ToxML demonstrating the Background information:

```
<Background>
  <StudyReportNumber>AA65WK.123.BTL</StudyReportNumber>
  <StudyTitle>In vivo Mouse Micronucleus Assay Following Subcutaneous
  Administration of ALX-0600</StudyTitle>
  <References>
    <Reference>http://www.accessdata.fda.gov/drugsatfda_docs/
    nda/2012/203441Orig1s000PharmR.pdf#page=176</Reference>
  </References>
  <GLPQACompliance>GLP compliant</GLPQACompliance>
  <Confidentiality>Non-confidential</Confidentiality>
  <StudyStartDate>20020925</StudyStartDate>
</Background>
```

The Study Call and Study Call Criteria are two of the most important data fields captured at the study level. The Study Call Criteria defines the particular set of rules or methodology used in determining what value was assigned in the Study Call field. A typical Study Call Criteria for a Salmonella only Ames assay is: "Criteria for a positive response were: for the tester strains TA98 and TA100, a 2-fold increase over the mean revertants per plate of the appropriate vehicle control; for tester strains TA1535, TA1537, and TA1538, at least a 3-fold increase over the mean revertants per plate of the appropriate vehicle control." If these conditions were met, the Study Call value would be "Positive." In the context of ToxML this portion would look as follows:

```
<StudyCallCriteria>Criteria for a positive response were: for the tester
strains TA98 and TA100, a 2-fold increase over the mean revertants per plate
of the appropriate vehicle control; for tester strains TA1535, TA1537, and
TA1538, at least a 3-fold increase over the mean revertants per plate of the
appropriate vehicle control.</StudyCallCriteria>
<StudyCalls>
```

```
    <Call>Positive</Call>
  </StudyCalls>
```

The test level encompasses information and details about the test system such as species, strain, sex and metabolic activation system; test conditions such as assay technique, solvent vehicle, and experiment id; negative and positive test control information such as the control compound name and the treatment results for the controls; and, importantly, the test results.

Test levels in ToxML are split into their own separate tests based primarily on any difference in either the test system or test conditions. *In vitro* assays typically utilise multiple strains within a given species. For example, a bacterial mutation study using the TA98, TA100, TA1535 and TA1537 strains of salmonella with and without a metabolic activation system should have eight tests defined in a ToxML record. The same principal holds true for *in vivo* assays where parameters like sex vary and cause the tests to be grouped accordingly. A single test record sample demonstrating the test system, test conditions and test results is shown next:

```
<Test>
  <TestSystem>
    <Species>Salmonella typhimurium</Species>
    <Strain>TA100</Strain>
    <MetabolicActivation>Present</MetabolicActivation>
    <MetabolicActivationSystem>
      <SystemType>S9</SystemType>
      <Species>Syrian hamster (mesocricetus auratus)</Species>
      <Sex>Male</Sex>
      <Tissue>Liver</Tissue>
      <FinalConcentration>10%</FinalConcentration>
    </MetabolicActivationSystem>
  </TestSystem>
  <TestCondition>
    <SolventVehicle>
      <Substance>
        <Substance>Water</Substance>
      </Substance>
    </SolventVehicle>
    <ExperimentId>3</ExperimentId>
    <TestId>3</TestId>
    <AssayTechnique>Preincubation</AssayTechnique>
  </TestCondition>
  <TestResults>
    <TestCalls>
      <Call>Weakly positive</Call>
    </TestCalls>
  </TestResults>
</Test>
```

The treatment level is comprised of one treatment group for each dosage being tested within a given test record. The treatment groups contain information about the dosage regimens, subject counts (for *in vivo* studies) and the treatment group results including any comments, observations, assay-specific results, and treatment-related outcomes or effects. The data fields at this level naturally contain the most variation across similar *in vitro* and *in vivo* assays, making the disparate fields available only within the context of their respective assays.

The dosage regimen, as defined in ToxML, is the most homologous data of the treatment groups across toxicological assays. Typically, a dosage regimen will contain a dose value with units, a dose duration (the period of time over which the dose was administered) and a treatment time (the time period for which the subject was exposed to the substance). A treatment group for an *in vitro* assay where cells were treated with 12.5 $\mu g\ kg^{-1}$ of a test substance continuously exposed for 72 hours would look like the following:

```
<TreatmentGroup>
  <DosageRegimen>
    <Dose>
      <Value>12.5</Value>
      <Units>micro-g kg⁻¹</Units>
    </Dose>
    <TreatmentTime>
      <Value>72.0</Value>
      <Units>hour</Units>
    </TreatmentTime>
    <DoseDuration>
      <Value>72.0</Value>
      <Units>hour</Units>
    </DoseDuration>
  </DosageRegimen>
</TreatmentGroup>
```

Treatment results, as would be expected, is one area where data fields can, and often do, have more differences than most other areas within ToxML. Bacterial mutagenesis assay results are considerably different than *in vitro* chromosome aberration assay results as far as the detailed results that are recorded and reported (see also Chapter 2). The delineation of these disparate results is a naturally occurring artefact of the hierarchical structure of ToxML. A comparison of a bacterial mutation assay result and an *in vitro* chromosome aberration assay:

```
<TreatmentGroup>
  <Results>
```

```
      <RevertantCount>1126.0</RevertantCount>
      <RevertantCountSignificance>80.6+/−SEM</RevertantCount
Significance>
    </Results>
  </TreatmentGroup>
  <TreatmentGroup>
    <Results>
      <MitoticIndexMean>4.3</MitoticIndexMean>
      <TotalCellsScored>200</TotalCellsScored>
      <PercentStructureAberrationWithoutGaps>1.5</PercentStructure
      AberrationWithoutGaps>
      <PercentStructureAberrationWithGaps>4.0</PercentStructure
      AberrationWithGaps>
    </Results>
  </TreatmentGroup>
```


ToxML also has the capability to capture data down to the individual subject level within a given treatment group. These results sections are almost an exact parallel to the results captured at the group level adding in an identifier for the individual subject and omitting data fields that are based on populations greater than one. The following is an example of an *in vivo* micronucleus assay result for an individual subject including a few of the result data fields for the treatment group as a whole (notably the PCENCERatio, MNPCEPer1000PCE and MNPCESignificance data fields):


```
  <TreatmentGroup>
    <IndividualSubjectCount>6</IndividualSubjectCount>
    <Results>
      <PCENCERatio>0.48</PCENCERatio>
      <MNPCEPer1000PCE>0.035</MNPCEPer1000PCE>
      <MNPCESignificance>Not significant</MNPCESignificance>
      <IndividualResults>
        <IndividualResult>
          <SubjectId>2595</SubjectId>
          <MNPCE>
            <MNPCECount>1</MNPCECount>
            <TotalPCECount>2000</TotalPCECount>
          </MNPCE>
          <PercentPCE>
            <PCENCERatio>0.42</PCENCERatio>
          </PercentPCE>
        </IndividualResult>
      </IndividualResults>
    </Results>
  </TreatmentGroup>
```

A complete example of an *in vivo* micronucleus study has been provided in the Appendix.

4.2.3 SEND

SEND is the Standard for Exchange of Nonclinical Data with regulatory agencies. It is an implementation of the Clinical Data Interchange Standards Consortium (CDISC) Standard Data Tabulation Model (SDTM)[14] for representing nonclinical data. The SDTM defines a standard structure for study data tabulations that specifies how testing results from studies can be aggregated together into a set of self-contained data files and modelled according to three general observation classes: Interventions, Events, and Findings. Interventions are investigational, therapeutic, and other treatments administered as specified in the protocol. Events are planned protocol milestones. Findings are the observations resulting from tests usually at specific time points.

SDTM provides a standardised table model for submitting study data tabulations to regulatory authorities consistent with current industry practices and regulatory requirements, which specify the use of the SAS Version 5 transport format.[15] However, efforts are currently underway to replace the transport format with an XML format. SDTM was originally developed to handle submissions of human clinical trial data.[16]

SEND is an implementation of STDM that includes those parts relevant to nonclinical studies and supplemented with domains and variables that are specific to non-human studies.[17] As a standard, it defines a common semantic model for describing toxicological data endpoints (called "data domains") for single-dose toxicity, repeat-dose toxicity and carcinogenicity studies. It uses a controlled vocabulary that contract research organisations (CROs), vendors, companies and regulators can adopt to facilitate the exchange of data. Development of the standard was driven by the need to move towards the electronic submission of data to regulatory agencies. With compliance for electronic submission required by the U.S. FDA by the end of 2016, usage of SEND as a communication standard has been implemented by several vendors (such as Instem[18] and PointCross[19]) who have developed tools to support reading, writing, managing and analysing data in SEND datasets.

The conceptual organisation of the SEND format for a non-clinical *in vivo* study is presented in Figure 4.4. The relationships are similar to those in other representations such as ToxML. However, in contrast, the study data is tabulated by animal subject and there is an emphasis on traceability, that is, maintaining the representation of raw data as it was collected as well as after data standardisation. The approach also differs as SEND uses a data

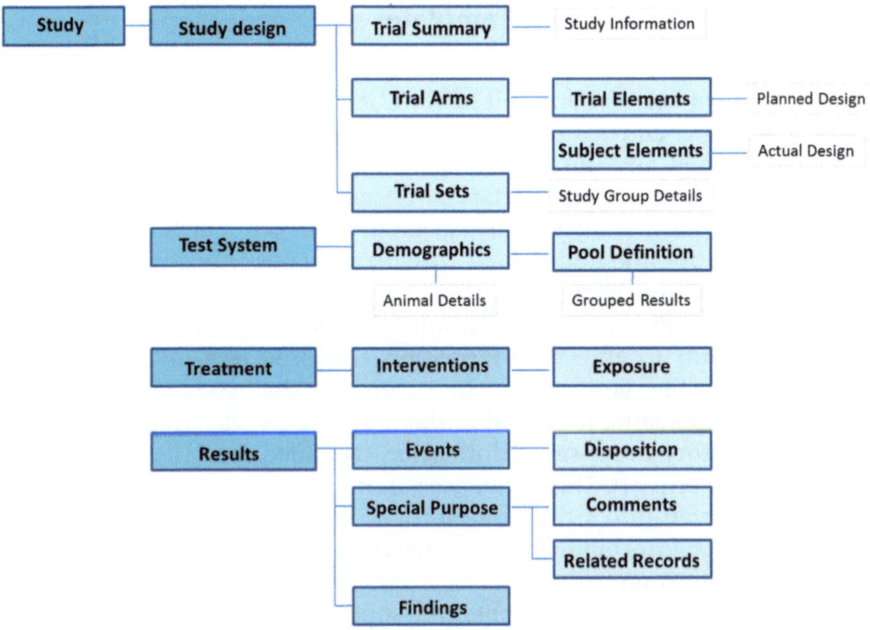

Figure 4.4 SEND format conceptual organisation.

model built around conducting a non-clinical trial of animal subjects *versus* a collection of toxicity studies for assessing one or more drugs or chemicals. SEND places more emphasis on how a study was planned and how the study was actually conducted. SEND datasets are organised as a set of relational tables where the hierarchical relationships are implemented through a set of foreign keys and unique identifiers. Identifiers used for unique record identification and relating records include study ids, domain ids, and animal subject ids.

The SEND model is rigid in its representation of specific types of endpoint data through its definition of domains. However, extending the model to include new study types is fairly straightforward and involves defining and adding a new domain as part of the Findings class.

The SEND controlled vocabulary is an important component of study data standardisation and is a critical component of achieving semantically-interoperable data exchange. Controlled terminology specifies the key concepts represented as preferred terms, definitions, synonyms, codes, and code systems. To ensure consistency, controlled terminology standards are maintained by the standards organisations (*i.e.*, external to the sponsor or applicant).

4.2.4 OECD Harmonised Templates

The Organisation for Economic Co-operation and Development (OECD) has developed a set of harmonised XML templates for transmission of dossier data on chemicals to European regulators, such as the European Chemicals Agency, and the OECD High Production Volume (HPV) Chemicals Program, reflecting properties or effects on human health and the environment.[20] Each study type is a separate, independent XML template that was developed and added for a specific regulatory purpose over time. Unlike other formats, these templates were not developed independently as a standard but as part of a software and database system called the International Uniform ChemicaL Information Database (IUCLID).[21] IUCLID was developed under contract for the European Commission and is freely distributed to sponsors. It is used in the preparation and transmission of data for regulatory purposes. Information compiled into IUCLID can be exchanged with other IUCLID databases or other databases supporting the format or exported into a reporting format. IUCLID is the recommended system to gather information within the OECD High Production Volume (HPV) Chemicals Program.

The harmonised templates contain a superset of all data required for reporting. The required fields for reporting vary by regulatory authority and country jurisdiction. The templates are designed for reporting study summary information and not full study reports. The templates include a controlled vocabulary of almost 10 000 terms to standardise study summary results and ensure exchangeability. Each template is structured into five main parts, featuring the data fields common to most templates. These include:

1) Administrative Data – record id, submitted substance id, date, author, history, study type, study summary, conclusions, executive summary, *etc.*
2) Data Source – reference, authors, source, testing laboratory, *etc.*
3) Materials and Methods – test guideline, GLP compliance, test material information, *etc.*
4) Results and Discussion – tables of results and their discussion
5) Overall remarks and attachments – attached background material and full study report.

Currently, templates are defined for 129 different study types ranging from *in vivo* and *in vitro* genetic toxicity and carcinogenicity to skin irritation and honey bee toxicity, as well as physico-chemical properties and including nanomaterial specific properties/endpoints. Most OHTs correspond to specific OECD Test Guidelines (see Chapter 2). OHT 201 captures studies of intermediate effects.

4.2.5 ISA-Tab

A number of alternative approaches for assessing toxicity that do not rely on traditional *in vivo* experiments are being developed. These methods incorporate several different technologies including: omics approaches (*e.g.*, transcriptomics, proteomics, metabolomics, *etc.*), high-content screening methods, bioreactors, cell biology, and computational toxicology. This bioscience information is being used to group compounds based on biological similarity for read-across purposes, identify biomarkers, support understanding mechanisms of action, predict hazards and establish exposure thresholds. The use of this information is severely limited without standards and ontologies to describe, annotate, organise, transfer and archive this information. This level of annotation ensures a minimum amount of information is used to describe the experiment and supports subsequent analysis of the data.

The ISA-Tab universal data format is a general framework to represent bioscience experiments, including the high-content screening studies, chemical analysis, and omics experiments.[22–24] It describes experiments including the study design, details of how individual assays are performed and links to external data files containing the results. The majority of terms used to describe the experiments are linked to ontology entries to support a common understanding of the meta data and its use in subsequent data analysis.

ISA-Tab models an experiment using four concepts: (1) investigations; (2) studies; (3) assays; (4) data:

- *Investigation*: an investigation describes the high-level concepts and is linked to one or more related studies;
- *Study*: the study ISA-Tab unit contains information on the subject under study (*e.g.*, a cell or tissue), its characteristics and any treatments applied;
- *Assay*: this ISA-tab unit describes tests performed on treated material described in the study unit, which produces qualitative or quantitative measurements;
- *Data*: datasets (*e.g.*, tab-separated spreadsheets) or external files in native formats (*e.g.*, Affymetrix CEL files) are linked to the experimental meta data described in the assay.

The following example illustrates how ISA-Tab can be used to represent transcriptomics data. The investigation contains general information on the experiment, including a title, abstract, publications, and authors. In this ISA-Tab example, there is only one study unit with some general information (identifier, title, abstract, *etc.*) and a single test or assay (transcriptomics) that was performed. The study unit also contains some information on the experimental design of the study, including the study factors (compound, dose, sample time point), the individual steps and protocols describing them (sample growth and treatment protocol, RNA extraction, *etc.*), as well as any study-related publications. The study design is outlined in a separate table that includes information on the study subject and characteristics (*e.g.*, organism, sex, organ, tissue, cell, and so on). The different values for each experimental factor are enumerated, with each row representing a unique treated sample. There are three study factors in this experiment (compound, dose, and sample time point) that were used as treatments performed in duplicate. The total number of entries in this study is 280 and each entry is named under the fields "Sample Name". For example, the "Sample Name" value SN301601603 represents the first replicate of a treatment of paracetamol at a dose of 200 micro-molar at a sample time point of 2 hours. The assay describes the sequence of steps with which the samples (described in the study table) were processed (the experimental graph). These steps are linked to individual protocols (described in the study unit). The content of the table is based on the template defined for the individual assay and may contain parameters such as "label". Each entry in the table will be based on the treated samples described in the study unit table. The processing of these samples is described, where a "Name" column is included at each step indicating how the different samples are combined. In this example, the "Data Transformation Name" is used to combine data generated from four samples in the data transformation step. The four rows are given the same name (*e.g.*, "RIF 2 h High *vs.* Control") describing how data from two samples treated with Rifampicin (RIF) and the two 2 hour control samples are combined. Where data is generated, it is linked to a file containing the raw data (which may be linked to ISA-Tab meta data (Figure 4.5)).

A series of tools are provided by the ISA group at the University of Oxford to support the community of users of ISA-Tab. This includes the following utilities:

- *ISAcreator configurator*: tools to support the generation of templates that describe experimental tests.
- *ISAcreator*: generates ISA-Tab formatted archives (a compressed collection of files containing all experimental meta data and linked files) using a desk-top tool enabling a user to enter the investigation, study, and assay level meta data and link to individual data files. The tool integrates with existing ontologies to provide semantic annotation of the fields and is customisable.
- *Validators*: tools to ensure an ISA-Tab record does not contain any errors.
- *Converters*: a series of converters are available to convert ISA-Tab records into formats for submission to public databases (such as ArrayExpress[25] and PRoteomics IDEntifications (PRIDE) database[26]) as well as to the Resource Description Framework (RDF).[27]
- *Analysis*: there are applications to perform data analysis such as a Risa[28] that parses ISA-Tab into the R data structures.[29]

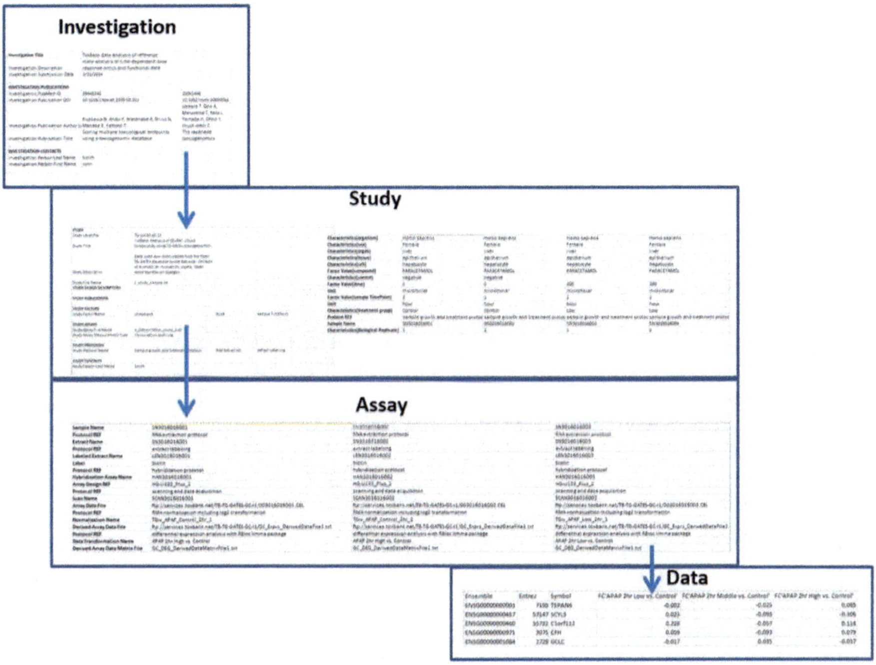

Figure 4.5 ISA-Tab format.

4.3 Databases and Data Sets of Toxicity Studies

Tables 4.1–4.4 list a series of toxicity databases, which are either general toxicity databases or that have been created focusing on toxicity data for reproductive/developmental studies, acute/chronic toxicity studies, and carcinogenicity/genetic toxicity studies. It is beyond the scope of this chapter to cover all databases in detail; however, for illustration, a number of examples are given in the following.

The Registry of Toxic Effects of Chemical Substances (RTECS®) database is a comprehensive collection of basic toxicity information including: prescription and non-prescription drugs, food additives, pesticides, fungicides, herbicides, solvents, diluents, chemical waste, reaction products of chemical waste and substances used in industrial and household situations.[30,31] The RTECS database covers six primary categories of toxicity data: acute toxicity, tumorigenicity, mutagenicity, skin and eye irritation, reproductive effects and multiple dose effects. In late 2011, Accelrys also began adding aquatic and *in vitro* toxicology data. It currently covers over 180 000 compounds with updates adding approximately 2000 new

Table 4.1 Summary of general toxicity databases (non-exhaustive list).

Database	Description	Link/reference
COSMOS	COSMOS Next Generation (COSMOS NG), resource for toxicological knowledge (COSMOS EU project)	https://ng.cosmosdb.eu
DrugMatrix	Reference database containing studies of the effects of these compounds at multiple time points and in multiple target organs. These studies included pharmacology, clinical chemistry, hematology, histology, body and organ weights, and clinical observations.	https://ntp.niehs.nih.gov/results/drugmatrix/index.html
DSSTox	Distributed Structure-Searchable Toxicity (DSSTox) Database Network (US EPA)	https://www.epa.gov/chemical-research/distributed-structure-searchable-toxicity-dsstox-database
ECHA	European Chemicals Agency database for REACH	https://echa.europa.eu/information-on-chemicals/registered-substances, https://echa.europa.eu/search-for-chemicals
IPCS INCHEM	Chemical safety information from intergovernmental organisations (International Programme on Chemical Safety (IPCS))	http://www.inchem.org
JECDB	Japanese Existing Chemical Data Base	http://dra4.nihs.go.jp/mhlw_data/jsp/SearchPageENG.jsp
Leadscope Toxicity Database	Toxicity study data	http://www.leadscope.com/toxicity_databases/
NTP CEBS	Chemical Effects in Biological Systems database (National Toxicology Program (NTP))	22 https://manticore.niehs.nih.gov/cebssearch

Pharma Pendium	Toxicity data from US FDA and EMA approval documents	http://www.pharmapendium.com
RTECS	Registry of Toxic Effects of Chemical Substances (National Institute for Occupational Safety and Health (NIOSH))	16 http://www.cdc.gov/niosh/rtecs/default.html
TOXNET	Toxicology Data Network (National Institutes of Health (NIH))	https://toxnet.nlm.nih.gov/help/newtoxnet/toxnetfs.html
TOXNET HSDB	Hazardous Substances Data Bank (National Institutes of Health (NIH))	https://toxnet.nlm.nih.gov/newtoxnet/hsdb.htm
ToxCast	Chemical screening data from the ToxCast project and Tox21 collaboration, accessible *via* iCSS ToxCast Dashboard (US EPA)	https://actor.epa.gov/dashboard/
ToxRefDB	Toxicity Reference Database containing *in vivo* toxicity studies (US EPA)	43 https://cfpub.epa.gov/si/si_public_record_report.cfm?Lab=NCCT&dirEntryId=227139, https://www.epa.gov/chemical-research/exploring-toxcast-data-downloadable-data
US FDA ICSAS	Informatics and Computational Safety Analysis Staff databases from the US FDA	http://www.accessdata.fda.gov/scripts/cder/ob/default.cfm
VITIC	Data from published and unpublished sources (Lhasa)	https://www.lhasalimited.org/products/vitic.htm

Table 4.2 Summary of reproductive and developmental toxicity databases (non-exhaustive list).

Database	Description	Link/reference
EDKB	Endocrine Disruptor Knowledge Base (US FDA, National Center for Toxicological Research (NCTR))	75 http://www.fda.gov/ScienceResearch/BioinformaticsTools/EndocrineDisruptorKnowledgebase/default.htm
ILSI	Developmental toxicity database (International Life Sciences Institute (ILSI))	76
Leadscope Drugs Reproductive and Developmental Database	Reproductive and developmental toxicity studies from the US FDA CDER (Center for Drug Evaluation and Research) product approval reviews.	http://www.leadscope.com/toxicity_databases/regulatory_databases/
Leadscope Food Safety Reproductive and Developmental Database	Reproductive and developmental studies from the US FDA CFSAN (Center for Food Safety and Applied Nutrition) reviews.	http://www.leadscope.com/toxicity_databases/regulatory_databases/
Mario Negri (CASEAR)	Mario Negri Institute for Pharmacological Research Developmental and reproductive toxicity database	https://qsartoolbox.org/resources/databases/
REPROTOX	Pregnancy, reproduction, and development data	77 http://www.reprotox.org/
Shepard's catalog	Teratogenic agents	78
TERIS	Teratogen Information System: Teratogenic exposures in pregnant women	79–84
TOXNET DART	Developmental and Reproductive Toxicology and Environmental Teratology Information (National Institutes of Health (NIH))	85 http://toxnet.nlm.nih.gov/newtoxnet/dart.htm
TOXNET LactMed	Drugs and Lactation Database. Possible adverse effects in the nursing infant (National Institutes of Health (NIH))	86 https://toxnet.nlm.nih.gov/newtoxnet/lactmed.htm
US FDA ICSAS	Informatics and Computational Safety Analysis Staff databases from the FDA	87 and 88

Table 4.3 Summary of acute and chronic toxicity databases (non-exhaustive list).

Database	Description	Link/reference
COSMOS	COSMOS Next Generation (COSMOS NG), resource for toxicological knowledge (COSMOS EU project)	https://ng.cosmosdb.eu
eTox	Drug safety database from the pharmaceutical industry legacy toxicology reports (Innovative Medicines Initiative (IMI) project eTox)	http://www.etoxproject.eu/
HESS	Hazard Evaluation Support System Integrated Platform (National Institute of Technology and Evaluation (NITE))	89 https://www.nite.go.jp/en/chem/qsar/hess-e.html
Leadscope Drugs Chronic and Subchronic Database	Chronic and subchronic toxicity database from the US FDA CDER (Center for Drug Evaluation and Research) product approval reviews.	http://www.leadscope.com/toxicity_databases/regulatory_databases/
Leadscope food safety acute toxicity database	Acute toxicity data from the US FDA CFSAN (Center for Food Safety and Applied Nutrition) PAFA (Priority-based Assessment of Food Additives) database.	http://www.leadscope.com/toxicity_databases/regulatory_databases/
Leadscope Food Safety Chronic and Subchronic Database	Chronic/subchronic toxicity database from the US FDA CFSAN (Center for Food Safety and Applied Nutrition) reviews.	http://www.leadscope.com/toxicity_databases/regulatory_databases/
Mario Negri	Mario Negri Institute for Pharmacological Research Repeated dose toxicity databases	https://services.toxbank.net/
RepDose	Repeated dose toxicity database (Fraunhofer Institute for Toxicology and Experimental Medicine)	https://repdose.item.fraunhofer.de

Table 4.4 Summary of carcinogenicity and genetic toxicity databases (non-exhaustive list).

Database	Description	Link/reference
TOXNET CPDB	Carcinogenicity Potency Database (National Institutes of Health (NIH))	https://toxnet.nlm.nih.gov/newtox-net/cpdb.htm
IARC	International Agency for Research on Cancer (IARC) and carcinogenicity classification	https://monographs.iarc.fr/agents-classified-by-the-iarc/
ISSCAN	Chemical carcinogens: structures and experimental data from Istituto Superiore di Sanita	90
TOXNET CCRIS	Chemical Carcinogenesis Research Information System. Chemical carcinogens: structures and experimental data. Developed by the National Cancer Institute (NCI).	91 https://toxnet.nlm.nih.gov/newtox-net/ccris.htm
Leadscope Drugs Genetox Database	Genetic toxicity database from the US FDA CDER (Center for Drug Evaluation and Research) product approval reviews.	http://www.leadscope.com/toxicity_databases/regulatory_databases/
Leadscope Food Safety Genetox Database	Genetic toxicity database from the US FDA CFSAN (Center for Food Safety and Applied Nutrition) reviews.	http://www.leadscope.com/toxicity_databases/regulatory_databases/
Leadscope SAR Genetox	Genetic toxicity database with validated structures to be used as a resource for preparing training sets for predictive systems creation.	http://www.leadscope.com/toxicity_databases/
Leadscope SAR Carcinogenicity	Carcinogenicity toxicity database with validated structures to be used as a resource for preparing training sets for predictive systems creation.	http://www.leadscope.com/toxicity_databases/
NTP	National Toxicology Program (NTP) genetic toxicology database	22
TOXNET GENE-TOX	Mutagenicity data from the US EPA	https://toxnet.nlm.nih.gov/newtox-net/genetox.htm
IRIS	Integrated Risk Information System. Data from the US EPA in support of human health risk assessment	https://www.epa.gov/iris

compounds per year. At this time, the database contains 279 237 acute toxicity studies for 154 702 chemicals, 8017 irritation studies for 4880 chemicals, 52 730 multiple dose studies for 13 953 chemicals, 46 385 RTECS mutation studies for 13 343 chemicals, 25 558 reproductive studies for 3724 chemicals, and 10 517 tumorigenic studies for 3724 chemicals. The high-level endpoint data includes LC50, LD50, LOAEL, TCLo, TD and TDLo values for select species and routes of administration and includes over 220 000 individual values.

The Leadscope Genetic Toxicity and Carcinogenicity Databases are two toxicity databases containing validated structures to be used as a resource for preparing training sets for predictive systems creation. The following sources were used: (1) the FDA Centre for Food Additives and Applied Nutrition (CFSAN) Food Additive Resource Management system (FARM) and Priority based Assessment of Food Additives (PAFA)[32]; (2) the US FDA's Center for Drug Evaluation and Research (CDER) Pharmacology Reviews based on the New Drug Approval (NDA) documents;[33,34] (3) the Chemical Carcinogenicity Research Information System (CCRIS);[35] (4) the National Toxicology Programs (NTP) genetic toxicology database;[36] (5) the Tokyo-Eiken database;[37] (6) and other publications. Since the CCRIS database no longer harvests public data, this database also contains entries from recent publications. The Leadscope Genetic Toxicity database includes 11 028 compounds with 179 732 test results from: CDER; CFSAN; NTP; CCRIS; and other primary data sources. The database provides overall compound calls for 46 genetic toxicity endpoints: 33 calls for bacterial mutagenesis; 4 calls for *in vitro* mammalian mutagenesis; 6 calls for *in vitro* chromosome aberration; 6 calls for *in vivo* micronucleus. The Leadscope Carcinogenicity database includes 3598 compounds with 11 538 test results from: CDER; CFSAN; NTP; CCRIS; and other primary data sources. Available endpoints with populations include: male rat (1774), female rat (1725), male mouse (1640) and female mouse (1675).

The RepDose (repeated dose toxicity) database was designed to support a variety of needs including assessing the threshold of toxicological concern concept,[38,39] read-across approaches or development of integrated testing strategies for risk assessment.[40–43] The RepDose database was developed at the Fraunhofer Institute of Toxicology and Experimental Medicine under the framework of the long-range research initiative of the Chemical Industry[44,45] and is continuously updated. One-third of the RepDose data are on-line.[46] The selection criteria adopted for the RepDose database includes both chemical classes and specific study types. The chemicals include mainly commercially available organics, some pesticides and a few drugs. On average 3–4 repeated dose studies following well-established guidelines, *i.e.*, OECD, are included per chemical. For each study, the design, scope and observed toxicological effects were recorded. Studies are extracted from literature, *e.g.*, NTP reports, MAK documentations, BUA reports, CICADs, EU risk assessment reports and peer-reviewed publications or are provided by companies. Non-guideline studies are also included when a study addresses a specific question. All studies are annotated with an indication of study quality.

The database focuses on studies with oral (gavage, feeding, and drinking water) or inhalation studies performed in rodents with exposure periods ranging from 14 days to lifetime. Some dog and dermal studies were added for certain scientific questions. RepDose uses a MySQL database and includes information on the tested chemical namely structural information and experimental physico–chemical data (*e.g.*, molecular weight, solubility in water, physical state, boiling point, dissociation constant) as well as the toxicological data. This toxicological information covers both the experimental design of the study as well as any observed effects. The experimental design provides details on the number of animals per dose group, the sex and the strain as well as the exposure protocol (duration, route, dose groups, and any post-exposure observation period). The observed effects are recorded for each target such as histopathological changes in critical organs, hematology, clinical chemistry and clinical symptoms with an indication of the Lowest Observed Effects Level (LOEL) per effect. An overall LOEL per study is also included. The database makes heavy use of standardised and curated glossaries for all fields in the database covering both the study design and the observed effects. The database currently includes in total about 880 chemicals and 3000 studies, 350 chemicals with 900 inhalation studies/700 compounds with 2000 oral studies.

COSMOS Next Generation (COSMOS NG)[47] is an openly available resource for toxicological knowledge, derived from the European Union and Cosmetics Europe co-funded COSMOS Project,[48] which was part of the SEURAT-1 (Safety Evaluation Ultimately Replacing Animal Testing) Research Initiative[49] focussed on the development of alternative methods for repeated dose toxicity assessment without the use of animal experiments. COSMOS NG provides access to a database as well as *in silico* profilers for toxicity and other associated chemistry-based predictive workflows and provides a platform to share high-quality chemistry and toxicity data. The database comprises a significant resource of toxicological and safety information pertinent to human health.[50] In 2019, the database in COSMOS NG has access to over 44 000 chemical structures and includes an inventory of cosmetics ingredients comprising more than 19 000 substances and 5500 individual chemical structures. Data are available for in excess of 12 500 toxicity studies for 1660 compounds across 27 endpoints relating to human health. There is an emphasis on repeated dose toxicity data, focusing on those for cosmetics ingredients. COSMOS NG also includes a downloadable, updated dataset for the development of Threshold of Toxicological Concern (TTC), which has been enriched with data for cosmetics ingredients.[51]

For further examples of toxicological databases, see also Chapter 2 on biological data sources and Chapter 7 for omics data repositories.

4.4 Process Management and Regulatory Compliance Databases

In December 2014, the US FDA issued a final Guidance for Standardised Study Data that should be provided as part of regulatory submissions in electronic format.[52] This guidance mandates that studies starting 24 months later (from December 2016) must use the appropriate FDA-supported standards, formats, and terminologies specified in the Catalog for NDA, Abbreviated New Drug Approval (ANDA), and some Biologics License Application (BLA) submissions. The Standard for Exchange of Nonclinical Data (SEND) describes the datasets and associated documents required for preclinical electronic submission.[5] Producers of SEND datasets include the submitting company as well as CROs generating test results.

Supporting regulatory submissions requires developing and maintaining compliant storage and workflow systems for testing data and supporting information for preparation of initial SEND submissions as well throughout the review process and afterwards for post-approval monitoring. These storage systems contain toxicity study data from CROs and other corporate sources, chemical representation, formulation, and synthetic preparation information, screening, omics, and other related safety information. As clinical trials are performed and additional approvals sought (NDA, ANDA) beyond the initial investigation approval (Investigative New Drug Approval), a long-term, compliant, central repository is needed for tracking regulatory submissions as well as supporting regulatory queries.

For toxicity information, regulatory storage functionality can be divided into two parts: that functionality needed to support the storage and maintenance of the data, and that required to generate electronic submissions and track them during the approval process. The storage of toxicity data requires accepting and/or manipulating testing data from possibly several CROs and/or internal sources and integrating the results from studies and tests from multiple sources into submission sets for specific compounds. This may be in either a GLP or non-GLP environment, but should support 21 CFR Part 11 compliance.[53] More sophisticated databases may support process management, from study design and protocol management to resource management and scheduling technician duties such as data collection and management of animal room activities (*e.g.*, Instem Provantis[6]).

The toxicity database may be linked or integrated with a chemical compound database whose focus is to store and maintain chemical information such as stereochemical structure representation, formulation, source, purity and relationships to other chemicals. More sophisticated chemical databases may support chemical registration, track inventory and generate labels for safe handling (*e.g.*, Instem Provantis[6]).

Reporting capabilities of a regulatory database are required for both internal communications as well as communications with regulatory agencies.

Internal reports support safety review, regulatory queries and decision support. Capabilities include reporting individual animal studies, group summary tables, incidence analysis by animal group, symptom and severity, time to response, mass location diagrams, and raw data and audit trail reports.

Analytical tools for assessment of test data are required for safety assessment and decision making. Capabilities should include the statistical analysis features toxicologists expect, such as definition of trial sets, group averages, assessment of outliers and exclusions. Additionally, features for a regulatory compliant database should include types of analyses that agencies might perform (for example, creation of trial sets that are different from the original data and analyses involving different outliers and exclusions) (*e.g.*, Instem Provantis,[6] PointCross DSIMS[7]).

Informatics tools may also be applied to regulatory compliance databases for data mining existing data warehouses. Big data techniques may analyse trends across related compounds, studies, animals, *etc.* Informatics tools may also be employed to support quality control through identification of data and information errors or to support research and development efforts.

The driving force for employing regulatory databases is not a move towards a paperless work environment but a pragmatic need to support electronic submissions to regulatory agencies. Most important is the capability to create and validate SEND documents. SEND dataset creation requires taking electronic data from multiple sources and converting it into the SEND equivalent. Terminology must be mapped into the SEND controlled vocabulary from that used by CROs or sponsors or other companies. Data sets must be organised, standardised, and integrated together into a SEND dataset corresponding to a SEND study design. These data set definitions and vocabulary mappings may require significant intellectual analysis and time investment and may have to be performed for each data source. There are several vendors with products for creating SEND datasets (Instem Submit[6] and PointCross DSIMS[7]).

Rules have been defined for validating SEND datasets.[54] SEND dataset generators include comprehensive checks of the datasets to ensure they conform to the rules of the current SEND Implementation Guide. Company-specific rules can be added for SEND dataset checking to support enhanced checking of SEND submissions and to ensure that additional variables are provided for internal data warehouse population in sponsor organisations. Viewing and editing of SEND documents is required for quality control prior to submission. Users need to thoroughly review their datasets, checking the translations used to generate the SEND dataset and to address issues found by automated checkers.

Process management involves management and creation of automated workflows for retrieving and processing SEND datasets and associated documents and files. This management extends to submission, review tracking, handling regulatory inquires, version control supporting changes, audit trail logs, and long-term compliant storage of the datasets.

4.5 Alternative Approaches Databases

4.5.1 ToxBank

ToxBank was one of the projects in the SEURAT-1 cluster co-funded by the European Union and Cosmetics Europe.[55,56] It was initiated in part to ensure data and protocols generated as part of long-term European research initiatives into alternative methods are made available to the global scientific community. The majority of other projects in SEURAT-1 were focussed on the development of new methods, which include mode-of-action-driven *in vitro* methods based on cell lines of human origin, bioreactor development, the identification of biomarkers, and computation models to support hazard and risk assessment.

The ToxBank data warehouse is a publicly available on-line system designed to support the diverse and geographically distributed SEURAT-1 cluster. Partners in the SEURAT-1 initiative were generating a wide variety of experimental data and protocols, including transcriptomics, proteomics, metabolomics, epigenetics, high-content screening, chemical analysis, and so on. This data is being supplemented with data generated from other groups, such as Tox21[57] or the TG-GATES project[58.59] The system was designed to allow investigators to upload protocols, reports and data and well as to search, analyse and retrieve this information. This process includes appropriate investigation and protocol registration and version control and also allows investigators to set and modify access controls to the information over time.

The ToxBank data warehouse uses the open ISA-Tab universal data exchange format to represent, import and access any formatted investigation results. The ISA-Tab is generated using a version of ISAcreator customised for use with the SEURAT-1 project. The software supports the generation of a complete ISA-Tab archive that describes the experimental design, the assays performed and links to the generated data. Forms are available to describe the three primary units of an ISA-Tab investigation: investigation, study, and assay.

The ISAcreator software has been customised to allow for additional annotation of investigations with registered SERUAT-1 authors and projects as well as specific keywords used to organise information across the project. This customisation also links protocols uploaded to the warehouse to each step of the investigation. To support subsequent searching of the ISA-Tab meta data through the ToxBank data warehouse, a number of specific fields were required to be completed and linked to ontologies. These include cells or cell lines as well as treated chemicals. Since the warehouse provides the opportunity to perform substructure, similarity and exact match searching on the chemicals treated in the experiment, the compound field needs to be linked to chemical ontologies such as the OpenTox[60] and CheMBL[61] ontologies.

Investigations are also annotated with a SEURAT-1 customised ontology that supports searching and organisation of information across the project. Once the ISA-Tab formatted investigation is prepared within the ISAcreator software and validated to ensure it does not contain any errors, an ISA-Tab archive can be generated. This archive is then uploaded to the ToxBank data warehouse (see Figure 4.6). The ISA-Tab investigation is stored using the Resource Description Framework (RDF), which supports linked resource and semantic web approaches.

The ISA-Tab standard provides a mechanism for standardising and organising meta information about the experiment. There is a limited format for representing the experimental data; so, to support both searching of the information and integration of analysis of the data, the underlying processed (endpoint) data has been standardised for use in the SEURAT-1 project. Most experiments generate some type of processed data. For example, transcriptomics generates fold changes for individual genes (generated from the treated sample and the control) and high-throughput screening generates AC50 values for individual compounds tested in a particular assay (generated from an analysis of the concentration response). The processed data is standardised to support searching; for example, to identify all investigations where the fold change for a particular gene is greater than 1.5. It is also helpful when analysing the data, such as for identifying enriched biological pathways (for a particular treatment or dose) based on important genes identified from a transcriptomics experiment.

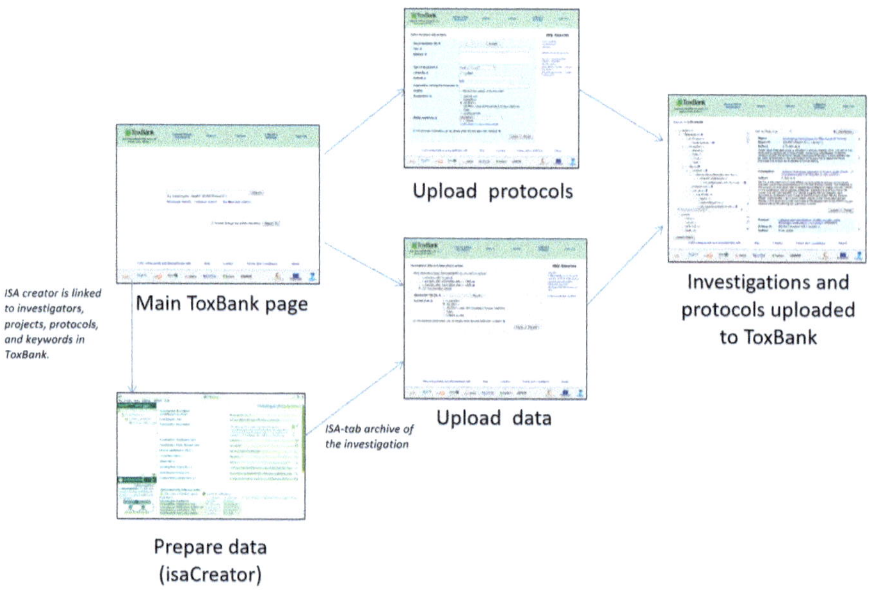

Figure 4.6 How data is organised and uploaded to the ToxBank data warehouse.

4.5.2 ToxCast/DSSTox/Tox21

ToxCast is a multi-year research project being undertaken by the US Environmental Protection Agency.[62] As part of this effort, approximately 700 cell- or protein-based high-throughput screening assays are being run against over two thousand chemicals. These assays probe toxicity-related signaling pathways and are being investigated to help prioritise chemicals of concern for further testing in programs such as the US EPA's Endocrine Disruption Screening Program or the Toxic Substance Control Act. Similar data are generated as part of the Tox21 project, which includes participation from the National Institute of Environmental Health Sciences/National Toxicology Program, the National Institutes of Health/National Chemical Genomics Center (NCGC), the Environmental Protection Agency, and the Food and Drug Administration. This group is generating screening information on close to 10 000 chemicals. The ToxCast and Tox21 data along with information on animal studies and other related data resources is being made available through the US EPA computational toxicology research web site.[63] The data has been processed, annotated and organised and made available in multiple formats, including the iCSS ToxCast Dashboard.[64] The data format includes information on the chemical structures, annotations of the assay, raw and processed data (as well as the step used to process the data) and a database of related *in vivo* data called ToxRefDB.[65]

The website contains details of the chemicals tested and provides information concerning the chemical structures, and Quality Control (QC) grade and description.[66] The chemical library currently consists of 1860 unique chemical substances that were screened as part of the ToxCast Phase II library screening plus chemicals from the full Tox21 library, totalling 8599 substances. The chemical structure records include the standard fields used for all DSSTox-related files, including names and structure IDs, information on the tested form, information on the chemical structure (*e.g.*, SMILES string) and properties (*e.g.*, molecular weight). An SD file containing all chemical structures is also provided. The QC fields consist of a grade (*e.g.*, Pass, Pass Molecular Weight, Caution, and Not determined) as well as information on the purity, whether the molecular weight was confirmed and status information on the sample, such as whether it has been withdrawn.

Assay description files are also available for download for the ToxCast and Tox21 data[42] to support interpretation of the individual assays and results. It also supports integrating this data with investigations performed elsewhere. This semantic annotation is based on the BioAssay Ontology (BAO)[67] expanded with additional information to consistently describe the ToxCast assay parameters. This annotation includes descriptions of the assay design types and readout information. It also annotates specific technology targets (such as mRNA) and intended targets (such as enzyme activity).

The data generated from the ToxCast and Tox21 groups is collected by the US EPA and processed in six steps to generate the final data files, once they are converted to a common format for each assay supplier: step 1: assigns replicate and concentration indexes; step 2: removes or defines individual data values based upon the well quality; step 3: applies specific normalisation to the assay endpoint; step 4: uses different functions (constant, hill, and gain-loss) to model the concentration response; step 5: after an analysis of different concentration response models, identifies the winning model and determines the point of departure; step 6: highlights concentration series that may be erroneous or anomalies. The original data, the data derived at each step and the final processed data are made available as an SQL database along with a pipeline utilising the R statistical package[14] to process the data.[42]

4.5.3 PubChem

PubChem is an online database that includes chemical and biological activity data developed by the National Institutes of Health (NIH) as part of the Molecular Libraries and Imaging component.[68] The database currently contains over 68 000 000 (non-unique) compounds and over 1 000 000 bioassays. It is organised into three databases: PubChem Substance (includes information on chemical samples and data depositor as well as links to the biological activities), PubChem Compound (contains unique chemical structures), and PubChem BioAssay (including conditions and readout). As a partner on the Tox21 project, the NCGC is uploading Tox21 assay results to PubChem.

4.5.4 NIEHS/CEBS

The Chemical Effects in Biological Systems (CEBS) database is an online database developed and housed at the National Institute of Environmental Health Sciences (NIEHS) as an integrated repository for toxicogenomics data.[69] It is a public resource of environmental health data housing depositions of data from academic, industrial and governmental laboratories. CEBS displays data in the context of biology and study design, and permits data integration across studies. Published data may be downloaded from the site while unpublished data is available with permission from the data author.

More genomics database resources are listed in Chapter 7.

4.6 Integrated Approaches

The benefit of using a harmonised approach to organising toxicology data in a database is to be able to rapidly search and easily analyse toxicological data across different chemicals, different study types and different sources of information. This includes having access to high-level summaries as well as details of the study design and experimental data. This integrated approach ensures all relevant data is retrieved and reports are generated in a consistent manner. It also allows the data to be easily analysed and computational models built to support hazard and risk assessment. A number of systems have been developed to facilitate this type of analysis, including Leadscope,[70] VITIC, the US EPA Aggregated Computational Toxicology Resource (ACToR),[71] DSSTox,[11,44] and TOXNET[72] (see also Chapter 2 and Chapter 5 on integrated technologies). Overarching portals such as the OECD eChemPortal[73] allow simultaneous parallel searches for a chemical across different databases. ELIXIR[74] is an intergovernmental organisation coordinating a "distributed infrastructure for life-science information", including the ELIXIR Data Platform comprising Core Data Resources and Deposition Databases.

Databases of toxicity information frequently contain a variety of different types of data and sources of information. The compound record is the unique point for association of all test data and a unique key is usually assigned to a compound as a reference point. Examples of unique keys include CAS (Chemical Abstracts Service) Registry Numbers, InChIs (see Chapter 3), EC numbers, or individual database keys such as Leadscope numbers. As new compounds are added to the database, often they will be registered by structure and class information to assign a unique record for them. This registration process frequently involves comparing the structure against existing database structures to determine prior existence and to avoid duplicates in the database. This is frequently a problem when combining datasets from different sources such as adding public data to a corporate database where the conventions for representing the structure of the compound may be different. Sometimes the problem of avoiding duplicates cannot be solved solely through compound structure registration.

There may be several types of structural forms in the database: salts, free-acids and bases, stereoisomers and mixtures. The formulation used for toxicity testing may or may not be recorded as part of the database record. Different salt forms may have been tested for the same compound. The stereochemistry of the compound may not have been fully known when it was initially tested, but became elucidated over time. Since test data may be associated with different structural forms and at different points in time, it is necessary to integrate this information so that everything known about a compound can be easily referenced and retrieved together.

How to handle the integration of structural information is one design criterion of the database. All structural forms of a compound may be organised under a single record with a unique key. Alternatively, relationships between different structural forms may be defined explicitly, through named relationships or a systematic unique key assignment where a component of the key (*e.g.*, a suffix) indicates a different form (*e.g.*, a salt code). Alternatively, the relationships between different structural forms could be dynamic, whereby structure searching (exact, family, or similarity searching) could aggregate toxicity data for the test compound or a group of related structures.

Retrieval and display of all known testing data for a compound can provide a toxicological profile of the compound over several endpoints. Compound records may be retrieved by unique key, chemical names, structure searches or searching by data identifiers such as study numbers. Examination of all data for a compound may help highlight the areas of toxicological concern, such as reproductive birth defects and the exposure levels required to see toxicity effects. It may help isolate translation effects between species and genders that can only be identified by examination of multiple test results at the same time. It may help determine the specificity of the toxic effect, such as a positive result only being seen when an assay considers metabolic activation. It may indicate the relative sensitivity of different assays to the compound class of interest. Lastly, it may indicate apparent discrepancies in the data, possibly indicating that a particular test result may be suspicious or that the test subject or assay was not appropriate for the compound. Retrieval and display of the complete study information including test source, experimental protocol and conditions along with result findings can help verify suspicions or confirm conclusions.

Retrieval of compounds related to the test compound can help support the toxicological profile of the test compound. Similarity structure searching and compound clustering can help construct a toxicological profile by compound class providing further support to the endpoint study results for the test compound. Results can be considered across a range of endpoints, or used to create an overall endpoint call for the compound when multiple endpoint studies exist.

Multiple study results for a test compound frequently occurs and integration of this data must be supported. Merging of companies has resulted in the corresponding merging of their test data. Incorporation of compound libraries, such as catalogues and public databases, requires integration. Data integration requires both integrated data retrieval and data assessment. It requires more than retrieving multiple silos of information. It requires organisation of results across different studies and at different levels (such as test and treatment levels) in order to assess why duplicate studies may have different results.

Frequently, it is important to determine an overall endpoint call from duplicate studies and a corresponding confidence level. This requires assessment of the test data and a defined approach towards grading the information. Some sources of information may be trusted over other sources (such as corporate data over public data). Sometimes the differences in the experiment will determine a preferred call, such as using results from a five-strain Ames assay study *versus* a two-strain mini Ames assay or using GLP *versus* non-GLP test results. Additionally, an algorithmic approach may be used, such as averaging results or using an "any positive is positive" approach. Defining the approach for calculating an overall endpoint call to resolve conflicting results is frequently based upon the "fit-for-purpose" of the call. For example, the rules for calculating an overall negative call when assessing drug safety may be more stringent than during the drug discovery process.

To illustrate a specific implementation of an integrated approach, the Leadscope system will be examined. Leadscope integrates many sources of toxicity data by converting the distributed data files and on-line databases into a well-structured XML (Extensible Markup Language) format known as ToxML (as discussed earlier in this chapter). The conversion process involves the mapping and normalisation of toxicological study information such as result findings and dosage regimens. The resulting ToxML document is then loaded into Leadscope's structure-searchable cheminformatics system, which can be accessed *via* Leadscope's data mining application or a collection of web services. Currently, this toxicity database contains over 182 000 chemical structures with toxicity data covering acute, multiple dose studies (including chronic and subchronic), carcinogenicity, genetic toxicity, reproductive, developmental, tumorigenic and irritation, all collected from numerous sources including RTECS; NTP; DSSTox; CPDB, *etc.* Proprietary structures and toxicity data can be loaded in a similar manner and accessed behind a company's firewall to provide seamless access to public and proprietary information.

Leadscope databases represent toxicology data in a compound–study–test–treatment level hierarchy that is described and supported by the ToxML model of toxicology data. Different salts, mixtures, and free-forms of the compound are maintained as separate database records but are linked together in the database. A compound registration system determines unique structural records and assigns them a Leadscope number as a unique key. Compounds retrieved from external sources are registered by assignment of a Leadscope number to avoid record duplication and to enable integration of results from different sources under a single compound record.

Toxicology data for each compound is organised by study type. Studies within each study type are organised by toxicological endpoint. Studies from different sources are displayed as integrated results in a single spreadsheet. An example of integrated study results for acetaminophen is illustrated in Figure 4.7.

Test results for each study or set of studies may be explored by drilling down into the next level of the compound–study–test–treatment hierarchy of information. This next example illustrates how a slice of test-level information in the hierarchy is presented with common data fields in a spread-sheet for acetaminophen in Figure 4.8. The test data is that for the study selected in Figure 4.7.

Leadscope also provides a general study spreadsheet where the user has complete control over the data presented. The user can select the level of information in the hierarchy to display as well as individual data fields. This allows comparisons of data across multiple compounds, study types and

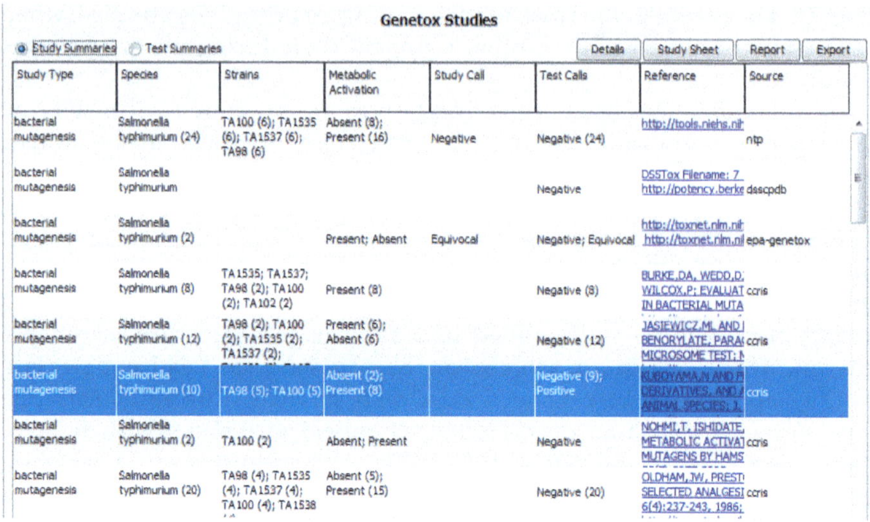

Figure 4.7 Integration of study data from several sources.

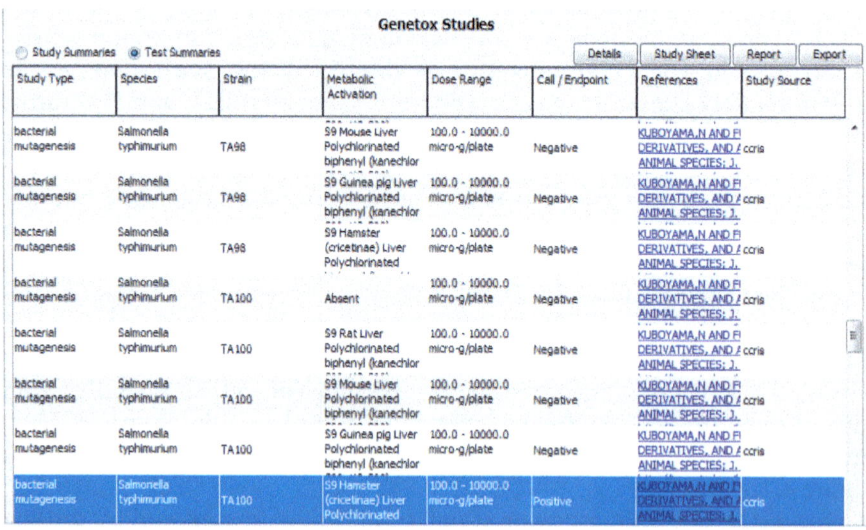

Figure 4.8 Test data presented for a single study.

endpoints as well as studies for a common endpoint. An example illustrating multiple study type results for several compounds integrated from several sources is presented in Figure 4.9.

When integrating data from several sources, Leadscope uses grading algorithms to assign overall endpoint calls for a compound. While these

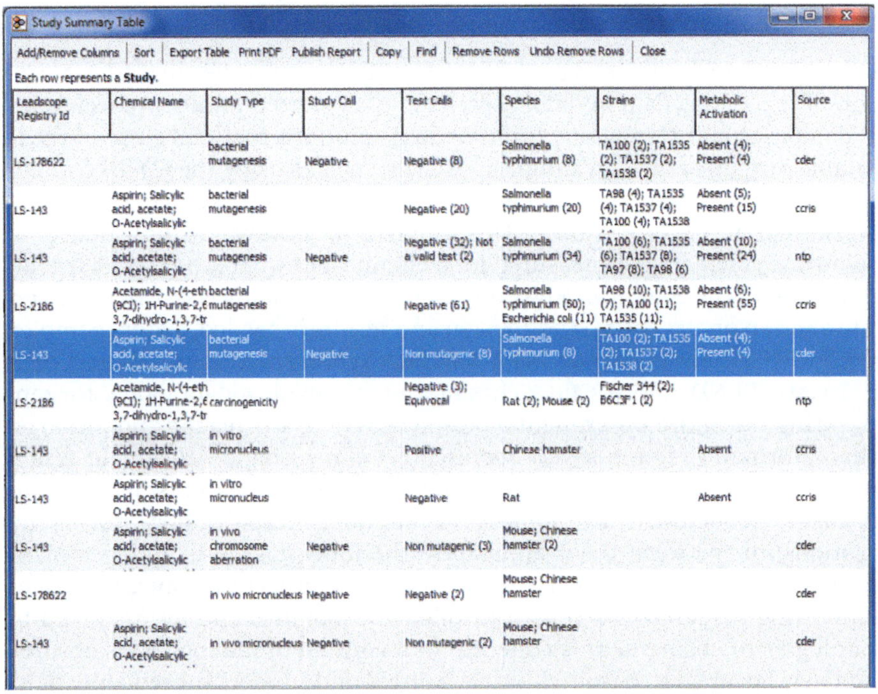
Study Summary Table

Add/Remove Columns | Sort | Export Table Print PDF Publish Report | Copy | Find | Remove Rows Undo Remove Rows | Close

Each row represents a **Study**.

Leadscope Registry Id	Chemical Name	Study Type	Study Call	Test Calls	Species	Strains	Metabolic Activation	Source
LS-178622		bacterial mutagenesis	Negative	Negative (8)	Salmonella typhimurium (8)	TA100 (2); TA1535 (2); TA1537 (2); TA1538 (2)	Absent (4); Present (4)	cder
LS-143	Aspirin; Salicylic acid, acetate; O-Acetylsalicylic	bacterial mutagenesis		Negative (20)	Salmonella typhimurium (20)	TA98 (4); TA1535 (4); TA1537 (4); TA100 (4); TA1538	Absent (5); Present (15)	ccris
LS-143	Aspirin; Salicylic acid, acetate; O-Acetylsalicylic	bacterial mutagenesis	Negative	Negative (32); Not a valid test (2)	Salmonella typhimurium (34)	TA100 (6); TA1535 (6); TA1537 (8); TA97 (8); TA98 (6)	Absent (10); Present (24)	ntp
LS-2186	Acetamide, N-(4-eth (9CI); 1H-Purine-2,6 3,7-dihydro-1,3,7-tr	bacterial mutagenesis		Negative (61)	Salmonella typhimurium (50); Escherichia coli (11)	TA98 (10); TA1538 (7); TA100 (11); TA1535 (11);	Absent (6); Present (55)	ccris
LS-143	Aspirin; Salicylic acid, acetate; O-Acetylsalicylic	bacterial mutagenesis	Negative	Non mutagenic (8)	Salmonella typhimurium (8)	TA100 (2); TA1535 (2); TA1537 (2); TA1538 (2)	Absent (4); Present (4)	cder
LS-2186	Acetamide, N-(4-eth (9CI); 1H-Purine-2,6 3,7-dihydro-1,3,7-tr	carcinogenicity		Negative (3); Equivocal	Rat (2); Mouse (2)	Fischer 344 (2); B6C3F1 (2)		ntp
LS-143	Aspirin; Salicylic acid, acetate; O-Acetylsalicylic	in vitro micronucleus		Positive	Chinese hamster		Absent	ccris
LS-143	Aspirin; Salicylic acid, acetate; O-Acetylsalicylic	in vitro micronucleus		Negative	Rat		Absent	ccris
LS-143	Aspirin; Salicylic acid, acetate; O-Acetylsalicylic	in vivo chromosome aberration	Negative	Non mutagenic (3)	Mouse; Chinese hamster (2)			cder
LS-178622		in vivo micronucleus	Negative	Negative (2)	Mouse; Chinese hamster			cder
LS-143	Aspirin; Salicylic acid, acetate; O-Acetylsalicylic	in vivo micronucleus	Negative	Non mutagenic (2)	Mouse; Chinese hamster			cder

Figure 4.9 Multiple study type results presented for similar compounds.

algorithms differ by endpoint, they consider the quality of the study source (*e.g.*, NTP), the quality of the study data (*e.g.*, strains tested), conflicts between test data in a given study, conflicts between different study results and results from different structural forms (*e.g.*, salts). The purpose of the endpoint determines whether results may be averaged or only the most significant findings considered. Examples of endpoints graded by Leadscope include: bacterial mutation, Salmonella, *E. coli*, rodent carcinogenicity, and liver toxicity.

Leadscope provides a searching capability supporting the retrieval of integrated results. This includes structure search capabilities (exact, family, substructure, similarity, stereochemical, name, id) combined with searching by data source, and the presence of particular study type or endpoint data. These different functions are combined using Boolean logic operators to construct a query operating on the database or pre-defined or user-defined subsets of data. These searches are coupled with different ways to view, filter and export the data as well as generate reports. This information is linked to data mining and model building software to extract knowledge from the chemical and toxicological information.

4.7 Conclusions

This chapter has described a number of existing approaches to the organisation of toxicology data in databases, as well as given a number of examples of databases and standalone or integrated solutions that make use of these database organisations. A simple approach, such as using the SD file format, provides a convenient data exchange format where there is a one-to-one relationship to the tested chemical. This supports the case where individual endpoints are calculated or measured for a chemical or where there is an overall study call for the chemical. However, when a toxicity study is well defined and information on the entire study design and results is needed, a more complex structured format incorporating a semantic representation is required such as ToxML, SEND, and IUCLID. These formats provide a data model for representing the entire experimental design as well as the results and findings. This information is inherently hierarchical and a single flat tabular representation (such as an SD file) is limited in the types of relationships that can be easily encoded. The flat tabular information is incapable of capturing the relationships between the overall study, the individual tests and treatment groups that make up a toxicity study. All these approaches require the use of supporting resources, such as controlled vocabularies and terms linked to ontologies to ensure data is collected in a consistent manner. This ensures data may be integrated from multiple sources and support subsequent analysis and exploration especially when looking across multiple studies and endpoints. The choice of which standard to use for organising toxicity databases also depends on the tools available to support the harvesting of the data in the desired format, tools for parsing and validating the data, as well as its intended use, such as if the data is generated for a regulatory submission or for research purposes. How to organise the related chemical structure information is also an important consideration in the development of such databases, including how to register unique chemical structures and how to link different but related chemical products, such as different salt forms.

The emergence of new techniques for assessing hazard and risk has meant that data is being generated that does not adhere to the more rigid study guidelines associated with studies that would be accepted by regulatory agencies. This includes techniques such as transcriptomics studies or high content screening. Approaches to representing this type of information are becoming increasingly important, such as ISA-Tab. These approaches also emphasise the importance of linking experimental meta data to generally accepted ontologies, such as the BioAssay Ontology (BAO), to annotate the information semantically in order to support interpretation of the results and any subsequent integrated analysis of the data. The format of the individual data results is often not standardised and it will be important to standardise this data over time. It is also becoming increasingly important to link the experimental steps to data processing protocols which may have been encoded as a pipeline.

In the future, it will become increasingly important to search and analyse large databases of historical toxicity studies alongside other types of data such as omics, high-content screening and high-throughput screening, when performing hazard or risk assessments. These diverse experiments highlight the importance of developing standards for integrating these different silos of information to support viewing and analysing the data holistically. However, the challenge in the future will be to overcome the inherent inertia that accompanies adoptions in and changes to standards for organising toxicity databases. The problem is confounded by the rapid increase of the volume and variety of the data produced – major characteristics of big data – and the cost and resources required for the on-going activities to generate high-quality databases that adhere to a data standard and also contain the latest information. These high-quality databases are the key to effective support for *in silico* approaches to safety assessment. The databases underpin all aspects of *in silico* assessment and hence the quality of these databases drives the quality of any *in silico* safety assessment.

Appendix

Complete *In vivo* Micronucleus Study Record.


```
<?xml version=“1.0” encoding=“UTF-8” standalone=“yes”?>
<Compounds>
  <Compound version=“3.0.19”>
    <Ids>
      <Id type=“cas”>7783-90-6</Id>
      <Id type=“authority”>7783-90-6</Id>
      <Id type=“reg”>LS-1735</Id>
    </Ids>
    <Names>
      <Name type=“chemName”>Silver Chloride</Name>
      <Name type=“synonym”>EK2004-0092</Name>
    </Names>
    <ToxicityStudies>
      <InVivoMicronucleusStudies>
        <Study>
          <StudyType>in vivo micronucleus</StudyType>
          <StudyCalls>
            <Call>Positive</Call>
          </StudyCalls>
```

<StudyCallCriteria>Evaluation Criteria: A positive result was defined as a statistically significant increase in the number of micronucleated PCEs for at least one of the dosing levels and a statistically significant dose-related

response. A negative result was defined as any deviation from the above criterion. In addition to statistical analysis, the final evaluation included biological relevance consideration.</StudyCallCriteria>

<StudyComments>Micronucleus Assay: Male mice in groups of 6 per dose or controls were harvested 24 hours after dosing (exceptions: 6 additional animals for 125 mg kg^{-1} and vehicle control dose level for harvest after 48 hours as well, and 3 additional animals for 125 mg kg^{-1} to ensure that 6 animals survived to harvest). Signs of toxicity included rough hair coat and/or hunched posture in all mice at 62.5 mg kg^{-1} and 125 mg kg^{-1} dose groups one hour post dose. At the 125 mg kg^{-1} dose group at the 48 hour time plot there was a statistically significant decrease in PCE : NCE ratio compared to vehicle control, indicating that the test article or its metabolites is cytotoxic to the target tissue (bone marrow). The vehicle control, corn oil, induced appropriate percent micronucleated PCEs when compared to historical control data. The positive control, cyclophosphamide, induced a statistically significant increase in micronucleated PCEs as compared to the vehicle control (mean and standard error=1.23 +/− 0.24%). The test article tested negative under study condition in the micronucleus assay.

Conclusions: In the range finding study, a maximum tolerated dose of 125 mg kg^{-1} was established based on clinical observations. No differences were observed between sexes and only males were used in the micronucleus assay. The test article induced signs of clinical toxicity, rough hair and/or hunched posture, at 62.5 and 125 mg kg^{-1} of silver chloride. There was statistically significant decrease in the PCE : NCE ratio for the 125 mg kg^{-1} dose group, indicating that the test article was cytotoxic to the target issue- bone marrow. The test article did not induce statistically significant increases in micronucleated PCEs at any dose level when compared to vehicle control responses.

Note 1: More animals were used in the study than could be accounted for in the various tables. Samples were based on at least 2000 PCEs/animal. The PCE : NCE ratio was determined by scoring the number of PCEs and NCEs observed while scoring at least 500 erythrocytes per animal.

Note 2: The study report indicates that the study sponsor, Eastman Kodak Company, was responsible for determination and documentation of identity, strength, purity, stability, and uniformity of the test article and the determination of stability, homogeneity, and concentration of the dosing preparations. If correct documentation with test substance purity, storage stability, and dose preparations were performed by the sponsor, then the study is considered to be Acceptable-Guideline.</StudyComments>

<StudyCallComments>Analytical Analysis: The micronuclei frequency was determined by analysing the number of Micronucleated PCEs from at least 2000 PCEs/animal. The PCE : NCE ratio was determined by scoring the number of PCEs and NCEs observed while scoring at least 500 erthrocytes per animal. The staining procedure permitted the differentiation, the color, of PCEs and NCEs.</StudyCallComments>

```
<EntryInformation>
  <Owner>Dave</Owner>
  <OwnerEmail>dave@somecompany.com</OwnerEmail>
  <CreationDate>20140818</CreationDate>
  <ModificationDate>20140818</ModificationDate>
</EntryInformation>
<Background>
  <StudyReportNumber>6132-202</StudyReportNumber>
  <BatchLotNumber>12-424</BatchLotNumber>
  <StudyTitle>Erexson, G.L. (2004). In vivo Mouse Micronu-
  cleus Assay.</StudyTitle>
  <Guideline>OPPTS 870.5395; OECD 474</Guideline>
  <PerformingLaboratory>Covance Laboratories Inc. (Vienna,
  Virginia)</PerformingLaboratory>
  <Duration>
    <Value>1.0</Value>
    <Units>day</Units>
  </Duration>
  <Reviewer>Steven L. Malish, PhD</Reviewer>
  <GLPQACompliance>QA & amp; GLP compliant</
  GLPQACompliance>
  <Confidentiality>Confidential</Confidentiality>
  <ReportDate>20050727</ReportDate>
  <StudyStartDate>20050105</StudyStartDate>
  <StudyEndDate>20050105</StudyEndDate>
  <ReviewDate>20050727</ReviewDate>
  <PhysicalCharacteristics>
    <PercentPurity>>99.5 a.i.</PercentPurity>
    <PhysicalState>Powder</PhysicalState>
    <Color>White</Color>
  </PhysicalCharacteristics>
</Background>
<Tests>
  <Test>
    <TestSystem>
      <Species>Mouse</Species>
      <Strain>Crl:CD-1 (ICR) BR</Strain>
      <TargetCell>Bone marrow</TargetCell>
      <Sex>Male</Sex>
      <Supplier>Charels River Laboratories, Raleigh, NC</
      Supplier>
      <InitialAge>8 Weeks</InitialAge>
      <InitialWeight>29.9–36.3 g</InitialWeight>
    </TestSystem>
    <TestCondition>
      <SolventVehicle>
```

<Substance>
<Substance>Corn oil</Substance>
<Quantity>
<Value>20.0</Value>
<Units>mL kg^{-1}</Units>
</Quantity>
</Substance>
</SolventVehicle>
<RegulatoryTestType>Regulatory original</RegulatoryTestType>
<RouteOfExposure>Injection – Intraperitoneal</RouteOfExposure>
<StainName>May-Gruenwald + Giemsa</StainName>
</TestCondition>
<NegativeTestControls>
<TestControl>
<ControlCompoundName>Corn oil</ControlCompoundName>
<TreatmentGroup>
<DosageRegimen>
<DoseVolume>20 mL</DoseVolume>
</DosageRegimen>
<ControlType>Negative</ControlType>
<Results>
<PCENCERatio>0.53</PCENCERatio>
<MNPCEPer1000PCE>0.015</MNPCEPer1000PCE>
<IndividualResults>
<IndividualResult>
<SubjectId>2602</SubjectId>
<MNPCE>
<MNPCECount>1</MNPCECount>
<TotalPCECount>2000</TotalPCECount>
</MNPCE>
<PercentPCE>
<PCENCERatio>0.47</PCENCERatio>
</PercentPCE>
</IndividualResult>
<IndividualResult>
<SubjectId>2606</SubjectId>
<MNPCE>
<MNPCECount>0</MNPCECount>
<TotalPCECount>2000</TotalPCECount>
</MNPCE>
<PercentPCE>
<PCENCERatio>0.78</PCENCERatio>

```
            </PercentPCE>
          </IndividualResult>
          <IndividualResult>
            <SubjectId>2624</SubjectId>
            <MNPCE>
              <MNPCECount>0</MNPCECount>
              <TotalPCECount>2000</TotalPCECount>
            </MNPCE>
            <PercentPCE>
              <PCENCERatio>0.38</PCENCERatio>
            </PercentPCE>
          </IndividualResult>
          <IndividualResult>
                <SubjectId>2628</SubjectId>
            <MNPCE>
              <MNPCECount>0</MNPCECount>
              <TotalPCECount>2000</TotalPCECount>
            </MNPCE>
            <PercentPCE>
              <PCENCERatio>0.48</PCENCERatio>
            </PercentPCE>
          </IndividualResult>
          <IndividualResult>
                <SubjectId>2632</SubjectId>
            <MNPCE>
              <MNPCECount>2</MNPCECount>
              <TotalPCECount>2000</TotalPCECount>
            </MNPCE>
            <PercentPCE>
              <PCENCERatio>0.56</PCENCERatio>
            </PercentPCE>
          </IndividualResult>
        </IndividualResults>
      </Results>
    </TreatmentGroup>
  </TestControl>
        </NegativeTestControls>
        <PositiveTestControls>
          <TestControl>
          <ControlCompoundName>Cyclo-
          phosphamidemonohydrate</Control
          CompoundName>
          <TreatmentGroup>
            <DosageRegimen>
                <Dose>
                  <Value>80.0</Value>
```

```
            <Units>mg kg−1</Units>
          </Dose>
        </DosageRegimen>
        <ControlType>Positive</ControlType>
        <Results>
          <PCENCERatio>0.6</PCENCERatio>
      <MNPCEPer1000PCE>0.615</MNPCEPer
      1000PCE>
      <IndividualResults>
        <IndividualResult>
      <SubjectId>2600</SubjectId>
      <MNPCE>
        <MNPCECount>21</MNPCECount>
        <TotalPCECount>2000</TotalPCECount>
      </MNPCE>
      <PercentPCE>
        <PCENCERatio>0.63</PCENCERatio>
      </PercentPCE>
    </IndividualResult>
    <IndividualResult>
      <SubjectId>2601</SubjectId>
      <MNPCE>
        <MNPCECount>39</MNPCECount>
        <TotalPCECount>2000</TotalPCECount>
      </MNPCE>
      <PercentPCE>
        <PCENCERatio>0.49</PCENCERatio>
      </PercentPCE>
    </IndividualResult>
    <IndividualResult>
      <SubjectId>2603</SubjectId>
      <MNPCE>
        <MNPCECount>32</MNPCECount>
        <TotalPCECount>2000</TotalPCECount>
      </MNPCE>
      <PercentPCE>
        <PCENCERatio>0.54</PCENCERatio>
      </PercentPCE>
    </IndividualResult>
    <IndividualResult>
      <SubjectId>2610</SubjectId>
      <MNPCE>
        <MNPCECount>14</MNPCECount>
        <TotalPCECount>2000</TotalPCECount>
      </MNPCE>
      <PercentPCE>
```

```xml
        <PCENCERatio>0.81</PCENCERatio>
      </PercentPCE>
    </IndividualResult>
    <IndividualResult>
      <SubjectId>2613</SubjectId>
      <MNPCE>
        <MNPCECount>17</MNPCECount>
        <TotalPCECount>2000</TotalPCECount>
      </MNPCE>
      <PercentPCE>
        <PCENCERatio>0.53</PCENCERatio>
      </PercentPCE>
    </IndividualResult>
    </IndividualResults>
  </Results>
  </TreatmentGroup>
</TestControl>
  </PositiveTestControls>
  <Treatments>
    <TreatmentGroup>
      <DosageRegimen>
        <Dose>
          <Value>31.25</Value>
          <Units>mg kg⁻¹</Units>
        </Dose>
        <HarvestTime>
            <Value>24.0</Value>
            <Units>hour</Units>
          </HarvestTime>
        </DosageRegimen>
        <IndividualSubjectCount>6</Individual
        SubjectCount>
      <Results>
        <PCENCERatio>0.48</PCENCERatio>
        <MNPCEPer1000PCE>0.035</MNPCE
        Per1000PCE>
        <MNPCESignificance>Not significant</
        MNPCESignificance>
        <IndividualResults>
          <IndividualResult>
            <SubjectId>2595</SubjectId>
            <MNPCE>
              <MNPCECount>1</MNPCECount>
              <TotalPCECount>2000</Total
              PCECount>
            </MNPCE>
```

<PercentPCE>
<PCENCERatio>0.42</PCENCE
Ratio>
</PercentPCE>
</IndividualResult>
<IndividualResult>
<SubjectId>2596</SubjectId>
<MNPCE>
<MNPCECount>0</MNPCECount>
<TotalPCECount>2000</Total
PCECount>
</MNPCE>
<PercentPCE>
<PCENCERatio>0.43</PCENCE
Ratio>
</PercentPCE>
</IndividualResult>
<IndividualResult>
<SubjectId>2609</SubjectId>
<MNPCE>
<MNPCECount>1</MNPCE
Count>
<TotalPCECount>2000</
TotalPCECount>
</MNPCE>
<PercentPCE>
<PCENCERatio>0.33</
PCENCERatio>
</PercentPCE>
</IndividualResult>
<IndividualResult>
<SubjectId>2620</SubjectId>
<MNPCE>
<MNPCECount>1</MNPCE
Count>
<TotalPCECount>2000</
TotalPCECount>
</MNPCE>
<PercentPCE>
<PCENCERatio>0.62</
PCENCERatio>
</PercentPCE>
</IndividualResult>
<IndividualResult>
<SubjectId>2633</SubjectId>
<MNPCE>

```
                <MNPCECount>4</MNPCE
                Count>
                <TotalPCECount>2000</
                TotalPCECount>
              </MNPCE>
              <PercentPCE>
              <PCENCERatio>0.61</
              PCENCERatio>
            </PercentPCE>
          </IndividualResult>
        </IndividualResults>
      </Results>
    </TreatmentGroup>
  <TreatmentGroup>
    <DosageRegimen>
      <Dose>
        <Value>62.5</Value>
        <Units>mg kg^-1</Units>
      </Dose>
      <HarvestTime>
        <Value>24.0</Value>
        <Units>hour</Units>
      </HarvestTime>
    </DosageRegimen>
    <IndividualSubjectCount>6</Individual
    SubjectCount>
    <Results>
      <PCENCERatio>0.49</PCENCERatio>
      <MNPCEPer1000PCE>0.0050</
      MNPCEPer1000PCE>
      <MNPCESignificance>Not significant</
      MNPCESignificance>
      <IndividualResults>
        <IndividualResult>
          <SubjectId>2593</SubjectId>
          <MNPCE>
            <MNPCECount>1</MNPCE
            Count>
            <TotalPCECount>2000</Total
            PCECount>
          </MNPCE>
          <PercentPCE>
            <PCENCERatio>0.47</PCENCE
            Ratio>
          </PercentPCE>
        </IndividualResult>
```

```
<IndividualResult>
  <SubjectId>2597</SubjectId>
  <MNPCE>
    <MNPCECount>0</MNPCE
    Count>
    <TotalPCECount>2000</Total
    PCECount>
  </MNPCE>
  <PercentPCE>
    <PCENCERatio>0.58</PCENCE
    Ratio>
  </PercentPCE>
</IndividualResult>
<IndividualResult>
  <SubjectId>2598</SubjectId>
  <MNPCE>
    <MNPCECount>4</MNPCE
    Count>
    <TotalPCECount>2000</Total
    PCECount>
  </MNPCE>
  <PercentPCE>
    <PCENCERatio>0.52</PCENCE
    Ratio>
  </PercentPCE>
</IndividualResult>
<IndividualResult>
  <SubjectId>2608</SubjectId>
  <MNPCE>
    <MNPCECount>3</MNPCE
    Count>
    <TotalPCECount>2000</Total
    PCECount>
  </MNPCE>
  <PercentPCE>
    <PCENCERatio>0.42</PCENCE
    Ratio>
  </PercentPCE>
</IndividualResult>
<IndividualResult>
  <SubjectId>2615</SubjectId>
  <MNPCE>
    <MNPCECount>2</MNPCE
    Count>
    <TotalPCECount>2000</Total
    PCECount>
```

</MNPCE>
<PercentPCE>
<PCENCERatio>0.47</PCENCERatio>
</PercentPCE>
</IndividualResult>
</IndividualResults>
</Results>
</TreatmentGroup>
<TreatmentGroup>
<DosageRegimen>
<Dose>
<Value>125.0</Value>
<Units>mg kg^{-1}</Units>
</Dose>
<HarvestTime>
<Value>24.0</Value>
<Units>hour</Units>
</HarvestTime>
</DosageRegimen>
<IndividualSubjectCount>15</IndividualSubjectCount>
<Results>
<PCENCERatio>0.44</PCENCERatio>
<MNPCEPer1000PCE>0.035</MNPCEPer1000PCE>
<MNPCESignificance>Not significant</MNPCESignificance>
<IndividualResults>
<IndividualResult>
<SubjectId>2614</SubjectId>
<MNPCE>
<MNPCECount>2</MNPCECount>
<TotalPCE-Count>2000</TotalPCECount>
</MNPCE>
<PercentPCE>
<PCENCERatio>0.62</PCENCERatio>
</PercentPCE>
</IndividualResult>
<IndividualResult>

<SubjectId>2616</SubjectId>
<MNPCE>
<MNPCECount>1</MNPCECount>
<TotalPCECount>2000</TotalPCECount>
</MNPCE>
<PercentPCE>
<PCENCERatio>0.42</PCENCERatio>
</PercentPCE>
</IndividualResult>
<IndividualResult>
<SubjectId>2618</SubjectId>
<MNPCE>
<MNPCECount>0</MNPCECount>
<TotalPCECount>2000</TotalPCECount>
</MNPCE>
<PercentPCE>
<PCENCERatio>0.36</PCENCERatio>
</PercentPCE>
</IndividualResult>
<IndividualResult>
<SubjectId>2619</SubjectId>
<MNPCE>
<MNPCECount>2</MNPCECount>
<TotalPCECount>2000</TotalPCECount>
</MNPCE>
<PercentPCE>
<PCENCERatio>0.36</PCENCERatio>
</PercentPCE>
</IndividualResult>
<IndividualResult>
<SubjectId>2626</SubjectId>
<MNPCE>
<MNPCECount>2</MNPCECount>
<TotalPCECount>2000</TotalPCECount>

</MNPCE>
<PercentPCE>
<PCENCERatio>0.45</
PCENCERatio>
</PercentPCE>
</IndividualResult>
</IndividualResults>
</Results>
</TreatmentGroup>
<TreatmentGroup>
<DosageRegimen>
<Dose>
<Value>125.0</Value>
<Units>mg kg^{-1}</Units>
</Dose>
<HarvestTime>
<Value>48.0</Value>
<Units>hour</Units>
</HarvestTime>
</DosageRegimen>
<IndividualSubjectCount>15</
IndividualSubjectCount>
<Results>
<PCENCERatio>0.3</
PCENCERatio>
<MNPCEPer1000PCE>0.025</
MNPCEPer1000PCE>
<MNPCESignificance>Not
significant</MNPCE
Significance>
<IndividualResults>
<IndividualResult>
<SubjectId>2599</
SubjectId>
<MNPCE>
<MNPCECount>0</
MNPCECount>
<TotalPCE-
Count>2000</
TotalPCECount>
</MNPCE>
<PercentPCE>
<PCENCERatio>0.4</
PCENCERatio>
</PercentPCE>

```
</IndividualResult>
<IndividualResult>
  <SubjectId>2611</
  SubjectId>
  <MNPCE>
    <MNPCECount>3</
MNPCECount>
    <TotalPCE-
    Count>2000</
    TotalPCECount>
  </MNPCE>
  <PercentPCE>
    <PCENCERatio>0.26</
    PCENCERatio>
</PercentPCE>
</IndividualResult>
<IndividualResult>
  <SubjectId>2612</
  SubjectId>
  <MNPCE>
    <MNPCECount>0</
    MNPCECount>
    <TotalPCECount>2000</
    TotalPCECount>
  </MNPCE>
  <PercentPCE>
    <PCENCERatio>0.38</
    PCENCERatio>
  </PercentPCE>
</IndividualResult>
  <IndividualResult>
    <SubjectId>2617</
    SubjectId>
    <MNPCE>
      <MNPCECount>1</
      MNPCECount>
      <TotalPCE-
      Count>2000</
      TotalPCECount>
    </MNPCE>
    <PercentPCE>
      <PCENCERatio>0.32</
      PCENCERatio>
    </PercentPCE>
```

</IndividualResult>
<IndividualResult>
<SubjectId>2631</SubjectId>
<MNPCE>
<MNPCECount>1</MNPCECount>
<TotalPCECount>2000</TotalPCECount>
</MNPCE>
<PercentPCE>
<PCENCERatio>0.16</PCENCERatio>
</PercentPCE>
</IndividualResult>
</IndividualResults>
</Results>
</TreatmentGroup>
</Treatments>
</Test>
</Tests>
</Study>
</InVivoMicronucleusStudies>
</ToxicityStudies>
<Structure>
<Molfile>
<Header>
<Line1>LS-1735</Line1>
<Line2>CDK 8/18/2014,6:57</Line2>
</Header>
<ConnectionTable><![CDATA[2 1 0 0 0 0 0 0 0 0999 V2000
0.5000 0.0000 0.0000 Cl 0 0 0 0 0 0 0 0 0 0 0 0
−0.5000 0.0000 0.0000 Ag 0 0 0 0 0 0 0 0 0 0 0 0
1 2 1 0 0 0 0
M END
]]></ConnectionTable>
<VerificationLevel>0</VerificationLevel>
</Molfile>
</Structure>
</Compound>
</Compounds>

References

1. K. van Leeuwen, T. W. Schultz, T. Henry, R. Diderich and G. D. Veith, Using chemical categories to fill data gaps in hazard assessment, *SAR QSAR Environ. Res.*, 2009, **20**, 207–220.
2. S. Dimitrov and O. Mekenyan, An introduction to read-across for the prediction of effects of chemicals, in *In Silico Toxicology: Principles and Applications*, ed. M. T. D. Cronin and J. C. Madden, The Royal Society of Chemistry, Cambridge, 2010, pp. 372–384.
3. M. T. D. Cronin, An introduction to chemical grouping, categories and read across to predict toxicity, in *Chemical Toxicity Prediction: Category Formation and Read-Across Applications*, ed. M. T. D. Cronin, J. C. Madden, S. J. Enoch and D. W. Roberts, The Royal Society of Chemistry, Cambridge, 2013, pp. 1–29.
4. M. T. D. Cronin, Evaluation of categories and read-across for toxicity prediction allowing for regulatory acceptance, in *Chemical Toxicity Prediction: Category Formation and Read-Across Applications*, ed. M. T. D. Cronin, J. C. Madden, S. J. Enoch and D. W. Roberts, The Royal Society of Chemistry, Cambridge, 2013, pp. 155–167.
5. G. J. Myatt and K. P. Cross, *In silico* Solutions for Predicting Efficacy and Toxicity, in *Human-based Systems for Translation Research*, ed. R. Coleman, The Royal Society of Chemistry, 2015, pp. 194–218.
6. R. Kroes, A. G. Renwick, M. Cheeseman, J. Kleiner, I. Mangelsdorf, A. Piersma, B. Schilter, J. Schlatter, F. van Schothorst, J. G. Vos and G. Wurtzen, Structure-based thresholds of toxicological concern (TTC): guidance for application to substances present at low levels in the diet, *Food Chem. Toxicol.*, 2004, **42**, 65–83.
7. G. Daston, D. J. Knight, M. Schwarz, T. Gocht, R. S. Thomas, C. Mahony and M. Whelan, SEURAT: Safety Evaluation Ultimately Replacing Animal Testing—Recommendations for future research in the field of predictive toxicology, *Arch. Toxicol.*, 2015, **89**(1), 15–23.
8. MySQL 5.6 Reference Manual, http://dev.mysql.com/doc/refman/5.6/en/.
9. A. Dalby, J. G. Nourse, W. G. Hounshell, A. K. I. Gushurst, D. L. Grier, B. A. Leland and J. Laufer, Description of several chemical structure file formats used by computer programs developed at Molecular Design Limited, *J. Chem. Inf. Comput. Sci.*, 1992, **32**(3), 244–255.
10. https://www.3dsbiovia.com/products/collaborative-science/biovia-draw/ctfile-no-fee.html.
11. C. L. Russom, S. P. Bradbury, S. J. Broderius, D. E. Hammermeister and R. A. Drummond, Predicting modes of action from chemical structure: Acute toxicity in the fathead minnow (Pimephales promelas), *Environ. Toxicol. Chem.*, 1997, **16**(5), 948–967.
12. A. M. Richard and C. R. Williams, Distributed structure-searchable toxicity (DSSTox) public database network: a proposal, *Mutat. Res.*, 2002, **499**(1), 27–52.

13. M. Ali, M. Patel, D. Wilkinson, P. Judson, K. Cross and D. Bower, ToxML, a data exchange standard with content controlled vocabulary used to build better (Q)SAR models, *SAR QSAR Environ. Res.*, 2013, **24**(6), 429–438.
14. http://www.cdisc.org/sdtm.
15. http://www.fda.gov/downloads/ForIndustry/DataStandards/StudyDataStandards/UCM312964.pdf.
16. http://www.lexjansen.com/pharmasug/2004/FDACompliance/FC08.pdf.
17. http://www.fda.gov/downloads/Drugs/GuidanceComplianceRegulatoryInformation/Guidances/UCM292334.pdf.
18. http://www.instem.com/.
19. http://www.pointcrosslifesciences.com/.
20. http://www.oecd.org/ehs/templates/.
21. https://iuclid6.echa.europa.eu/.
22. http://isa-tools.org/.
23. P. Rocca-Serra, M. Brandizi, E. Maguire, N. Sklyar, C. Taylor, K. Begley, D. Field, S. Harris, W. Hide, O. Hofmann, S. Neumann, P. Sterk, W. Tong and S.-A. Sansone, ISA software suite: supporting standards-compliant experimental annotation and enabling curation at the community level, *Bioinformatics*, 2010, **26**(18), 2354–2356.
24. S. A. Sansone, P. Rocca-Serra, D. E. Maguire, C. Taylor, O. Hofmann, H. Fang, S. Neumann, W. Tong, L. Amaral-Zettler, K. Begley, T. Booth, L. Bougueleret, G. Burns, B. Chapman, T. Clark, L. A. Coleman, J. Copeland, S. Das, A. de Daruvar, P. de Matos, I. Dix, S. Edmunds, C. T. Evelo, M. J. Forster, P. Gaudet, J. Gilbert, C. Goble, J. L. Griffin, D. Jacob, J. Kleinjans, L. Harland, K. Haug, H. Hermjakob, S. J. Ho Sui, A. Laederach, S. Liang, S. Marshall, A. McGrath, E. Merrill, D. Reilly, M. Roux, C. E. Shamu, C. A. Shang, C. Steinbeck, A. Trefethen, B. Williams-Jones, K. Wolstencroft, I. Xenarios and W. Hide, Toward interoperable bioscience data, *Nat. Genet.*, 2012, **44**(2), 121–126.
25. http://www.ebi.ac.uk/arrayexpress/.
26. http://www.ebi.ac.uk/pride/archive/.
27. http://www.w3.org/RDF/.
28. A. González-Beltrán, S. Neumann, E. Maguire, P. Rocca-Serra and S. Sansone, The Risa R/Bioconductor package: integrative data analysis from experimental metadata and back again, *BMC Bioinf.*, 2014, **15** (Suppl 1), S11.
29. http://www.r-project.org/.
30. D. V. Sweet, V. P. Anderson and J. C. F. Fang, An overview of the Registry of Toxic Effects of Chemical Substances (RTECS): Critical information on chemical hazards, *Chem. Health Saf.*, 1999, **6**(6), 12–16.
31. U.S. Department of Health and Human Services, Public Health Service, Centers for Disease Control and Prevention, National Institute for Occupational Safety and Health, Registry of Toxic Effects of

Chemical Substances (RTECS®) Comprehensive Guide to the RTECS®, DHHS (NIOSH) Publication No. 97–119, 1997, http://www.cdc.gov/niosh/docs/97-119/pdfs/97-119.pdf.
32. R. D. Benz, Priority-based assessment of food additives database of the U.S. Food and Drug Administration Center for Food Safety and Applied Nutrition, *Environ. Health Perspect.*, 1992, **96**, 85–89.
33. US FDA CDER New&Generic Drugs Approvals: Drugs@FDA, http://www.accessdata.fda.gov/scripts/cder/drugsatfda/index.cfm.
34. US FDA CDER Drug and Biologic Approval Reports, http://www.fda.gov/cder/rdmt/.
35. CCRIS (Chemical Carcinogenesis Res. Information System): Developed and maintained by National Cancer Institute, http://toxnet.nlm.nih.gov/cgi-bin/sis/htmlgen?CCRIS.
36. R. W. Tennant, The genetic toxicity database of the National Toxicology Program: evaluation of the relationships between genetic toxicity and carcinogenicity, *Environ. Health Perspect.*, 1991, **96**, 47–51.
37. Tokyo-Eiken: Tokyo Metropolitan Institute of Public Health, http://www.tokyo-eiken.go.jp/henigen.
38. S. E. Escher, I. Tluczkiewicz, M. Batke, A. Bitsch, C. Melber, E. D. Kroese, H. E. Buist and I. Mangelsdorf, Evaluation of inhalation TTC values with the database RepDose, *Regul. Toxicol. Pharmacol.*, 2010, **58**(2), 259–274.
39. I. Tluczkiewicz, H. E. Buist, M. T. Martin, I. Mangelsdorf and S. E. Escher, Improvement of the Cramer classification for oral exposure using the database TTC RepDose – A strategy description, *Regul. Toxicol. Pharmacol.*, 2011, **61**, 340–350.
40. I. Tluczkiewicz, M. Batke, D. Kroese, H. Buist, T. Aldenberg, E. Pauné, H. Grimm, R. Kühne, G. Schüürmann, I. Mangelsdorf and S. E. Escher, The OSIRIS Weight of Evidence approach: ITS for the endpoints repeated-dose toxicity (RepDose ITS), *Regul. Toxicol. Pharmacol.*, 2013, **67**(2), 157–169.
41. S. E. Escher, M. Batke, S. Hoffmann-Doerr, H. Messinger and I. Mangelsdorf, Interspecies extrapolation based on the RepDose database—A probabilistic approach., *Toxicol. Lett.*, 2013, **218**(2), 159–165.
42. M. Batke, S. Escher, S. Hoffmann-Doerr, C. Melber, H. Messinger and I. Mangelsdorf, Evaluation of time extrapolation factors based on the database RepDose, *Toxicol. Lett.*, 2011, **205**(2), 122–129.
43. M. Batke, T. Aldenberg, S. Escher and I. Mangelsdorf, Relevance of non-guideline studies for risk assessment: the coverage model based on most frequent targets in repeated dose toxicity studies, *Toxicol. Lett.*, 2013, **218**(3), 293–298.
44. The European Chemical Industry Council Long Range Initiative, B1.1: "Identification of Structure Activity Relationship (SAR) Alerts (Combinatin of Functional Groups) for Substances with Low NOELs (RepDose Database)".
45. A. Bitsch, S. Jacobi, C. Melber, U. Wahnschaffe, N. Simetska and I. Mangelsdorf, REPDOSE: A database on repeated dose toxicity studies of

commercial chemicals – a multifunctional tool,, *Regul. Toxicol. Pharmacol.*, 2006, **46**, 202–210.
46. https://repdose.item.fraunhofer.de/.
47. Available from https://ng.cosmosdb.eu and/or https://cosmosdb.eu.
48. http://www.cosmostox.eu.
49. http://www.seurat-1.eu.
50. C. Yang, D. Hristozov, A. Tarkhov, T. Kleinöder, I. Boyer, M. Cronin, E. Fioravanzo, H. Kim, B. Heldreth, A. Mostrag-Szylchtying, J. Rathman, A. Richarz, C. Schwab, V. Vitcheva and A. Worth, COSMOS DB as an international share point for exchanging regulatory and toxicity data of cosmetics ingredients and related substances, *Toxicol. Lett.*, 2015, **238**(Suppl.), S382.
51. C. Yang, S. M. Barlow, K. L. Muldoon Jacobs, V. Vitcheva, A. R. Boobis, S. P. Felter, K. B. Arvidson, D. Keller, M. T. D. Cronin, S. Enoch, A. Worth and H. M. Hollnagel, Thresholds of Toxicological Concern for cosmetics-related substances: New database, thresholds, and enrichment of chemical space, *Food Chem. Toxicol.*, 2017, **109**, 170–193.
52. U.S. Department of Health and Human Services, Food and Drug Administration, Providing Regulatory Submissions in Electronic Format – Submissions Under Section 745A(a) of the Federal Food, Drug, and Cosmetic Act, Guidance for industry, 2014, http://www.fda.gov/downloads/Drugs/Guidances/UCM384686.pdf.
53. U.S. Department of Health and Human Services, Food and Drug Administration, Guidance for Industry Part 11, Electronic Records; Electronic Signatures – Scope and Application, 2003, http://www.fda.gov/downloads/RegulatoryInformation/Guidances/ucm125125.pdf.
54. http://www.cdisc.org/send.
55. P. Kohonen, E. Benfenati, D. Bower, R. Ceder, M. Crump, K. Cross, R. C. Grafström, L. Healy, C. Helma, N. Jeliazkova, V. Jeliazkov, S. Maggioni, S. Miller, G. Myatt, M. Rautenberg, G. Stacey, E. Willighagen, J. Wiseman and B. Hardy, The ToxBank Data Warehouse: Supporting the Replacement of *In Vivo* Repeated Dose Systemic Toxicity Testing, *Mol. Inf.*, 2013, **32**, 47–63.
56. P. Kohonen, R. Ceder, I. Smit, V. Hongisto, G. Myatt, B. Hardy, O. Spjuth and R. Grafström, Cancer biology, toxicology and alternative methods development go hand-in-hand, *Basic Clin. Pharmacol. Toxicol.*, 2014, **115**(1), 50–58.
57. NRC (National Research Council), *Science and Decisions: Advancing Risk Assessment*, Committee on Improving Risk Analysis Approaches Used by the U.S. EPA, National Academy Press, Washington, DC, 2009.
58. Y. Igarashi, N. Nakatsu, T. Yamashita, A. Ono, Y. Ohno, T. Urushidani and H. Yamada, Open TG-GATEs: a large-scale toxicogenomics database, *Nucleic Acids Res.*, 2015, **43**(Database issue), D921–D927.
59. http://toxico.nibiohn.go.jp/english/.
60. B. Hardy, *et al.*, Collaborative development of predictive toxicology applications, *J. Cheminf.*, 2010, **2**, 7.
61. E. Willighagen, The ChEMBL database as linked open data,, *J. Cheminf.*, 2013, **5**, 23.

62. D. J. Dix, K. A. Houck, M. T. Martin, A. M. Richard, R. W. Setzer and R. J. Kavlock, The ToxCast program for prioritizing toxicity testing of environmental chemicals, *Toxicol. Sci.*, 2007, **95**(1), 5–12.
63. US EPA, Toxicity Forecasting, Advancing the Next Generation of Chemical Evaluation, http://www.epa.gov/ncct/toxcast/data.html.
64. https://actor.epa.gov/dashboard/.
65. US EPA Toxicity Reference Database, https://cfpub.epa.gov/si/si_public_record_report.cfm?Lab=NCCT&dirEntryId=227139, https://www.epa.gov/chemical-research/exploring-toxcast-data-downloadable-data.
66. US EPA Distributed Structure-Searchable Toxicity (DSSTox) Database, http://www.epa.gov/ncct/dsstox/StandardChemFieldDefTable.html.
67. http://bioassayontology.org/.
68. S. Kim, J. Chen, T. Cheng, A. Gindulyte, J. He, S. He, Q. Li, B. A. Shoemaker, P. A. Thiessen, B. Yu, L. Zaslavsky, J. Zhang and E. E. Bolton, PubChem 2019 update: improved access to chemical data, *Nucleic Acids Res.*, 2019, **47**(D1), D1102–D1109.
69. M. Waters, S. Stasiewicz, B. A. Merrick, K. Tomer, P. Bushel, R. Paules, N. Stegman, G. Nehls, K. J. Yost, C. H. Johnson, S. F. Gustafson, S. Xirasagar, N. Xiao, C. C. Huang, P. Boyer, D. D. Chan, Q. Pan, H. Gong, J. Taylor, D. Choi, A. Rashid, A. Ahmed, R. Howle, J. Selkirk, R. Tennant and J. Fostel, CEBS – Chemical Effects in Biological Systems: a public data repository integrating study design and toxicity data with microarray and proteomics data, *Nucleic Acids Res.*, 2008, **36**(Database issue), D892–D900.
70. G. Roberts, *et al.*, LeadScope: Software for Exploring Large Sets of Screening Data, *J. Chem. Inf. Comput. Sci.*, 2000, **40**, 1302–1314.
71. R. S. Judson, *et al.*, Aggregating Data for Computational Toxicology Applications: The U.S., "Environmental Protection Agency (EPA) Aggregated Computational Toxicology Resource (ACToR) System", *Int. J. Mol. Sci.*, 2012, **13**(2), 1805–1831.
72. P. Wexler, TOXNET: An evolving web resource for toxicology and environmental health information, *Toxicology*, 2001, **157**(1–2), 3–10.
73. http://www.echemportal.org, participating databases: http://www.echemportal.org/echemportal/page.action?pageID=2.
74. https://elixir-europe.org.
75. D. Ding, L. Xu, H. Fang, H. Hong, R. Perkins, S. Harris, E. D. Bearden, L. Shi and W. Tong, The EDKB: an established knowledge base for endocrine disrupting chemicals, *BMC Bioinf.*, 2010, **11**(Suppl 6), S5.
76. E. Julien, C. C. Willhite, A. M. Richard, J. M. DeSesso and I. R. W. Group, Challenges in constructing statistically based structure-activity relationship models for developmental toxicity, *Birth Defects Res., Part A*, 2004, **70**(12), 902–911.
77. A. R. Scialli, A. Lione and G. K. B. Padgett, *Reproductive Effects of Chemical, Physical, and Biologic Agents: REPROTOX*, Johns Hopkins University Press, 1995.

78. T. H. Shepard, *Catalog of Teratogenic Agents*, The Johns Hopkins University Press, 10th edn, 2001.
79. R. L. Brent, The application of the principles of toxicology and teratology in evaluating the risks of new drugs for treatment of drug addiction in women of reproductive age, *NIDA Res. Monogr.*, 1995, **149**, 130–184.
80. R. L. Brent and D. A. Beckman, Environmental teratogens, *Bull. N. Y. Acad. Med.*, 1990, **66**(2), 123–163.
81. G. G. Briggs, R. K. Freeman and S. J. Yaffe, *Drugs in Pregnancy and Lactation: A Reference Guide to Fetal and Neonatal Risk*, Lippincott Williams & Wilkins, Philadelphia, Pa, 8th edn, 2008.
82. *Goodman and Gilman's the Pharmacological Basis of Therapeutics*, ed. L. L. Brunton, J. S. Lazo and K. L. Parker, The McGraw-Hill Companies, New York, 11th edn, 2006.
83. J. F. Cordero and G. P. Oakley Jr, Drug exposure during pregnancy: some epidemiologic considerations, *Clin. Obstet. Gynecol.*, 1983, **26**(2), 418–428.
84. J. F. Crow and C. Denniston, Mutation in human populations, *Adv. Hum. Genet.*, 1985, **14**, 59–123.
85. J. Z. Ji, L. Y. Han, C. W. Yap, L. Z. Sun, X. Chen and Y. Z. Chen, Drug Adverse Reaction Target Database (DART): Proteins Related to Adverse Drug Reactions, *Drug Saf.*, 2003, **26**(10), 685–690.
86. R. B. Fitzpatrick, LactMed: Drugs and Lactation database, *J. Electron. Resour. Med. Libr.*, 2007, **4**(1–2), 155–166.
87. E. J. Matthews, N. L. Kruhlak, R. Daniel Benz and J. F. Contrera, A comprehensive model for reproductive and developmental toxicity hazard identification: I. Development of a weight of evidence QSAR database, *Regul. Toxicol. Pharmacol.*, 2007, **47**(2), 115–135.
88. E. J. Matthews, N. L. Kruhlak, R. Daniel Benz, J. Ivanov, G. Klopman and J. F. Contrera, A comprehensive model for reproductive and developmental toxicity hazard identification: II. Construction of QSAR models to predict activities of untested chemicals, *Regul. Toxicol. Pharmacol.*, 2007, **47**(2), 136–155.
89. Y. Sakuratani, H. Q. Zhang, S. Nishikawa, K. Yamazaki, T. Yamada, J. Yamada, K. Gerova, G. Chankov, O. Mekenyan and M. Hayashi, Hazard Evaluation Support System (HESS) for predicting repeated dose toxicity using toxicological categories, *SAR QSAR Environ. Res.*, 2013, **24**(5), 351–363.
90. R. Benigni, C. Bossa, A. M. Richard and C. Yang, A novel approach: chemical relational databases, and the role of the ISSCAN database on assessing chemical carcinogenicity, *Ann. Ist. Super. Sanita*, 2008, **44**(1), 48–56.
91. R. R. Young, Genetic toxicology: web resources, *Toxicology*, 2002, **173**(1–2), 103–121.

CHAPTER 5

Making Big Data Available: Integrating Technologies for Toxicology Applications

NINA JELIAZKOVA*[a] AND VEDRIN JELIAZKOV[a]

[a]Ideaconsult Ltd, 4 A. Kanchev str., Sofia 1000, Bulgaria
*E-mail: jeliazkova.nina@gmail.com

5.1 Introduction

The most defining characteristics of toxicity big data are primarily neither volume nor velocity (streaming sources are not yet mainstream), but the variety (different forms of the data) and veracity (uncertainty of the data) are representative challenges of data management and analysis. Toxicities can be acute, chronic or sub-chronic, depending on the exposure extent (from single contact with a toxic substance to entire life exposure) (see Chapter 2). The classic way of testing toxicity is on whole animals or on extracts from organs (liver, *etc.*). Until recently, the publicly available toxicity data was mainly phenotypic (*e.g.*, acute toxicity on fish, salmonella mutagenicity, rodent carcinogenicity), the same phenotypic approach used before the adoption of the target-based approach for drug discovery in the mid-1980s. Representative data collections are, for example, US EPA DSSTox,[1] skin sensitisation data compilations,[2] the Fraunhofer repeated dose toxicity database,[3]

Issues in Toxicology No. 41
Big Data in Predictive Toxicology
Edited by Daniel Neagu and Andrea-Nicole Richarz
© The Royal Society of Chemistry 2020
Published by the Royal Society of Chemistry, www.rsc.org

the Benchmark mutagenicity dataset,[4] ISS toxicity datasets,[5] *etc.* These are small datasets, typically extracted from the literature, or generated and published in the framework of collaborative projects and published in the primary literature. These datasets consist of a few hundred to a few thousand chemical structures and are typically distributed as files in one of the established chemical formats (see Chapter 3). These and similar datasets were used as data sources for compiling a number of toxicity data collections, mostly using relational databases for data management and a web user interface or Application Programming Interface (API) for data access, for example the OpenTox database,[6] the US EPA databases ACToR[7] and ToxRefDB; OCHEM[8] and COSMOS DB[9] (see Chapter 4). At the same time, large data collections with bioactivity data for chemical compounds exist, mostly originating from drug discovery (*e.g.*, ChEMBL,[10] Open PHACTS); a recent summary is published in.[11] The largest publicly available data collection to date is PubChem.[12]

There is an increasing trend of combining and jointly analysing various sources of data (*e.g.*, cell-based viability assays, omics data, as well as clinical studies) that might relate to toxicity effects, as there is a possibility of finding common explanatory factors by exploiting the overlap in the perturbed pathways or gene–disease associations.[13] Toxicogenomics[14–16] (see Chapter 7) combines toxicology and genomics, metabolomics, proteomics, and comprises toxicogenomic databases (*e.g.*, Comparative Toxicogenomics Database,[17] Open TG-Gates,[18] Chemical Effects in Biological Systems (CEBS),[19] ToxCast, Tox21[20]) and generic bioinformatics analysis tools, as well as tools that help to relate adverse events to gene or protein functions.[21] The value of high-throughput screening for toxicology was first considered by the US "National Toxicology Program for the 21st Century"[22] in 2004, culminating with the large and multiphase ToxCast and Tox21 screening campaigns (latest data release – October 2015). Toxicity estimation is important in the context of drug discovery, in order to prevent drug side effects; but, also, in the context of the safe use of any chemical substance produced and included in, *e.g.*, food and everyday objects, as well as preventing environment pollution. Toxicogenomics and *in vitro* assay results in general are gaining acceptance by regulatory agencies in Europe.[23] The high-throughput screening findings are still difficult to relate to phenotypic observations; partly for this reason, frameworks such as Adverse Outcome Pathways (AOPs), as promoted by the OECD,[24] are becoming popular, even if they offer a simplified representation of biological events leading to adverse effects (see also Chapter 9). Integration with pathway databases (ConsensusPathDB,[25] KEGG, Reactome,[26] WikiPathways[27]) for the purpose of pathway enrichment and biological network analysis[28] is also being investigated.

Hundreds of bioinformatics-related databases are published annually by the Nucleic Acids Research (NAR) Database journal (currently just one of the journals collecting bioinformatics databases). The NAR online Molecular Biology Database Collection contains 1685 databases as of 2016. By considering these biological databases within the area of toxicology, the volume factor becomes substantial: "biological data volumes will soon rival those produced by astronomical observation".[29] The European Molecular Biology Laboratory (EMBL) European Bioinformatics Institute (EBI) maintains the largest bioinformatics data resources in Europe (sequences, proteins, chemicals, pathways, ontologies and literature) with 75 petabytes as of December 2015.

The ultimate goal of integrating all these disparate resources would be to enable multifaceted analysis of all aspects of chemistry and biology and contribute to the understanding of different aspects of toxicology. While most of the current efforts are focused on data annotation and linking, the integration approaches of the future will increasingly rely on modern data analytics and machine learning methods.

5.2 Data Integration Approaches

The integration of heterogeneous information is an ongoing challenge in many areas, but especially in life sciences, due to the increasing amount and diversity of the generated data. The first methodology for the integration challenge (and the primary solution until the late 1990s) was building data warehouses and custom solutions, designed to manage all the information under a central authority and a single database schema. The integration approach that was followed and widely used in enterprises, though still providing a single schema view (mediator schema), is designed to query many independent data sources on the fly (virtual integration). The data warehousing approach utilises Extract-Transform-Load (ETL) processes to access, convert and load the content of different data sources into the same physical database. The user can query all loaded data resources at the same time and receive the results. The virtual integration requires sending queries to the different sources at their original locations and presenting the results in one unified view as soon as they are received. Both approaches may implement the same technologies for data storage (*e.g.*, relational, NoSQL) and data access. The ETL processes may involve connecting to remote resources *via* web services and the virtual integration could consist of wrappers connecting to local or remote databases and combining the results only when presented to the user. The choice of the data integration strategy depends on the answer to the following two main questions: first, how to establish the correspondence between entities in different databases; and second, how to integrate query results. Matching entities from different data sources could be done either by converting the data sources into a common data model, or keeping the data models distinct and establishing equivalence between the separate entities. Equivalence in the latter case can be established through the use of shared ontology identifiers to annotate the data. A particular challenge in matching entities from different sources across either entity matching approach is the need to define the identity criteria for entities of that type. All query integration approaches require entity matching. This can be done during the conversion to a common data model (index time merging) or on the fly (query time merging, federated search), as well as through a hybrid approach.[30]

In the context of data for toxicology, the major entities to be mapped are the chemical substances and structures and the results of toxicity testing. The state-of-the-art approach to match chemical structures across databases is to use the IUPAC International Chemical Identifier (InChI).[31] The matching rule is "if the two search results returned include one and the same InChI, then the results are for the same chemical structure". For example, UniChem[32] utilises standard InChIs linking chemical structures between different chemistry resources. In reality, the use of InChIs presents a number of challenges, since InChIs, as it is the case for most other chemical identifiers, can differ for the same chemical structure, depending on the options used for its generation and the specific rules for chemical structures standardisation used by the given resource. There are typically differences of the standardisation rules between publicly accessible and industry databases, motivated by different usage patterns.

The publicly accessible chemical databases generally associate chemical substances with single structures and do not provide easily accessible data on substance identity and composition. Cefic's Long-Range Research Initiative (LRI) AMBIT software[33] was extended to enable the representation of chemical substances in real industry conditions by supporting complex compositions (including impurities, additives, UVCB – Chemical substances with unknown or variable composition, complex reaction products and biological materials).[34] The AMBIT tool[34,35] is available online and is loaded with non-confidential REACH data supplied by the European Chemicals Agency (ECHA). The data model also supports nanomaterials, as implemented in the eNanoMapper project,[36] and is briefly described in Chapter 3.

Representing experimental data is even more challenging. The most generic model for representing biological experiments is the ISA (Investigation-Study-Assay) framework, which is aimed at interoperable bioscience data description[37] and is described in Section 5.5. ISA relies on ontologies to represent entities of chemical structures, gene and protein identifiers, endpoints and assays. While repositories of genes and proteins as Entrez and Uniprot exist, there are semantic challenges with the mapping (*e.g.*, the Entrez gene id is species specific), which need to be resolved depending on the context. Integrating transcriptomic and metabolomic data is particularly challenging, due to the specifics of the analytical technologies and study designs.[38] For example, if chiral metabolites are not distinguished by the analytical platform, mapping to the existing entries in the pathway databases is not trivial. Identifier mapping frameworks[39,40] are developed to facilitate linking between different databases.

Multiple schema matching and schema mapping techniques (machine learning, semantic, probabilistic, similarity, aggregation, domain specific, semi-automatic) have been proposed to identify the correspondence between attributes of independent data schemas.[41] At the dawn of the big data era, the "no-schema" approach was introduced as more efficient time wise and more appropriate in certain cases, as the classic "schema-first" approach requires substantial effort to elaborate the data model upfront, and furthermore, no global schema may exist, and the integrator may not have control of the data sources. This vision was reflected in the "dataspaces" architecture,[42] where a dataspace can be seen as a set of distributed, heterogeneous and autonomous data sources, providing partially unstructured data. Each data source is designed and built independently to address a specific need and may provide implicit or explicit connections to other data sources. Integration in dataspaces is regarded as dynamically evolving ("Pay-as-you-go" integration) and provides increasingly better semantic consistency and richer query facilities. Currently, we find implementation of this vision in linked data[43] and data lakes.[44] A data lake is a repository of raw data in its original formats, able to store vast amount of data, which serve as input for processing and analysis tools, as well as ingestion into structured databases. The storage implementation can vary from files on Hadoop file systems, to combination of NoSQL, relational and semantic databases.

There are a number of completed and ongoing efforts to integrate publicly available as well as confidential toxicological data resources. OpenTox[45] was one of the first attempts to provide programmatic access to toxicology data and models, and, importantly, was designed as a set of distributed services,[33] providing a common data model and representational state transfer (REST) API, featuring diverse backend solutions and programming languages. Open PHACTS[46] also identifies all objects *via* Uniform Resource Identifiers (URIs), uses a single triple store and provides an API, hiding the SPARQL endpoint for ensuring user friendliness and performance. As multiple Resource Description Framework (RDF) data models co-exists in the Open PHACTS triple store, and each RDF data model has its own URI pattern for identifying objects; Open PHACTS uses a dedicated Identity Mapping Service (IMS) to establish the mapping between URIs obtained from different sources. Toxygates[47] combines accessing metadata through remote SPARQL and NoSQL solutions for storing and querying 800M data points for 170 chemicals. EBIs RDF platform[48] uses triple stores and offers a federated SPARQL endpoint. The ToxBank data warehouse[49] from the SEURAT-1 research initiative[50] offers REST API and was initially developed to store RDF representation of ISA-Tab[37] archives into a single triple store; it was later extended to support unstructured data upload as well as a virtual integration with COSMOS DB *via* REST API. The recently released US EPA Computational toxicology dashboard[51] is a large effort of organising the publicly available data with a user friendly interface. The integration of public and private data is of significant importance for the industry, and various solutions are actively developed: eTox[52] workflows,[53,54] Syngenta federated search,[55] Clariant CompTox suite,[56] Roche graph based record linkage[57] and others. Marti-Solano *et al.*[58] have reviewed the integration challenges in the context of pharmaceutical research and development. Ontologies and standards are highlighted as critical for data integration within this context.

5.3 Ontologies and Shared Vocabulary

An informal definition of ontology in this context is a "shared representation of a domain", which is computer readable. This definition of applied ontologies differs from the ontology in the original philosophical sense:[59] they are mostly used as hierarchical vocabularies rather than an axiom-based formal ontology. The most widely used case is annotation and usage of formal reasoning, while exploiting constraints defined in ontologies is rare. The W3C SKOS (Simple Knowledge Organization System)[60] data model is designed to meet this major use case and supports representation of vocabularies, classification schemes and taxonomies. Technically, ontologies are represented as collections of related entities; each entity could be described as a collection of triples (or statements), where each triple consists of a subject, a predicate, and an object. The predicate (also called a property) denotes the relationship between two objects. The W3C Resource Description Framework provides means to formally describe the triples. The RDF defines a data model (how the data is organised), instead of specifying a data format (how the data is written into a file). Once a data model is defined, it can be serialised into different formats, such as RDF/XML, N3, TURTLE and, lately, JSON-LD. The objects are modelled in RDF as Classes (rdf:Class), and Classes have specific Instances. Relationships are modelled with Properties (rdf:Property). The Web Ontology Language OWL is built on top of RDF, and, compared to RDF, imposes restrictions on what is allowed to be represented. Because of such restrictions, the OWL subsets OWL-Lite and OWL-DL (Description Logic) allow the performance of automated machine reasoning, useful for detecting data quality issues and inferring new facts such as interactions and mechanisms.[61] In OWL, there are Object properties and Data properties (owl:Property, which is a subclass of rdf:Property). An Object property specifies a relation between Instances, while a Data property specifies a relation between an Instance and a simple data value (string, integer, *etc.*). Properties cannot be used as Classes and *vice versa*.

Linked data representation is considered a state-of-the-art technology for data integration in life sciences, and there are a number of implemented solutions (SADI,[62] OpenTox,[33] Bio2RDF,[63] Open PHACTS, EBI-RDF[64]). The EBI-RDF platform uses a different approach, maintaining multiple RDF repositories for different resources and allowing federated searches across all of them. It is mandatory for all entities in the EBI-RDF platform to be assigned equivalent identifiers *via* the http://identifiers.org registry, which is essentially implementing the mapping between the distributed resources. PubChem and ChEMBL databases provide downloadable RDF dumps, which is a reliable option for large data transfers and allows for setting up local copies of the data repositories. However, accurate knowledge representation and annotation are labour intensive and rely on human curators. As an example, the latest release of the ChEBI (Chemical Entities of Biological Interest) ontology[65] has 50 504 entities; this is much less than the 90 million chemical structures in public databases such as PubChem.

The Bioportal[66] contains over 500 biomedical ontologies. There are multiple and overlapping ontologies describing the same or similar entities and the choice of an ontological term is not trivial.[67] The ontologies are not static, new versions are released at regular intervals, different for the different ontologies (from yearly to monthly or even on a daily basis). A number of ontology integration efforts exist, resulting in the compilation of "integrated data sources" as the Unified Medical Language System or the eNanoMapper ontology for nanomaterial safety.[68] Tools enabling semi-automatic matching of ontologies are actively being developed.[69–71] An overview of the ontology and ontology mapping evolution and related challenges is given in.[72]

5.4 Web Services and Cloud Technology

The cloud technology is becoming increasingly attractive to life sciences. Web platforms integrate data storage, cheminformatics and data analysis tools. Short overviews of web services advertised by open source and commercial cheminformatics packages can be found in.[73–76] Initially, commercial cheminformatics packages provided only demo web applications with partial or restricted functionality. Desktop tools have been traditionally easier to develop and, for the target users, confined to a company intranet environment, whereas web-only tools seemed to be of less value. This is changing with the adoption of cloud computing and mobile devices and the increasing number of chemical information resources offering REST API access to data (such as PubChem REST,[77] EPA ACToR,[78] Ensembl REST API,[79] LINCS API,[80] EBI RDF Platform,[81] Proteomics Standard Initiative Common QUery InterfaCe,[82] OpenPhacts API[83]) or data and analysis tools (GenomeSpace,[84] EBI web services, Embassy cloud[29]). This explosion of web API resources in life sciences is increasingly recognised by publishers, for example the Nucleic Acids Research (NAR) database journal has introduced a special section for Web services since 2015.

There are a number of public cloud providers, as well as cloud solutions that can be deployed as a private cloud (within a single organisation). Private clouds are an attractive solution for storing sensitive data. Hybrid clouds can integrate two or more (private or public) clouds.

5.5 Standards

Establishing a common Application Programming Interface for a given type of resource facilitates their integration, because it alleviates the need for schema matching at the backend. Essentially, the API defines a common schema, interaction rules, and ensures all resources of the same kind are compliant with the API. An example of the implementation of this approach in the genomics field is the Global Alliance for Genomics and Health (GA4GH) Data Working Group.[85] The Google Genomics API[86] implements the GA4GH API. An example of an interoperable REST API outside of the biomedical domain is OData (Open Data Protocol),[87] an industry standard known as the "SQL of the web" that provides an uniform way to query data.

The opportunity in definition and usage of standards in toxicogenomics for ensuring data interoperability was recognised early on[88] and extensions of the MIAME (Minimum information about a microarray experiment) have been developed. The most generic model for representing biological experiments currently is ISA-Tab, which aims to provide an interoperable bioscience data description.[37] The universality is achieved by modelling the experimental graph, explicitly describing all steps in an experiment and recording details of the input and output nodes (*i.e.*, how samples are processed). ISA (Investigation-Study-Assay) includes three major layers: Investigation, Study and Assay where the actual experimental readouts are stored in an additional data layer. The format specification and variety of tools of the open source ISA framework are supported by the ISA team at the University of Oxford e-Research Centre.

Since the ISA version 1 published in 2008, the ISA platform implementation relies on tab delimited text files and the file format representation is known as ISA-Tab. The ISA team started migration in 2015 to a new JSON format, instead of tab delimited files as a more flexible and efficient way to store the experimental meta data. Currently, there are ISA v1 and ISA v2 JSON versions. The ISA v2 release is based solely on JSON, where the backward compatibility with the ISA v1 and ISA-Tab format is preserved *via* ISA-tools project utilities. The ISA team develops and maintains documentation and tools, supporting the new ISA-JSON formats. The ISA-JSON specification is based on a set of JSON schemas used to validate the syntactical and logical correctness of the ISA-JSON information. It distinguishes between core components and extensions; for example, the support for nanomaterials is considered an extension. The JSON schemas describe various components of the three layers of the ISA platform (*i.e.*, investigation, study, assay, sample, source, ontology annotation, *etc.*).The ISA-JSON specification was officially released at the end of October 2016.[89]

The default approach for representation of chemical compounds in ISA-Tab is an ontology entry, which typically points to a single chemical structure. This is insufficient for describing substances of complex compositions such as nanomaterials, hence a material file was introduced to address this need in ISA-Tab-Nano.[90] The ISA-Tab-Nano project is an effort of the National Cancer Institute (NCI), National Cancer Informatics Program (NCIP) and Nanotechnology Informatics Working Group (US Nano WG).

Developing and adhering to standards would be the ideal approach to integration, but in reality, especially in life sciences, there are multiple standard schemas, rather than a single one, due to different research needs, competing groups, new kinds of data and existing business models.

5.6 Integrative Data Analysis

Data integration has been an established field of research in computer science since the 1980s; theoretical research and implementations in practice are widely available, including textbooks.[30] However, the term "data integration", especially in bioinformatics, is often used in a different sense, namely as the joint analysis of different types of biological data and building a combined model taking into account all datasets at the same time. Some authors argue that data integration in life sciences is different than IT integration, and again redefine the data integration as a modelling problem. Following this view, "data integration" becomes "integrative data analysis" or "network integration", and the challenges outlined are in fact computational and machine learning challenges.[91–94] Typical examples are:

- the selection of the appropriate machine leaning methods to use
- how to analyse network or graph data
- how to combine results from different methods
- how to handle noisy, incomplete and unbalanced data
- how to learn efficiently when the assumption for identically and independently distributed data does not hold.[95]

Machine learning methods are now being successfully applied in many domains and various data analytics platforms (*e.g.*, MapReduce[96] platforms such as Apache Hadoop, Apache Spark), generic workflow management systems such as KNIME[97] or bioinformatics-specific ones like Galaxy or Embassy.[98,99] An important challenge of big data analysis is the distributed nature of the data; scalability, ownership and security issues arise, requiring close collaboration with systems and database researchers in order to make use of the most appropriate data storage, computational and visualisation platforms.

5.7 Data Sharing Infrastructure

The data resources at the European Bioinformatics Institute (EMBL-EBI), the "home for big data in biology",[100] are exemplary of the power of data sharing.[29] Data resources for all types of biological data are actively maintained and integrated using semantic technologies (ontologies, RDF, federated SPARQL), standards (ISA) and fast search engines,[101] and provide access *via* web browsers and application programming interfaces. The services are part of the European research infrastructure for life sciences ELIXIR,[102] coordinating the implementation of distributed solutions for biological data management (see Chapter 4). ELIXIR promotes the FAIR principles, defining that the data resources and tools should be Findable, Accessible, Interoperable and Reusable[103] by third parties. Large-scale data analysis is available *via* the Embassy cloud and the European Grid Infrastructure (EGI) federation.

There are a number of initiatives and resources worldwide aimed at supporting data sharing, such as the Research Data Alliance,[104] the Data Commons catalog,[105] and the National Institutes of Health (NIH) Data Commons.[106] BioSharing[107,108] is a curated resource, launched in 2011 and implementing the FAIR principles. It currently contains 671 data standards, 838 databases, and 87 policies in the life, environmental and biomedical sciences. Generic file sharing platforms such as FigShare,[109] DataDryad[110] as well as Wikidata[111] and Wikipedia[112] are used for sharing, *e.g.*, chemical structures, genes and experimental data.

The European Commission is promoting open science data, supported by a freely accessible infrastructure: OpenAire[113] integrates institutional repositories and also provides the Zenodo[114] repository to upload research output (datasets and publications) files up to 50 GB. Zenodo is hosted at CERN, funded by the EU and CERN,[115] and provides integration with DropBox and GitHub. Users can define collections and communities, and configure the uploaded files for restricted access and embargo periods.

While Zenodo serves mainly archival purposes, the pan-European collaborative data infrastructure (EUDAT CDI)[116] provides generic data services such as storage and computing services to European researchers and research communities, and offers a joint metadata service, integrating metadata from different communities into easily searchable and open catalogues. There are a number of services implementing cloud facilities:

- B2ACCESS: authentication and Authorisation platform, identity provider, implemented by Unity IDM[117]
- B2DROP offering cloud services using owncloud[118]
- B2SHARE providing file sharing
- B2STAG: file transfer services, based on the iRods data management system[119] and GridFTP[120]
- B2SAFE providing replication and data management policies
- B2FIND implementing metadata searches, and finally
- BHOST allowing custom applications to be integrated within the EUDAT infrastructure.

5.8 Conclusions

The quest for a proper data integration architecture in the current age of big data is ongoing. Loosely coupled systems, allowing end users to share and integrate data without understanding data query languages or schema mapping, are increasingly considered more appropriate and are actively being developed.

The big data in toxicology are characterised by their variety, and the integration with omics data types adds a significant volume. The domain knowledge is represented by biomedical ontologies, which also play a significant role in the process of linking between independent data resources. Semantic technologies for integration are state of the art, but other storage and inference solutions are gaining popularity due to improved performance. Integration of data analysis tools is critical for making the data available for modern, complex, multidisciplinary applications.

References

1. A. M. Richard, C. Yang and R. S. Judson, Toxicity Data Informatics: Supporting a New Paradigm for Toxicity Prediction, *Toxicol. Mech. Methods*, 2008, **18**(2–3), 103–118.
2. P. S. Kern, G. F. Gerberick, C. A. Ryan, I. Kimber, A. Aptula and D. A. Basketter, Local lymph node data for the evaluation of skin sensitization alternatives: a second compilation, *Dermatitis*, 2010, **21**(1), 8–32.
3. A. Bitsch, S. Jacobi, C. Melber, U. Wahnschaffe, N. Simetska and I. Mangelsdorf, REPDOSE: A database on repeated dose toxicity studies of commercial chemicals–A multifunctional tool, *Regul. Toxicol. Pharmacol.*, 2006, **46**(3), 202–210.
4. K. Hansen, S. Mika, T. Schroeter, A. Sutter, A. ter Laak, T. Steger-Hartmann, N. Heinrich and K.-R. Müller, Benchmark Data Set for in Silico Prediction of Ames Mutagenicity,*J. Chem. Inf. Model.*, 2009, **49**(9), 2077–2081.
5. R. Benigni, C. Bossa, A. M. Richard and C. Yang, A novel approach: chemical relational databases, and the role of the ISSCAN database on assessing chemical carcinogenicity, *Ann. Ist. Super. Sanita*, 2008, **44**(1), 48–56.
6. R. Benigni, O. Tcheremenskaia, V. Jeliazkov, N. Jeliazkova, I. Nikolova, S. E. Escher, M. Batke, T. Baier, B. Hardy and R. Affentranger, OpenTox deliverable D3.4. Final database with additional content, 2011.
7. R. S. Judson, M. T. Martin, P. Egeghy, S. Gangwal, D. M. Reif, P. Kothiya, M. Wolf, T. Cathey, T. Transue, D. Smith, J. Vail, A. Frame, S. Mosher, E. A. C. Hubal and A. M. Richard, Aggregating Data for Computational Toxicology Applications: The U.S. Environmental Protection Agency (EPA) Aggregated Computational Toxicology Resource (ACToR) System, *Int. J. Mol. Sci.*, 2012, **13**(12), 1805–1831.
8. I. Sushko, S. Novotarskyi, R. Körner, A. K. Pandey, M. Rupp, W. Teetz, S. Brandmaier, A. Abdelaziz, V. V. Prokopenko and V. Y. Tanchuk, *et al.*, Online chemical modeling environment (OCHEM): web platform for data storage, model development and publishing of chemical information.,*J. Comput.-Aided Mol. Des.*, 2011, **25**(6), 533–554.
9. COSMOS DB, http://www.cosmostox.eu/what/COSMOSdb/, http://cosmosdb.eu/ accessed April 28, 2017.

10. A. Gaulton, L. J. Bellis, A. P. Bento, J. Chambers, M. Davies, A. Hersey, Y. Light, S. McGlinchey, D. Michalovich, B. Al-Lazikani and J. P. Overington, ChEMBL: a large-scale bioactivity database for drug discovery, *Nucleic Acids Res.*, 2011, **40**(Database issue), D1100–D1107.
11. S. Ekins, A. M. Clark, C. Southan, B. A. Bunin and A. J. Williams, Chapter 16. Small-molecule Bioactivity Databases, in *High Throughput Screening Methods*, ed. J. A. Bittker and N. T. Ross, Royal Society of Chemistry, Cambridge, 2017, pp. 344–371.
12. S. Kim, P. A. Thiessen, E. E. Bolton, J. Chen, G. Fu, A. Gindulyte, L. Han, J. He, S. He, B. A. Shoemaker, J. Wang, B. Yu, J. Zhang and S. H. Bryant, PubChem substance and compound databases, *Nucleic Acids Res.*, 2016, **44**(D1), D1202–D1213.
13. J. Piñero, À. Bravo, N. Queralt-Rosinach, A. Gutiérrez-Sacristán, J. Deu-Pons, E. Centeno, J. García-García, F. Sanz and L. I. Furlong, DisGeNET: a comprehensive platform integrating information on human disease-associated genes and variants, *Nucleic Acids Res.*, 2017, **45**(D1), D833–D839.
14. C. A. Afshari, H. K. Hamadeh and P. R. Bushel, The evolution of bioinformatics in toxicology: Advancing toxicogenomics, *Toxicol. Sci.*, 2011, **120**(Suppl. 1), S225–S237.
15. S. C. Sahu, *Toxicogenomics: A Powerful Tool for Toxicity Assessment*, John Wiley and Sons, 2008.
16. L. Suter, L. E. Babiss and E. B. Wheeldon, Toxicogenomics in predictive toxicology in drug development, *Chem. Biol.*, 2004, **11**(2), 161–171.
17. A. P. Davis, C. J. Grondin, R. J. Johnson, D. Sciaky, B. L. King, R. McMorran, J. Wiegers, T. C. Wiegers and C. J. Mattingly, The Comparative Toxicogenomics Database: update 2017, *Nucleic Acids Res.*, 2017, **45**(D1), D972–D978.
18. Y. Igarashi, N. Nakatsu, T. Yamashita, A. Ono, Y. Ohno, T. Urushidani and H. Yamada, Open TG-GATEs: a large-scale toxicogenomics database, *Nucleic Acids Res.*, 2015, **43**(D1), D921–D927.
19. S. Xirasagar, S. F. Gustafson, C. C. Huang, Q. Pan, J. Fostel, P. Boyer, B. A. Merrick, K. B. Tomer, D. D. Chan, K. J. Yost, D. Choi, N. Xiao, S. Stasiewicz, P. Bushel and M. D. Waters, Chemical effects in biological systems (CEBS) object model for toxicology data, SysTox-OM: Design and application, *Bioinformatics*, 2006, **22**(7), 874–882.
20. A. M. Richard, R. S. Judson, K. A. Houck, C. M. Grulke, P. Volarath, I. Thillainadarajah, C. Yang, J. Rathman, M. T. Martin, J. F. Wambaugh, T. B. Knudsen, J. Kancherla, K. Mansouri, G. Patlewicz, A. J. Williams, S. B. Little, K. M. Crofton and R. S. Thomas, ToxCast Chemical Landscape: Paving the Road to 21st Century Toxicology, *Chem. Res. Toxicol.*, 2016, **29**(8), 1225–1251.
21. ToxReporter: viewing the genome through the eyes of a toxicologist, Database, 2016.
22. National Toxicology program, https://ntp.niehs.nih.gov/.

23. W. T. M. Jansen and J. H. R. H. M. Schretlen, Toxicogenomics Implementation Strategies, in *Toxicogenomics-based Cellular Models*, ed. J. Kleinjans, Elsevier, 2014, pp. 321–336.
24. OECD, Adverse Outcome Pathways, Molecular Screening and Toxicogenomics, http://www.oecd.org/chemicalsafety/testing/adverse-outcome-pathways-molecular-screening-and-toxicogenomics.htm accessed July 24, 2017.
25. A. Kamburov, K. Pentchev, H. Galicka, C. Wierling, H. Lehrach and R. Herwig, ConsensusPathDB: toward a more complete picture of cell biology, *Nucleic Acids Res.*, 2011, **39**(Database issue), D712–D717.
26. G. Joshi-Tope, M. Gillespie, I. Vastrik, P. D'Eustachio, E. Schmidt, B. de Bono, B. Jassal, G. R. Gopinath, G. R. Wu, L. Matthews, S. Lewis, E. Birney and L. Stein, Reactome: a knowledgebase of biological pathways, *Nucleic Acids Res.*, 2005, **33**(Database issue), D428–D432.
27. M. Kutmon, A. Riutta, N. Nunes, K. Hanspers, E. L. Willighagen, A. Bohler, J. Mélius, A. Waagmeester, S. R. Sinha, R. Miller, S. L. Coort, E. Cirillo, B. Smeets, C. T. Evelo and A. R. Pico, WikiPathways: capturing the full diversity of pathway knowledge, *Nucleic Acids Res.*, 2016, **44**(D1), D488–D494.
28. K. Wanichthanarak, J. F. Fahrmann and D. Grapov, Genomic, Proteomic, and Metabolomic Data Integration Strategies, *Biomarker Insights*, 2015, **10**(Suppl. 4), 1–6.
29. C. E. Cook, M. T. Bergman, R. D. Finn, G. Cochrane, E. Birney and R. Apweiler, The European Bioinformatics Institute in 2016: Data growth and integration, *Nucleic Acids Res.*, 2016, **44**(D1), D20–D26.
30. A. Doan, A. Halevy and Z. Ives, *Principles of Data Integration*, Elsevier, 2012.
31. InChI Trust, Trust http://www.inchi-trust.org/.
32. J. Chambers, M. Davies, A. Gaulton, A. Hersey, S. Velankar, R. Petryszak, J. Hastings, L. Bellis, S. McGlinchey and J. P. Overington, UniChem: a unified chemical structure cross-referencing and identifier tracking system, *J. Cheminf.*, 2013, **5**(1), 3.
33. N. Jeliazkova and V. Jeliazkov, AMBIT RESTful web services: an implementation of the OpenTox application programming interface, *J. Cheminf.*, 2011, **3**(1), 18.
34. N. Jeliazkova, V. Koch, Q. Li, U. Jensch, J. S. Reigl, R. Kreiling, I. Georgiev and B. Hubesch, Linking LRI AMBIT chemoinformatic system with the IUCLID substance database to support read-across of substance endpoint data and category formation, *Toxicol. Lett.*, 2016, **258**, S114–S115.
35. LRI AMBIT tool for read across, https://ambitlri.ideaconsult.net/ accessed April 28, 2017.
36. N. Jeliazkova, C. Chomenidis, P. Doganis, B. Fadeel, R. Grafström, B. Hardy, J. Hastings, M. Hegi, V. Jeliazkov, N. Kochev, P. Kohonen, C. R. Munteanu, H. Sarimveis, B. Smeets, P. Sopasakis, G. Tsiliki, D. Vorgrimmler and E. Willighagen, The eNanoMapper database for nanomaterial safety information, *Beilstein J. Nanotechnol.*, 2015, **6**, 1609–1634.

37. S.-A. Sansone, P. Rocca-Serra, D. Field, E. Maguire, C. Taylor, O. Hofmann, H. Fang, S. Neumann, W. Tong and L. Amaral-Zettler, *et al.*, Toward interoperable bioscience data, *Nat. Genet.*, 2012, **44**(2), 121–126.
38. R. Cavill, D. Jennen, J. Kleinjans and J. J. Briedé, Transcriptomic and metabolomic data integration, *Briefings Bioinf.*, 2016, **17**(5), 891–901.
39. C. Batchelor, C. Y. A. Brenninkmeijer, C. Chichester, M. Davies, D. Digles, I. Dunlop, C. T. Evelo, A. Gaulton, C. Goble and A. J. G. Gray, *et al.*, Scientific Lenses to Support Multiple Views over Linked Chemistry Data, in *Lecture Notes in Computer Science (including subseries Lecture Notes in Artificial Intelligence and Lecture Notes in Bioinformatics)*, Springer Verlag, 2014, vol. 8796, pp. 98–113.
40. M. P. van Iersel, A. R. Pico, T. Kelder, J. Gao, I. Ho, K. Hanspers, B. R. Conklin and C. T. Evelo, The BridgeDb framework: standardized access to gene, protein and metabolite identifier mapping services, *BMC Bioinf.*, 2010, **11**(1), 5.
41. Z. Bellahsene, A. Bonifati and E. Rahm, *Schema Matching and Mapping*, Springer, Berlin Heidelberg, 2011.
42. M. Franklin, A. Halevy and D. Maier, From databases to dataspaces: a new abstraction for information management, *SIGMOD Rec.*, 2005, **34**(4), 27–33.
43. C. Bizer, Interlinking Scientific Data on a Global Scale, *Data Sci. J.*, 2013, **12**, GRDI6–GRDI12.
44. N. Miloslavskaya and A. Tolstoy, Big Data, Fast Data and Data Lake Concepts, *Procedia Comput. Sci.*, 2016, **88**, 300–305.
45. B. Hardy, N. Douglas, C. Helma, M. Rautenberg, N. Jeliazkova, V. Jeliazkov, I. Nikolova, R. Benigni, O. Tcheremenskaia and S. Kramer, *et al.*, Collaborative development of predictive toxicology applications, *J. Cheminf.*, 2010, **2**(1), 7.
46. P. Groth, A. Loizou, A. J. G. Gray, C. Goble, L. Harland and S. Pettifer, API-centric Linked Data integration: The Open PHACTS Discovery Platform case study, *J. Web Semant.*, 2014, **29**, 12–18.
47. J. Nystrom-Persson, Y. Igarashi, M. Ito, M. Morita, N. Nakatsu, H. Yamada and K. Mizuguchi, Toxygates: Interactive toxicity analysis on a hybrid microarray and linked data platform, *Bioinformatics*, 2013, **29**(23), 3080–3086.
48. S. Jupp, J. Malone, J. Bolleman, M. Brandizi, M. Davies, L. Garcia, A. Gaulton, S. Gehant, C. Laibe, N. Redaschi, S. M. Wimalaratne, M. Martin, N. Le Novère, H. Parkinson, E. Birney and A. M. Jenkinson, The EBI RDF platform: Linked open data for the life sciences, *Bioinformatics*, 2014, **30**(9), 1338–1339.
49. P. Kohonen, E. Benfenati, D. Bower, R. Ceder, M. Crump, K. Cross, R. C. Grafström, L. Healy, C. Helma and N. Jeliazkova, *et al.*, The ToxBank Data Warehouse: Supporting the Replacement of In Vivo Repeated Dose Systemic Toxicity Testing, *Mol. Inf.*, 2013, **32**(1), 47–63.

50. SEURAT-1, Towards the Replacement of in vivo Repeated Dose Systemic Toxicity Testing, http://www.seurat-1.eu/.
51. US EPA, Chemistry Dashboard, https://www.epa.gov/chemical-research/chemistry-dashboard accessed April 28, 2017.
52. e-tox project, http://www.etoxproject.eu/ accessed April 27, 2017.
53. M. Cases, K. Briggs, T. Steger-Hartmann, F. Pognan, P. Marc, T. Kleinöder, C. Schwab, M. Pastor, J. Wichard and F. Sanz, The eTOX Data-Sharing Project to Advance in Silico Drug-Induced Toxicity Prediction, *Int. J. Mol. Sci.*, 2014, **15**(11), 21136–21154.
54. P. Carrió, O. López, F. Sanz and M. Pastor, eTOXlab, an open source modeling framework for implementing predictive models in production environments, *J. Cheminf.*, 2015, **7**(1), 8.
55. T. P. Eyres, Extracting more value from data silos: using the semantic web to link chemistry and biology for innovation, *EMBnet. J.*, 2013, **19**(B), 36.
56. Q. Li, V. Koch, U. Jensch, J. Schneider-Reigl, N. Jeliazkova, B. Hubesch and R. Kreiling, Development of the Clariant CompTox Suite: A chemoinformatic system to support read-across and category formation within the scope of safety assessments of substances, *Toxicol. Lett.*, 2015, **238**(2), S172–S173.
57. S. Croset, J. Rupp and M. Romacker, Flexible data integration and curation using a graph-based approach, *Bioinformatics*, 2016, **32**(6), 918–925.
58. M. Marti-Solano, E. Birney, A. Bril, O. Della Pasqua, H. Kitano, B. Mons, I. Xenarios and F. Sanz, Integrative knowledge management to enhance pharmaceutical R&D, *Nat. Rev. Drug Discovery*, 2014, **13**(4), 239–240.
59. J. DiFrisco, G. Pontarotti, F. Boem, G. Schlaepfer, E. Sokolowska and E. Fernández-Labandera, Ontological Issues in the Life Sciences, *Biol. Theory*, 2015, **10**(2), 176–181.
60. SKOS Simple Knowledge Organization System, https://www.w3.org/2004/02/skos/ accessed April 28, 2017.
61. M. Herrero-Zazo, I. Segura-Bedmar, J. Hastings and P. Martínez, DINTO: Using OWL Ontologies and SWRL Rules to Infer Drug–Drug Interactions and Their Mechanisms, *J. Chem. Inf. Model.*, 2015, **55**(8), 1698–1707.
62. L. L. Chepelev and M. Dumontier, Semantic Web integration of Cheminformatics resources with the SADI framework, *J. Cheminf.*, 2011, **3**(1), 16.
63. J. Lee, M.-D. Pham, J. Lee, W.-S. Han, H. Cho, H. Yu and J.-H. Lee, Processing SPARQL queries with regular expressions in RDF databases, *BMC Bioinf.*, 2011, **12**(Suppl. 2), S6.
64. M. Barros and F. M. Couto, Knowledge Representation and Management: a Linked Data Perspective, *Yearb. Med. Inform.*, 2016, 25(1), 178–183.

65. P. de Matos, R. Alcántara, A. Dekker, M. Ennis, J. Hastings, K. Haug, I. Spiteri, S. Turner and C. Steinbeck, Chemical Entities of Biological Interest: an update, *Nucleic Acids Res.*, 2010, **38**(Database issue), D249–D254.
66. BioPortal, http://bioportal.bioontology.org/ accessed April 28, 2017.
67. J. Malone, R. Stevens, S. Jupp, T. Hancocks, H. Parkinson and C. Brooksbank, Ten Simple Rules for Selecting a Bio-ontology, *PLoS Comput. Biol.*, 2016, **12**(2), e1004743.
68. J. Hastings, N. Jeliazkova, G. Owen, G. Tsiliki, C. R. Munteanu, C. Steinbeck and E. Willighagen, eNanoMapper: harnessing ontologies to enable data integration for nanomaterial risk assessment, *J. Biomed. Semant.*, 2015, **6**, 10.
69. E. Rahm, Towards Large-Scale Schema and Ontology Matching, in *Schema Matching and Mapping*, ed. Z. Bellahsene, A. Bonifati and E. Rahm, Springer, Berlin Heidelberg, 2011, pp. 3–27.
70. P. Shvaiko and J. Euzenat, Ontology Matching: State of the Art and Future Challenges, *IEEE Trans. Knowl. Data Eng.*, 2013, **25**(1), 158–176.
71. L. Otero-Cerdeira, F. J. Rodríguez-Martínez and A. Gómez-Rodríguez, Ontology matching: A literature review, *Expert Syst. Appl.*, 2015, **42**(2), 949–971.
72. A. Groß, C. Pruski and E. Rahm, Evolution of biomedical ontologies and mappings: Overview of recent approaches, *Comput. Struct. Biotechnol. J.*, 2016, **14**, 333–340.
73. N. Jeliazkova, Web tools for predictive toxicology model building, *Expert Opin. Drug Metab. Toxicol.*, 2012, **8**(7), 791–801.
74. J. G. Frey and C. L. Bird, Cheminformatics and the Semantic Web: adding value with linked data and enhanced provenance, *Wiley Interdiscip. Rev.: Comput. Mol. Sci.*, 2013, **3**(5), 465–481.
75. J. G. Frey and C. L. Bird, Web-based services for drug design and discovery, *Expert Opin. Drug Discovery*, 2011, **6**(9), 885–895.
76. I. V. Tetko, U. Maran and A. Tropsha, Public (Q)SAR Services, Integrated Modeling Environments, and Model Repositories on the Web: State of the Art and Perspectives for Future Development, *Mol. Inf.*, 2016, **36**(3), DOI: 10.1002/minf.201600082.
77. PubChem REST tutorial.
78. US EPA ACToR Web Services, http://actorws.epa.gov/actorws/.
79. A. Yates, K. Beal, S. Keenan, W. McLaren, M. Pignatelli, G. R. S. Ritchie, M. Ruffier, K. Taylor, A. Vullo and P. Flicek, The Ensembl REST API: Ensembl Data for Any Language, *Bioinformatics*, 2015, **31**(1), 143–145.
80. CLUE API, https://clue.io/api accessed April 28, 2017.
81. S. Jupp, J. Malone, J. Bolleman, M. Brandizi, M. Davies, L. Garcia, A. Gaulton, S. Gehant, C. Laibe, N. Redaschi, S. M. Wimalaratne, M.

Martin, N. Le Novère, H. Parkinson, E. Birney and A. M. Jenkinson, The EBI RDF platform: linked open data for the life sciences, *Bioinformatics*, 2014, **30**(9), 1338–1339.
82. N. del-Toro, M. Dumousseau, S. Orchard, R. C. Jimenez, E. Galeota, G. Launay, J. Goll, K. Breuer, K. Ono, L. Salwinski and H. Hermjakob, A new reference implementation of the PSICQUIC web service, *Nucleic Acids Res.*, 2013, **41**(Web Server issue), W601–W606.
83. Open PHACTS Discovery Platform API, https://dev.openphacts.org/ accessed April 28, 2017.
84. GenomeSpace, http://www.genomespace.org/ accessed April 28, 2017.
85. Global Alliance for Genomics and Health, http://genomicsandhealth.org/.
86. Google Genomics API, https://cloud.google.com/genomics/.
87. OData (Open Data Protocol), http://www.odata.org/.
88. W. B. Mattes, S. D. Pettit, S. A. Sansone, P. R. Bushel and M. D. Waters, Database development in toxicogenomics: Issues and efforts, *Environ. Health Perspect.*, 2004, **112**(4), 495–505.
89. S.-A. Sansone, P. Rocca-Serra, A. Gonzalez-Beltran, D. Johnson and ISA community, ISA Model and Serialization Specifications 1.0, ZENODO, 2016, DOI: 10.5281/zenodo.163640.
90. D. G. Thomas, S. Gaheen, S. L. Harper, M. Fritts, F. Klaessig, E. Hahn-Dantona, D. Paik, S. Pan, G. A. Stafford, E. T. Freund, J. D. Klemm and N. A. Baker, ISA-TAB-Nano: A Specification for Sharing Nanomaterial Research Data in Spreadsheet-based Format, *BMC Biotechnol.*, 2013, **13**(1), 2.
91. V. Gligorijević and N. Pržulj, Methods for biological data integration: perspectives and challenges, *J. R. Soc., Interface*, 2015, **12**(112), 20150571.
92. A. R. Joyce and B. Ø. Palsson, The model organism as a system: integrating 'omics' data sets, *Nat. Rev. Mol. Cell Biol.*, 2006, **7**(3), 198–210.
93. D. Gomez-Cabrero, I. Abugessaisa, D. Maier, A. Teschendorff, M. Merkenschlager, A. Gisel, E. Ballestar, E. Bongcam-Rudloff, A. Conesa and J. Tegnér, Data integration in the era of omics: current and future challenges, *BMC Syst. Biol.*, 2014, **8**(Suppl. 2), I1.
94. Y. Li, F.-X. Wu and A. Ngom, A review on machine learning principles for multi-view biological data integration, *Briefings Bioinf.*, 2016, bbw113.
95. Z.-H. Zhou, N. V. Chawla, Y. Jin and G. J. Williams, Big Data Opportunities and Challenges: Discussions from Data Analytics Perspectives [Discussion Forum], *IEEE Comput. Intell. Mag.*, 2014, **9**(4), 62–74.
96. J. Dean and S. Ghemawat, MapReduce: simplified data processing on large clusters, *Commun. ACM*, 2008, **51**(1), 107.
97. KNIME, https://www.knime.org accessed April 28, 2017.

98. M. I. Jordan and T. M. Mitchell, Machine learning: Trends, perspectives, and prospects, *Science*, 2015, **349**(6245), 255–260.
99. H. Kashyap, H. A. Ahmed, N. Hoque, S. Roy and D. K. Bhattacharyya, Big Data Analytics in Bioinformatics: A Machine Learning Perspective, 2015.
100. EMBL-EBI services, http://www.ebi.ac.uk/services accessed April 28, 2017.
101. S. Squizzato, Y. M. Park, N. Buso, T. Gur, A. Cowley, W. Li, M. Uludag, S. Pundir, J. A. Cham, H. McWilliam and R. Lopez, The EBI Search engine: providing search and retrieval functionality for biological data from EMBL-EBI, *Nucleic Acids Res.*, 2015, **43**(W1), W585–W588.
102. ELIXIR: A distributed infrastructure for life-science information, http://www.elixir-europe.org.
103. M. D. Wilkinson, M. Dumontier, I. J. Aalbersberg, G. Appleton, M. Axton, A. Baak, N. Blomberg, J.-W. Boiten, L. B. da Silva Santos and P. E. Bourne, *et al.*, The FAIR Guiding Principles for scientific data management and stewardship, *Sci. Data*, 2016, **3**, 160018.
104. A. Treloar, The Research Data Alliance: globally co-ordinated action against barriers to data publishing and sharing, *Learn. Publ.*, 2014, **27**(5), 9–13.
105. Database Commons a catalog of biological databases, http://databasecommons.org/index.jsp accessed April 28, 2017.
106. The NIH Commons, https://datascience.nih.gov/commons accessed April 28, 2017.
107. BioSharing, https://biosharing.org/ accessed April 28, 2017.
108. P. McQuilton, A. Gonzalez-Beltran, P. Rocca-Serra, M. Thurston, A. Lister, E. Maguire and S.-A. Sansone, BioSharing: curated and crowd-sourced metadata standards, databases and data policies in the life sciences, *Database*, 2016, **2016**, baw075.
109. figshare, https://figshare.com/ accessed April 28, 2017.
110. Dryad Digital Repository, http://datadryad.org/ accessed April 28, 2017.
111. S. Burgstaller-Muehlbacher, A. Waagmeester, E. Mitraka, J. Turner, T. Putman, J. Leong, C. Naik, P. Pavlidis, L. Schriml, B. M. Good and A. I. Su, Wikidata as a semantic framework for the Gene Wiki initiative, *Database*, 2016, **2016**, baw015.
112. P. Ertl, L. Patiny, T. Sander, C. Rufener and M. Zasso, Wikipedia Chemical Structure Explorer: substructure and similarity searching of molecules from Wikipedia, *J. Cheminf.*, 2015, **7**(1), 10.
113. OpenAIRE, https://www.openaire.eu/ accessed April 28, 2017.
114. Zenodo, https://zenodo.org/ accessed April 28, 2017.
115. EU FP7 project OpenAIREplus, http://cordis.europa.eu/project/rcn/100079_en.html accessed April 28, 2017.

116. EUDAT Collaborative Data Infrastructure, https://www.eudat.eu/ accessed April 28, 2017.
117. UNITY Identity Manager, http://www.unity-idm.eu/ accessed April 28, 2017.
118. ownCloud, https://owncloud.org/ accessed April 28, 2017.
119. iRODS Open Source Data Management Software, https://irods.org/ accessed April 28, 2017.
120. GridFTP, http://toolkit.globus.org/toolkit/docs/latest-stable/gridftp/ accessed April 28, 2017.

CHAPTER 6

Storing and Using Qualitative and Quantitative Structure–Activity Relationships in the Era of Toxicological and Chemical Data Expansion

SULEV SILD[a], GEVEN PIIR[a], DANIEL NEAGU[b] AND UKO MARAN*[a]

[a]University of Tartu, Institute of Chemistry, Ravila 14A, Tartu 50411, Estonia; [b]University of Bradford, Department of Computer Science, Richmond Road, BD7 1DP Bradford, UK
*E-mail: uko.maran@ut.ee

6.1 Introduction

Continual advancements in toxicology and in chemistry-related sciences are producing complex experimental data in increasing volumes and require more and more sophisticated solutions to organise, integrate, analyse and interpret such data and also derivatives of these data. In general, the *volume* and *velocity* of the toxicology and chemistry data resources may not yet be a serious issue when considered within the context of Big Data. More important is how the *variety* and *complexity* of these data is managed and systematised, linked together to become a useful information resource, and finally

Issues in Toxicology No. 41
Big Data in Predictive Toxicology
Edited by Daniel Neagu and Andrea-Nicole Richarz
© The Royal Society of Chemistry 2020
Published by the Royal Society of Chemistry, www.rsc.org

abstracted through large-scale data analysis in the form of models to become a useful piece of knowledge that can support further decision making. While data organisation is an important topic, the knowledge organisation (the storage, accessing and consistent representation of predictive models) is the major issue.

Before looking into the organisation of predictive models, it is important to understand the size of the chemical space as we know it today. The Chemical Abstracts Service (CAS) databases contained, in the year 2016 (on 31 October), over 122 million organic and inorganic substances[1] that are reported in the scientific literature. The biggest data source of commercially available chemicals, the ZINC database, lists over 200 million substances, and as of October 2016, over 100 million substances are in stock, *i.e.*, readily available for purchase[2]. The PubChem database[3] and the ChEMBL database[4] contain over 90 million and almost 1.7 million compounds, respectively. The dynamics of the increase in the number of compounds speaks for itself. In the case of PubChem,[5] the number of compounds has grown almost five times in the last eight years. In the ChEMBL[6] database, the number of compounds has tripled in the last six years (Figure 6.1 and Table 6.1).

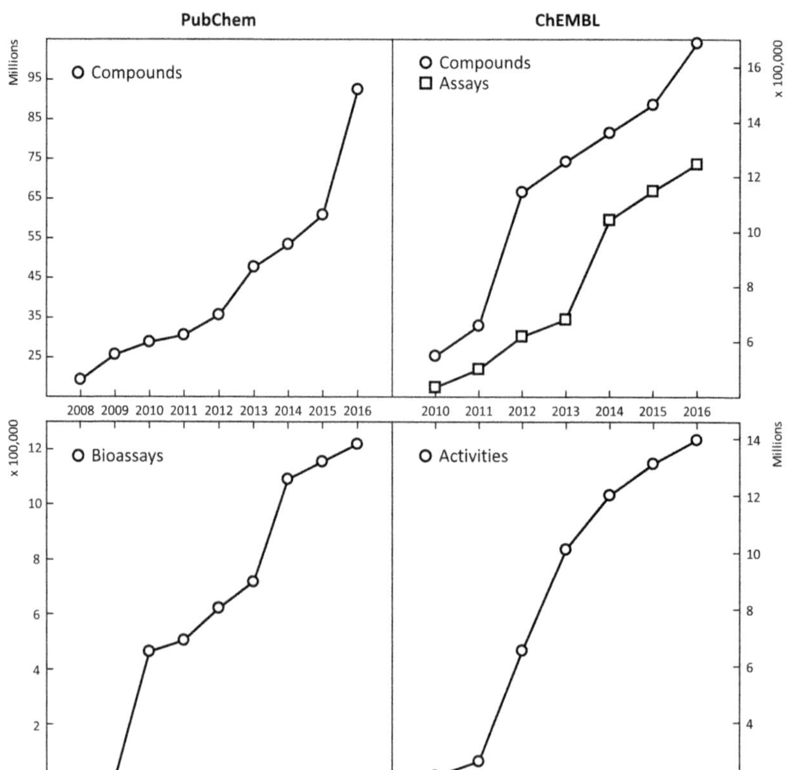

Figure 6.1 Increase in the number of compounds in the PubChem and ChEMBL databases and also related experimental data.

Table 6.1 Increase in the numbers of compounds, activities and assays in the PubChem and ChEMBL databases.

Year	PubChem		ChEMBL		
	Compounds	Bioassay	Compounds	Activities	Assays
2008	19 327 825	1197			
2009	25 668 433	1691			
2010	28 801 560	462 678	547 133	2 490 742	432 022
2011	30 455 811	504 222	658 075	3 030 317	499 867
2012	35 545 766	621 642	1 143 682	6 933 068	617 681
2013	47 757 896	717 429	1 254 575	10 509 572	679 259
2014	53 371 491	1 091 044	1 359 508	12 419 715	1 042 374
2015	60 774 418	1 154 363	1 463 270	13 520 737	1 148 942
2016[a]	92 365 853	1 218 719	1 686 695	14 371 219	1 246 132

[a]on 31 October 2016.

There are also many specialised databases in the area of toxicology. For example, ECOTOX[7] contains 14 656 compounds with 850 603 experimental data records collected from the literature (as of 31 October, 2016). The systematic data generation for predictive toxicology has received great attention in recent years, resulting in the ToxCast and Tox21 data compilations.[8,9] ToxCast includes 1800 carefully selected industrial and consumer products, food additives, and potentially "green" chemicals, which are experimentally measured with over 700 high-throughput assays covering a range of high-level cell responses and approximately 300 signalling pathways.[8] In the Tox21 initiative, a collection of 10 000 environmental chemicals and approved drugs have been tested for their potential to disrupt biological pathways that may result in toxicity using more than 100 assays.[9] One of the most recent chemical data compilations is the interactive Chemical Safety for Sustainability (iSSC) Chemistry Dashboard,[10] providing access to over 700 000 chemicals and millions of predicted physicochemical properties developed using machine-learning approaches. Those numbers are very large and become even larger if the virtual chemical universe is systematically explored, where over 166 billion molecules have been constructed by combining up to 17 atoms of C, N, O, S, and halogens, while considering valency rules and the stability of chemicals.[11]

This volume of data is huge within its own domain, but in the chemical world, the diversity and complexity of chemicals and their varied behaviours are even more important. When chemicals are put into different contexts, for instance, by linking with experimental data measured in organisms or in the environment, the amount of information grows immensely. For example, nearly 1.7 million compounds in the ChEMBL database have activities measured for over 14.3 million cases.[4] This view, however, is very rare because considering the over 1.24 million assays listed in ChEMBL, there is an average of 11.5 compounds per assay, and, for example, in the case of ECOTOX, there are on average 58 different experimental measurements per compound. Not all of this wealth of chemical data can be straightforwardly modelled, *i.e.*, abstracted. Therefore, when deriving abstractions from the information, *i.e.*, models, the size of the data set can often be large, but remains manageable. This data set is fit for the modelling conditions and is usually extracted from the wealth of chemical data. One should also keep in mind that our knowledge of creating predictive models today is not yet at the level where the full chemical universe could be analysed with a single model for different relevant endpoints or with a few clicks using off-the-shelf software.

In predictive toxicology and in the chemistry-related sciences, (Q)SARs form a general concept for modelling, covering both qualitative and quantitative approaches, for categorical and continuous experimental data, respectively. In these approaches, the chemical structure is typically represented numerically with molecular descriptors in three different ways: continuous values, categorical values (*e.g.*, indicator descriptors) and bit vectors (fingerprints). While the systematic representation of experimental and structural data is well established on the methodological side (see Chapter 3), the systematic mathematical representation of (Q)SARs in delivering final results is still under development. This is particularly due to the evolution of (Q)SARs from simple linear and multilinear relationships to complicated machine-learning models such as decision trees, artificial neural networks, support vector machines, Bayesian models, and ensemble models (*e.g.*, random forests). This evolution is particularly related to the recent growth and availability of data, *i.e.*, bigger data, that needs to be modelled for various toxicological endpoints and also the increased expectations of (Q)SAR users.

Since the seminal work of Hansch and Fujita,[12] many (Q)SAR models have been developed. A simple search (for "QS*R" query) in the Web of Science™[13] database gives over 21 000 publications starting from 1980. Not all of those publications include models, though. Despite this, there are a considerable number of models in publications that can be useful and potentially (re)used. However, our in-house analysis of publications (in 2013–2016) that include (Q)SAR models for different toxicity-related endpoints shows that only approximately 40% of these publications contain a sufficient amount of information for reusing the models. One can speculate that one possible reason for such a lack of reproducibility is that there was a lack of good reporting guidelines[14,15] and storage solutions[16] for predictive models. The information needed for reusing a predictive model could be expected to be found in the scientific publication and its supporting information, but this is not the current reality. Sometimes tiny but important bits of the information are left out, most likely unintentionally. Additionally, mathematically more complex models make it difficult to add the entire model representation to the publication, reducing the reusability. In addition, models developed with larger data sets are difficult to publish in a reproducible manner. Adding all such complex information to the supplementary materials section might be an answer, but even this requires new systematic and machine-readable approaches for the storage of predictive models, particularly within the emerging context of Big Data.

Developing an interoperable and extensible toxicological data and model governance is a data science necessity for a number of reasons. As expressed earlier, toxicity data include diverse chemical and biological information from measurements, (*in silico*) software outputs and (*in vivo* and *in vitro*) experiments that are distributed throughout a variety of data resources (see also Chapters 2 and 3). Different data formats, incomplete information about provenance and experiments, and computational errors increase the inconsistency of the collected information and make data governance challenging. Additionally, the quality of the chemical and toxicity data has a significant impact on the process of (Q)SAR model development, where it is crucial to ensure and record model accuracy and consistency. Automated data quality assessment and model applicability are still challenging in predictive toxicology: currently, this involves manual data curation and expert judgement. Moreover, good quality models usually represent a significant investment of time and knowledge. Once predictive models are built, it is important to contribute to the steps (model validation, model identification, model comparison and model ranking) simplifying the reuse of the model. This raises the need for toxicological (chemical) data and (Q)SAR model governance[17] that extends the general principles of data governance to the intelligence and knowledge involved in predictive toxicology and modelling tasks, and well-documented model reporting contributes a great deal to governance tasks.

Predictive models published in scientific articles are seldom used by anyone other than the original authors. When analysing models in the scientific literature, it is apparent that some published models are implemented in some standalone software or web application. This brings us to the concept where the predictive model can be observed as a commercial product, as the original authors usually see or expect to see some practical value in their model(s). This may motivate some model authors to keep important details about the models private or obscured (coded in software where they are not accessible). Many models are also derived from proprietary data that is commercially sensitive or for-profit purposes. Despite these practical considerations, one must keep in mind that our knowledge of the chemical universe and its toxic behaviours is yet very limited, and any abstraction from this universe in the form of models is subject to changes in time. This makes it very important to make published models easily accessible and (re)usable. This and the need for backtracking changes of the models bring us again to the question of the systematic storage of predictive models.

This chapter is about the technologies and approaches for storing predictive models such as (Q)SARs, with particular emphasis on predictive toxicology and on chemistry in general. First, we provide a summary of what constitutes a predictive model or, in other words, what must be present to make the model and its content usable and reproducible. This is followed by an analysis of existing file or data formats to collect and organise this information. Finally, solutions as to how these formats have been used in practice are described. Before going further, it is important to stress that we consider the predictive model as a whole body of immutable information that once derived must stay in this way, for one simple reason: once used for predictive purposes in decision support, one must be able to backtrack the model, understand how it was derived and evaluate the underlying data. Any alterations to the existing model will thus result in a new model.

6.2 Anatomy of Predictive Models

In the case of chemical structure and experimental (toxicological) data, it is very common to exchange such data between different software applications and services, so there are multiple well-established file formats for storing and sharing this information. The earlier examples of databases (CAS, ZINC, PubChem, ChEMBL, *etc.*) and many in-house databases clearly show this. Although the best practices for developing predictive models are well documented and published,[18,19] the storage of *in silico* models has not yet received much attention from scientific, industry and regulatory audiences. On the surface, model storage may appear to be a trivial issue, but in practice, it is a complex problem. The main question is what type of information must be considered when storing predictive models, or in other words, what are the skin, meat, and bones of predictive models? The skin of the predictive model is its mathematical representation. The meat of the model is the data from which the model was derived and validated, so that the model is basically a compact representation and replacement for the data. Finally, the bones of the model are the provenance information and metadata about the model and its components, which link the model and its applicability to the chemical information within the model domain. Consequently, to trust its usefulness, the model and its underlying data must be of a high quality and follow Linked Open Data concepts as much as possible.[20]

6.2.1 Mathematical Representation

On the surface, (Q)SAR models are essentially mathematical representations of the relationships between molecular descriptors and target properties. Publications of (Q)SAR models that do not make these mathematical relationships available are not that helpful to the scientific or user community. One can learn that somebody has developed a good predictive model and published a paper on a topic (toxicological endpoint) of interest, but without the mathematical representation, one can only read about the model, see some static figures, and obtain some quantitative information in the form of statistical parameters. Alternatively, one can try to redevelop the model, if the data (see next section) that the model was derived from is available. This may be a simple task if the algorithm used is simple enough, such as multiple linear regression analysis. However, redeveloping a model can change from simple to difficult amazingly quickly, for example, if the model has been developed with a stochastic algorithm, which uses random numbers at some point in the training phase. In such a case, one might obtain a similarly performing model, but replicating the original model is not possible. Therefore, depending on the model, different information is required for a complete mathematical representation. Table 6.2 includes the most commonly used model representations for (Q)SARs and lists the information needed to make use of these models. The list covers Multiple Linear Regression (MLR), Decision Trees (DT), Artificial Neural Networks (ANN) and Support Vector Machines (SVM). The common characteristic of all types of models is the independent variables (descriptors) section and how they are connected to the mathematical representation of the model. Speaking of mathematical representations, the most straightforwardly represented are the MLR models, where only coefficients for the descriptors and intercept are required. DT models require only a structure of the tree to be used and are therefore easy to reuse, too. In the case when multiple trees are used, such as in Random Forest models, a structure of all the trees is required. ANN and SVM models are more complex and require more information. In ANN models, one needs the structure of the network (*i.e.*, its topology of input, hidden and output layers), the number and weights of the neurons in the input and hidden layers, the number of hidden layers, the definition of the output layer (*e.g.*, 5-6-4-1), and also the type of activation function. For the SVM model, one needs to know what type of kernel basis function was used. Depending on the type of kernel, additional parameter values such as gamma, rho and the coefficient are needed, along with the list of support vectors for each descriptor and its coefficients.

6.2.2 Data Set Representation

The raw data that was used to derive the model is equally important because it provides a necessary context for verifying and understanding the model. The model can be used without the underlying data, but it is next to impossible to independently verify or validate the model without them. The data set

Table 6.2 Different model representations and the information required to reuse the listed models.

Model type	Information required
Multiple Linear Regression	• Descriptors (software + version) • Coefficient(s) for descriptor(s) and intercept (if used)
Decision Tree	• Descriptors (software + version) • Structure of the tree
Artificial Neural Network	• Descriptors (software + version) • Structure of network • Type of activation function for each layer • Weights for each neuron
Support Vector Machine	• Descriptors (software + version) • Type of kernel basis function and parameters • List of support vectors for each descriptor • List of coefficients

contains several interconnected components that form a complete data set. Each of those allows the verification of different steps in the model-building process. The first component is the representation of the molecules. There are several ways to represent molecules (chemical name, SMILES, InChI code, different file formats; see Chapter 3), but the most reliable from the point of view of the molecular descriptors is using the same file formats that were used in the modelling stages. If the descriptor calculation requires spatial coordinates, then the exact conformer of the chemical structure in the 3D representation must also be provided. The second component contains the values of the descriptors used. They are important when the recalculation of descriptors is attempted from the structures because it helps to verify the correctness and reproducibility of the descriptor calculation procedure. This information also allows checking if alternative descriptor calculation software is compatible with the used descriptors. The third and fourth components are the values of the modelled endpoint and the values of the predictions, respectively. The endpoint values are necessary for rebuilding the model, and the predicted values for checking the reproducibility of the model.

6.2.3 Model (and Data) Provenance and Metadata

The model provenance and metadata are important for the management of large (Q)SAR model collections and selecting appropriate models for decision making. The model provenance is mainly concerned with how the model developers arrived at the particular model. The model metadata is data (*i.e.*, information) about the model that facilitates the locating and processing of the model and related data. The provenance is closely related to the metadata, but they are not the same. The provenance records relevant data sources and processes involved in producing and delivering the model, while the metadata is used to represent the properties of the model. Because many of these properties are related to the provenance, the provenance is often represented as metadata. However, it is important to understand that not all metadata is related to the provenance.

The model provenance is the most important aspect for enabling the reuse of models. The provenance information for (Q)SAR models should include references to the data sources and all the data transformations that were performed on the raw data. This makes it possible to assess the quality of the experimental data and determine the reliability of the model. If data are combined from multiple sources, then each individual data point should also have appropriate references. Ideally, the model provenance should also capture the state of the system where the model was developed. This includes details about the handling of chemical structures (*e.g.*, representation of nitro groups and handling of salts), pre-processing steps (*e.g.*, structure minimisation or conformational space analysis), descriptor calculation and model development. For every calculation step, it is important to record the software and its version together with the setup used (*e.g.*, algorithm selection and parameter values). Often, it is sufficient to provide the provenance of the (Q)SAR model by a link to the related scientific publication, but this information is often incomplete and not machine-readable, which makes the reuse of models unnecessarily difficult. Section 6.4 of this chapter provides an overview of file formats for the storage of predictive models, where the model provenance is addressed in more detail. Related efforts include the development of the Chemical Information Ontology (CHEMINF) for chemical information entities, with a primary focus on data-driven research and the integration of calculated properties (descriptors).[21]

The main purpose of model metadata is to find its location and provide descriptive information for model discovery. Digital identifiers are essential metadata elements for the unique identification of models. If the identifier is permanent, such as a DOI (Digital Object Identifier), then it can be reliably used for locating models even if their physical location changes over time. Metadata also helps to organise large model collections to make the discovery of relevant models possible. Good-quality metadata provides the semantics of models and ensures that they can be effectively used. For example, it is easy to provide an overview of the models that are available for different endpoints and species. This can answer various questions, such as (1) who published it, (2) when was it published, (3) what are my rights for using this model, and (4) what type of model is it (*e.g.*, regression or classification). Section 6.5.1 of this chapter describes in more detail how different repositories organise models and their data.

6.3 Model Governance

Data governance is intrinsically related to the management of the quality, security and usage of digital resources: various frameworks for data governance have been described[22] to organise assets, generally by data principles, data quality, metadata, data access and the data lifecycle. Data governance is essentially a control framework for exercising control over the processes used within organisations to meet given standards and improve data quality. Because the characteristics of data and the needs of organisations are very different, it is difficult to establish a uniform data governance framework. Data governance in predictive toxicology (see Figure 6.2) was reviewed recently by Fu *et al.*,[15] and existing gaps in the development of current management frameworks according to data governance principles were identified. This section describes an extended data governance policy framework for models that supports efficient model management.[17]

Models are increasingly supporting decision-making processes. This means that a model has to be validated and accepted by regulatory bodies or management boards before it can be used within an organisation or in the regulatory context (*e.g.*, models in the environmental regulatory decision process).[23] Thus, a model can be considered as a valuable informational asset, and its proper usage and application makes it a candidate for governance.[24] Models can also represent a new source of risk for incorrect, inconsistent or

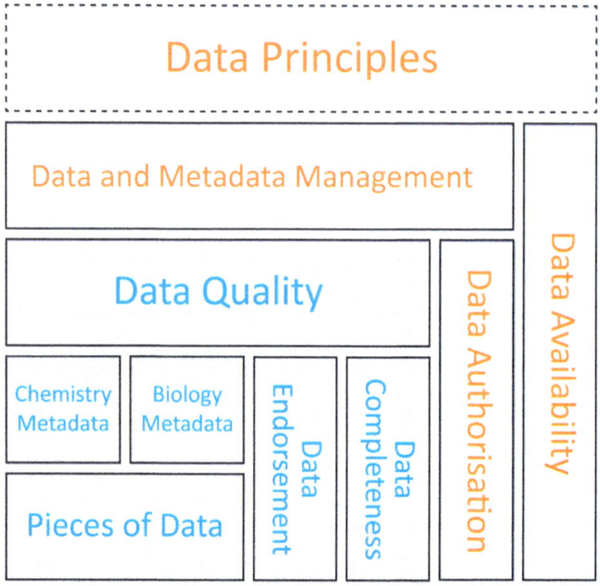

Figure 6.2 Decision domains of data governance for toxicological databases. Adapted from ref. 15, with permission from Springer Nature.

unsustainable decisions. To positively optimise such situations, model governance procedures should be strengthened through (a) assurance that the model is properly defined, described, stored and used and (b) the consistent recording of model improvements and updates to validate and maintain its effectiveness. These observations led the authors in ref. 17 to define Model Governance as a set of strategies that (1) help ensure that models achieve their purposes, (2) establish model reliability and its importance for organisational use, and (3) establish levels of controls and validation (see Figure 6.3).

Starting from the five OECD principles for model validation (embraced by regulatory bodies and widely implemented in predictive toxicology projects),[25] Palczewska *et al.* proposed in ref. 17 three groups of processes for model governance that build upon data governance procedures and extend them towards a data and model governance framework (see Figure 6.3).

According to this framework, a management board of specialist domain-knowledge experts defines the focus areas (*e.g.*, toxicity endpoints or chemical, physical or biological properties) for model development. Models developed from existing high-quality data already processed within the data governance framework are evaluated and stored in open or organisational inventories. Often, before the models find use as decision-supporting tools within an organisation, they are extensively evaluated and reviewed. For each model object, there are three model governance processes followed: model evaluation, model control, and model validation.[17] The stages of the proposed data and model governance framework[17] have recently been

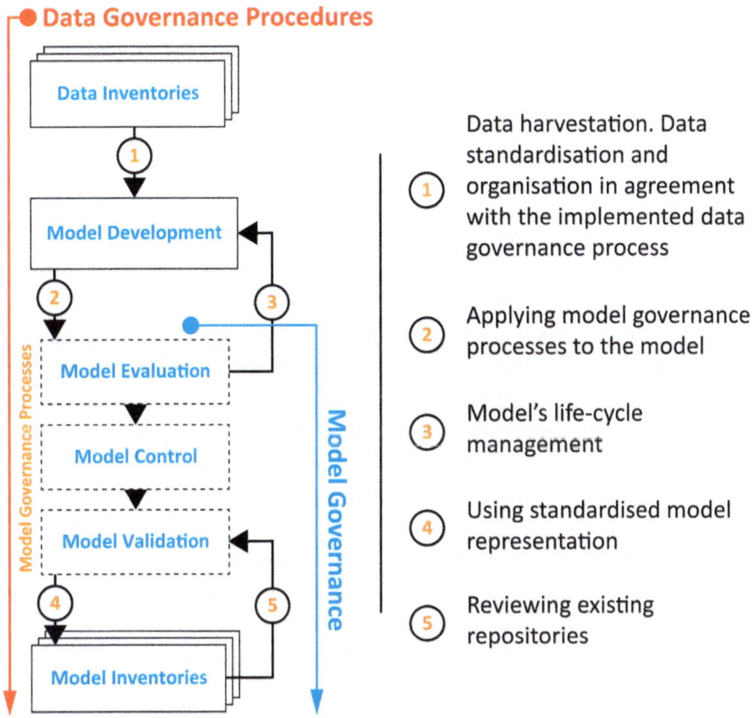

Figure 6.3 Data and model governance frameworks.

considered in a number of projects, *e.g.*, Syngenta M&DFARM[22] and EC FP7 and Cosmetics Europe's SEURAT-1 project COSMOS.[26] The COSMOS project applied these principles to data (managed in the COSMOS Database, https://cosmosdb.eu/)[27] and model[28] repositories that are managed in COSMOS Space and are freely available online.[26] Systematic representations of predictive models using various data and file formats are an essential part of model governance, as they allow easy access to the data and model mathematical representation and if available provide a systematic method for storage and use.

6.4 Data and File Formats for the Representation of Predictive Models

A predictive model can be represented to potential users in many different ways. Usually, modelling software packages have internal file formats to represent and archive predictive models. In that case, such models can be used only inside the corresponding software. Although the application of the model may be easy within the software, the wider usage of the models is restricted if the software is not freely available. Freely available alternatives such as Weka[29] or R[30] provide publicly accepted solutions, but those formats are still tailored to one specific software and tied to their environments. Additionally, they do not handle the descriptor calculation part of the model usage process. Another open source alternative is building models in the KNIME environment.[31] In this environment, the modeller can create a complete workflow and share it with users, whether directly or by applying it to the supplementary information of the publication such as in ref. 32. From the user's point of view, probably the simplest way to use a model is when it is built into some software. Many profit- and non-profit institutions have invested a great deal of time and money to develop predictive models and to integrate them with easily usable software (see Section 6.5.2 for a general overview). Depending on access policies, models can act as "black box" tools, making them less accessible and also less transparent for the user or for decision support. However, not all published models end up as local software resources. At best, the published information and knowledge is used as a basis for a new (improved) model. In the worst-case scenario, the model is going to be lost in the ever-growing amount of new information. This eventually leads to the need for data and/or file format(s) that allow the collection of information about descriptive and predictive models to make them usable in many instances. This section more thoroughly discusses efforts and respective data formats (see Table 6.3) that organise (Q)SAR model information in a more transparent and usable manner in publicly accessible formats.

6.4.1 QSAR Model Reporting Format

In 2004, in view of the regulatory use of predictive models for the risk assessment of chemicals and seeking a way to validate QSAR models, members of the Organisation for Economic Co-operation and Development (OECD)

Table 6.3 Areas covered by the data and file formats for the representation of predictive models.

	QMRF	QSAR-ML	PMML	QsarDB
Compounds	✓	✓[b]		✓
Properties	✓	✓		✓
Descriptors	✓	✓		✓
Models	✓[a]		✓	✓
Predictions	✓			✓
File	Single, with attachments	Single, linked with external resources	Single	Collection
Extensibility		✓	✓	✓
Machine readable		✓	✓	✓

[a]In practice, only MLR models are represented, although in a few cases, ANN models are attached as software-specific files.
[b]molecular structures are characterised with InChI codes and other appropriate attributes, and the actual representations of the 2D or 3D structures are stored externally.

formulated five principles[25] that QSAR models should follow in order to be eligible to be used in decision support: a defined endpoint, an unambiguous algorithm, a defined domain of applicability, appropriate measures of goodness-of-fit, robustness and predictivity, and, if possible, a mechanistic interpretation. To support the application of these principles, the QSAR Model Report Format (QMRF)[33] was proposed by the European Commission (EC) Joint Research Centre (JRC). The latest version of QMRF makes use of the XML file format and also provides a QMRF Editor[34] software for authoring QMRF documents.

QMRF sums up the essential information of the model and links it with the OECD principles in textual form. The content of the QMRF file is divided into ten chapters (see Figure 6.4). The first two chapters provide general information about the model (title, authors, software, publication, *etc.*). The information required to fulfil the OECD principles is covered in chapters three to eight of the QMRF file, with the exact descriptions provided in Figure 6.4. In those chapters, one can quite easily describe the modelling process and the results. Such a model

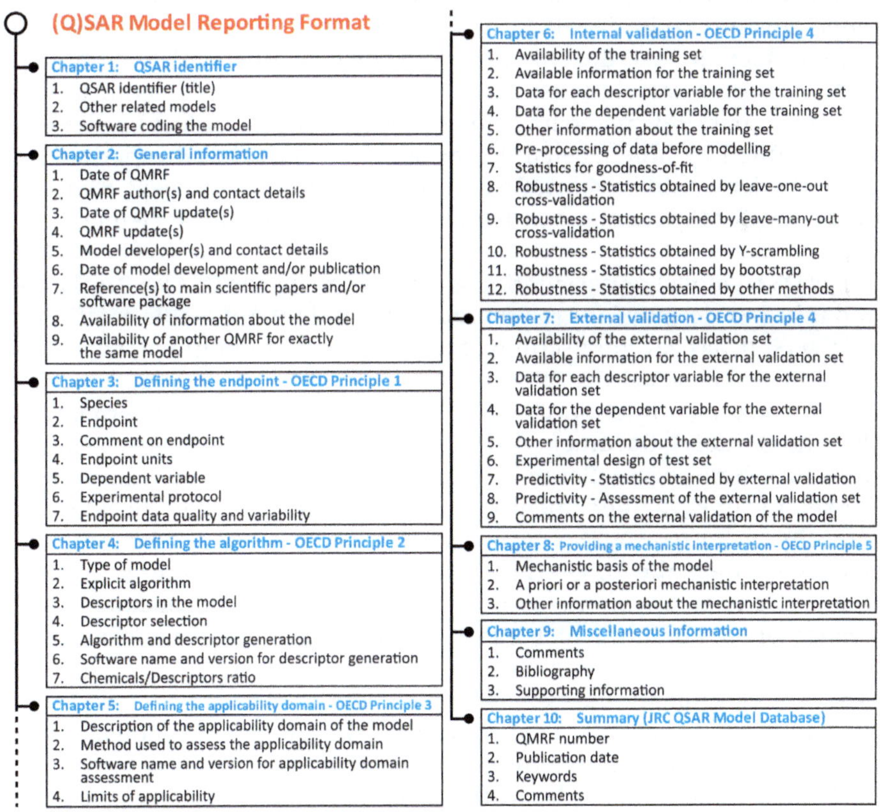

Figure 6.4 Schematic overview of the QMRF data format.

description very often repeats the information from the original scientific articles, and it is easily usable for simple (multiple) linear regression models. For more complicated mathematical definitions of models, this is not sufficient, and it is necessary to provide references to the original scientific document or provide this material as supporting information. A critical question remains: does the original paper or supporting information contain enough information for model usage? Chapter nine in the QMRF definition is used for miscellaneous (supporting) information, and chapter ten contains JRC QSAR Model Database (see Section 6.5.1.1) information and is filled by EC JRC when the submitted QMRF is included in the database after quality control, *i.e.*, of the clarity and completeness of the documentation. Table 6.3 summarises all model components that QMRF covers.

6.4.2 QSAR-ML

QSAR-ML is an open XML-based data format for the exchange of QSAR data sets.[35] Its XML schema is available at a dedicated homepage,[36] and Bioclipse software[37] provides a reference implementation and graphical tools. This format is aimed at the accurate description of QSAR data sets to reproduce their setup, enabling the validation of chemical structures and descriptor calculations. The QSAR-ML schema covers the following areas: data set metadata, molecular structures, molecular descriptors (defined by Blue Obelisk Descriptor Ontology)[38] and experimental property data. Table 6.4 briefly describes the most significant elements of the QSAR-ML schema, and Figure 6.5 shows how those elements are connected.

The QSAR model representation is not covered by the QSAR-ML schema, and it is only designed for the reproducible and interoperable creation of data sets. It has several useful features and some limitations. For example, molecular structures are characterised with InChI codes and the actual representations of 2D or 3D structures are stored externally, so that the QSAR-ML document only includes references to external files or links to web resources. Because the external data are dynamic and might change in time, the document can include checksums for checking the integrity of those resources. As checksums can only inform about the fact of changed data, the long-term storage of data sets still requires a proper independent archiving of external data. Each QSAR-ML document can include one endpoint, which can have multiple units. If multiple units are present, then the user must take care of translating the endpoint values into the same unit system. The metadata section can include literature references about the data sources. However, if the response values are compiled from multiple data sources then it is not easy to connect an individual response value with its original data source because it is not possible to assign literature references to individual response values. According to Table 6.3, QSAR-ML lacks model representation and also predicted values.

Table 6.4 Overview of major elements in the QSAR-ML data schema.

Element name	Element description
Metadata	Contains a list of literature references in BibTeXML format and also provides descriptive attributes about the data set, including its name, description, authors, source URL and license information.
Structurelist	Contains resources with molecular structures. Resources can be either paths to files or URLs and may also include checksum attributes for integrity checking. Resources contain one or more structure elements for defining the identity of the structure. Structure elements may also include annotations of errors/warnings.
Descriptorlist	Contains descriptors that can be uniquely identified *via* references to the Blue Obelisk Descriptor Ontology[38] and descriptor providers. Descriptors can also include parameters, given as key-value pairs, needed for descriptor calculations.
Descriptorproviders	Identifies software that was used for the calculation of descriptors, including its name, version, vendor and URL, and that provide implementations of descriptor algorithms.
Responselist	Contains values of a response variable or experimentally measured endpoint. Each response is referenced to a structure and response units (*e.g.*, LD50 or IC50), and its value may be given as a single value or an array of values.
Responseunit	Identifies units for response values and includes attributes for name, description, URL, *etc.*
Descriptorresultlist	Contains results of descriptor calculation, where each descriptor result includes a descriptor value and is linked to a descriptor and a structure.

6.4.3 The Predictive Model Markup Language

The Predictive Model Markup Language (PMML) standard is arguably the most interoperable representation of the statistical and data mining models currently available. This standard has been developed by the Data Mining Group consortium,[39] which includes representatives from major statistical software and information technology companies. Therefore, it has support at various degrees in many open source and proprietary software packages. Currently, this standard has reached a mature state that is the result of continuous development over more than a decade. As of October 2016, the latest version available is PMML v4.3. The standard is still evolving, and support for new model types is improving. For example, the latest release in August 2016 included new standards for representing probabilistic Gaussian process regression models and Bayesian network models. The PMML standard uses the XML mark-up language to represent models, and its structure is described as an XML Schema. It has a comprehensive specification covering different aspects of the representation of models. The specification can be divided into four major components (see Figure 6.6): data dictionary, data transformations, model descriptions, and model explanations.

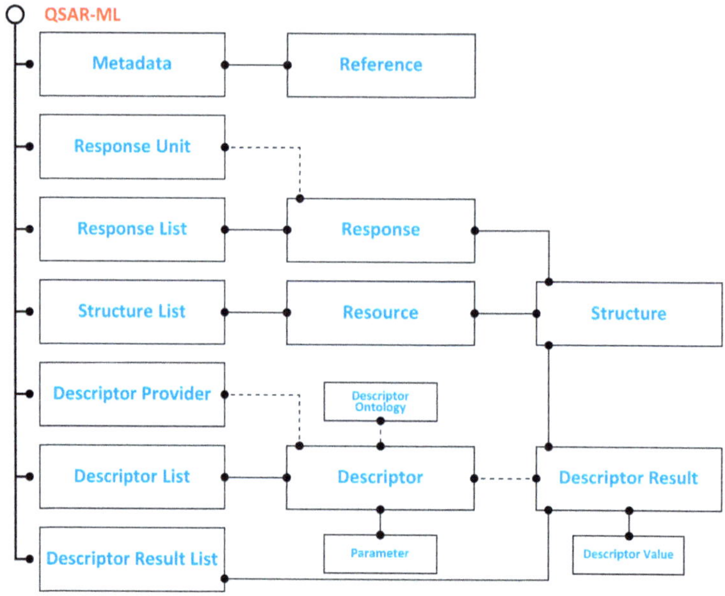

Figure 6.5 Schematic overview of the QSAR-ML data format.

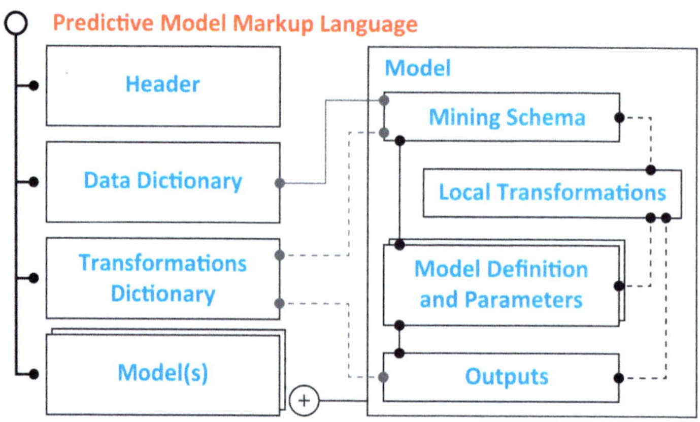

Figure 6.6 Schematic overview of the PMML data format.

The data dictionary is defined with a DataDictionary element, which contains a series of sub-elements that define data fields. Data fields are defined as continuous, categorical, or ordinal. Depending on this definition, further characterisation is possible, such as data type (*e.g.*, double or string), valid numerical ranges, and lists of valid, invalid, and missing values. This element can be viewed as metadata about the data set. In terms of (Q)SAR models, it defines descriptor (*e.g.*, descriptor ID, name, or data type) and target activity (*e.g.*, property ID or name) columns in the data set.

Data transformations describe data processing operations that are necessary for the usage of the model. These are usually simple functions that map input data to intermediate values that are easier to use in the specified models. The PMML specification defines various types of data transformations, including.

- Normalisation maps discrete or continuous values to numbers,
- Discretisation maps continuous values to discrete values,
- Value-mapping maps discrete values to discrete values,
- Built-in and user-defined functions that derive a new value by applying it to one or more parameters, and
- Aggregations that summarise a group of values (*e.g.*, min, max, or average).

Model representation is the most complex part of the specification and provides a comprehensive language for describing different models developed with various modelling techniques. PMML can represent both classification and regression models. The model definition starts with the mining schema element, which describes the purposes of the data fields and links them with the mathematical representation of the model. The mining schema also describes the handling of missing values, outlier treatment, and scaling of model outputs. The latest PMML specification includes 16 elements for different model types. Table 6.5 summarises the most relevant PMML elements for describing (Q)SAR models. Table 6.3, however, reveals that PMML is the ideal solution for a machine-readable mathematical representation of the predictive model, but it lacks at the same time a simple solution for storing all data related to the model.

6.4.4 QsarDB Archive Format

The QSAR DataBank or QsarDB archive format is designed for the seamless digital organisation and integration of full (Q)SAR model information: molecular and experimental data, mathematical representation of models and accompanying information on molecular structures, literature sources, *etc.*[14] The QsarDB archive format is based on the QsarDB data schema (see Section 6.5.1.2), which formalises (Q)SAR concepts, the data objects, dividing them into logical units and providing relationships between them. The QsarDB data format organises their presentation for computer systems and accommodates all the information necessary to fully reproduce the model. The QsarDB data schema defines five data registries for organising (Q)SAR related objects: compounds, properties, descriptors, models, and predictions (see Figure 6.7). Each object is essentially a container of data attributes, where some of the attributes are common for every type of object (*e.g.*, object's identifier, name, and description), while others are specific for certain types of objects. For example, the InChI code and CAS number attributes apply only for compound data. See Table 6.6 for additional details about QsarDB container objects.

Table 6.5 Overview of major elements in the PMML data schema.

Model element	Purpose
RegressionModel	This element is used to represent various regression (and classification) models, including multilinear regression, polynomial regression and logistic regression.
NeuralNetwork	This element is used for representing artificial neural networks (*e.g.*, back-propagation neural networks or radial basis function neural networks). Different network architectures can be created, and many activation functions for neurons are supported.
SupportVectorMachine-Model	This element describes binary and multi-class support vector machine models (SVM). Such models may use a large number of different kernels. Currently linear, polynomial, radial basis and sigmoid kernels are supported in the PMML specification.
NearestNeighborModel	This element represents k-Nearest Neighbours (k-NN) models, where the decision is made directly from the training data by finding the k most similar instances. This means that for the model evaluation, the entire data set must be available in the PMML document, which might cause difficulties for evaluating very large data sets. An alternative is to store the training data externally and reference it with the TableLocator element.
NaiveBayesModel	This element is used to represent naïve Bayes models.
ClusteringModel	This element represents centre-based and distribution-based cluster models (*e.g.*, k-means or Kohonen maps). This element can also be used as a part of a counter-propagation neural network model.
MiningModel	This element is for representing multiple models within one PMML document and makes it possible to implement various decision-making strategies. For example, it is possible to describe ensemble or consensus models (*e.g.*, Random Forests), where the output is calculated as the average or by majority voting from the outputs of included models. Another possibility is to use a decision tree to select a suitable model for the given input data. Multiple models could be combined, where additional models could be used for estimating missing values.

The content of each container can be extended with an unlimited set of data attachments, called cargos, which could be a model representation in PMML format, molecular structures in various formats, workflows, and so on. Related objects are linked with each other *via* identifiers through direct links at the registry level or indirect links at the cargo level. Some of the cargos are explicitly defined (*i.e.*, system cargos) in the QsarDB format. For example, every object could include a "bibtex" cargo containing bibliography

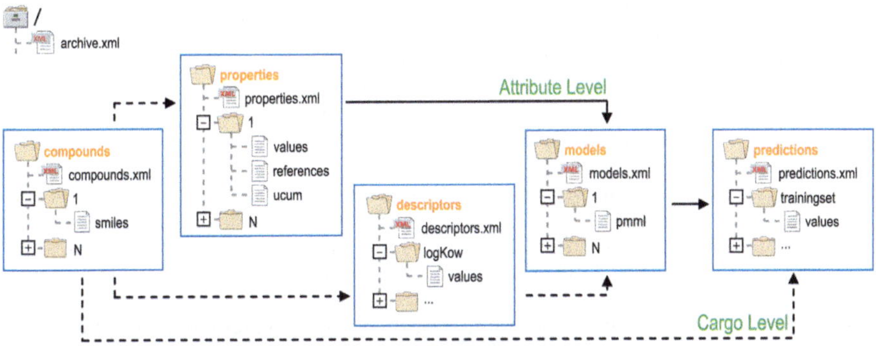

Figure 6.7 Components of the QsarDB data format with relationships at the Attribute and Cargo levels.

data in BibTeX format[40] with references to original sources. Similarly, property, descriptor and prediction objects have "value" cargos for storing their values in tab-separated value (TSV) format. These objects may also have "ucum" cargo in Unified Code for Units of Measure format[41] for defining their units. Each "value" cargo can be accompanied by "reference" cargo for linking every value with an appropriate literature reference. The QsarDB representation encourages the reuse of standard and well-established text-based formats (*e.g.*, TSV, SDF, or PMML files) and is designed to be extensible for different modelling approaches.

The QsarDB data format uses a so-called "collection of files approach" and introduces a naming scheme for grouping files systematically. This allows the use of multiple data formats that already exist, improves the extensibility of the data format and allows mixing them while making use of extensible mark-up language (XML) in the serialisation of the inner registries. The implementation of QsarDB data format separates molecular and experimental data and the mathematical representation of models into the five folders: compounds, properties, descriptors, models and predictions (Figure 6.7). All of them include a registry file for attribute values and cargo files that are both extensible and usually contain information in text format. Multiple storage options are available for QsarDB archives, including a ZIP (or some other) archive file, a directory tree, or a directory tree under the control of revision control software. For example, model developers could use commonly available revision control software to record changes performed inside the QsarDB archive directory and it will make it possible to check out older versions of the archive for testing and comparing differences between different versions. For easier creation of the QsarDB archive, both command line tools and a graphical editor are available at the QsarDB website.[42] The summary of model formats in Table 6.3 reveals that QsarDB includes all the necessary components to make (Q)SAR models transparent and reproducible.

Table 6.6 Overview of major objects in QsarDB data representation.

Compound	Represents a chemical compound that has a unique and well-defined chemical structure. Attributes for the characterisation of the compound include InChI code and CAS number. The molecular structure is given with one or more structure cargos (*e.g.*, SMILES or MOL file).
Property	Represents experimentally measured biological activities or physicochemical properties. Property can be characterised with endpoint (QMRF classification) and species attributes. A well-defined property should also contain a "ucum" cargo to define its unit.
Descriptor	Represents experimental or theoretically calculated data that are used as independent variables in QSAR models. Descriptor is characterised with an application attribute (*i.e.*, software name and version). If the descriptor is described in the Blue Obelisk Descriptor Ontology,[38] then additional information can be added with "bodo" cargo.
Model	Represents a mathematical relationship that relates a property to one or more descriptors. Model is linked to one property *via* the property ID attribute, while links to descriptors are defined indirectly inside the PMML cargo. Additional cargos might include workflows or applicability domain representation.
Prediction	Represents predicted values resulting from the application of the model. Prediction is linked to one model *via* the model ID attribute and is characterised with a prediction set type (*e.g.*, training, validation or testing set) and application attribute (*e.g.*, software and version used for the prediction).

6.5 Applications and Solutions for the Storage of Predictive Models

The following section describes how the file formats for the storage of predictive models described have been implemented in various web-based repository solutions and software packages. Among them, only the PMML format is not designed for (Q)SAR modelling. It has been implemented into major statistical software packages for keeping models internally or as a model exchange format. Other formats are better suited for (Q)SAR applications, but have a narrower software support. The QSAR-ML format in this list is special because it is not designed for the storage of models but is implemented in Bioclipse software for the reproducible representation of data sets and leaves the model representation to be handled as an exercise for the user. The QMRF and QsarDB data formats are not tied to a particular (Q)SAR model development software and are currently used as generic solutions for storing a large number of models in two web-based public repositories, QMRF Inventory and QsarDB Repository, respectively.

(Q)SAR models are conventionally implemented in standalone software or web applications either as services or as part of an integrated modelling environment. With the development of file and data formats for (Q)SAR models, repository approaches have also been evolved and have gained momentum. The main focus in the following subchapters is the implementation of the previously described file formats into repository solutions, followed by descriptions of integrated modelling environments and web solutions.

6.5.1 Repository Approaches

6.5.1.1 JRC QSAR Model Database

The JRC QSAR Model Database[43] is a collection of QMRF documents and associated data sets that was created to help identify (Q)SAR models for the purpose of hazard and risk assessment. As of October 2016, it contained 109 QMRF documents. Those documents are linked together with structure files, giving a full overview of the compounds that were used for modelling. All the QMRFs are searchable by the QMRF number, free text, endpoint, algorithm, software and authors and can be downloaded as PDF, HTML, XML or XLS files. Additionally, the chemical structures can be searched (CAS no., formula, chemical name, SMILES) either as an exact search or by similarity. The results show in which QMRF document the corresponding structure was used and whether it is part of the training or validation set. Unfortunately, there is no option to browse the database. The database contains QMRF documents for different linear and non-linear model representations for classification and regression models. Of those models, the most easily applicable models are linear regression models, where the exact equation is inside the document. For some non-linear models (*e.g.*, neural networks), an additional file with the model information is included, but not for all. This makes it difficult to use those particular models directly, and the interested party must contact the original author.

6.5.1.2 QsarDB Repository

The development of the QSAR DataBank archive format (see Section 6.4.4) and the necessity to make archives publicly available led to the development of a repository solution that allows easy access to model(s) in the QsarDB archives.[16,44] The QsarDB Repository allows uploading (Q)SAR models that are published in the scientific literature or well documented by other means. The repository organises uploaded data into Communities (*e.g.*, research groups and institutions) and Collections (*e.g.*, data grouped by topic), and it is open for contributions from the scientific community. The contributors can freely register to get access for uploading QsarDB archives to the repository. The uploaded archives will go through the submission workflow, which includes automated quality checks for the validation of the archive content.[16] In addition, moderators assigned to the Collections perform manual quality control of archives. It is a content-aware repository (a.k.a. a smart repository), which along with classical tasks of digital repository, *i.e.*, finding and downloading data, also offers integrated (Q)SAR services, such as the analysis and visualisation of data sets and models, access to (Q)SAR-specific metadata, and chemical search and prediction functionality. In this way, model developers can make their models and new knowledge easily available. Model users have easy access to full information about the (Q)SARs and can be more confident in making decisions based on the transparent representation of the models. QsarDB is also the first repository to provide digital identifiers for the unique identification of published (Q)SAR models. In particular, it assigns DOI numbers and HDL (Handle.Net Registry) identifiers to make (Q)SAR models citable in classical research publications. For example, several recent scientific publications have used[45,46] or published new (Q)SAR models[47–52] *via* the QsarDB repository. Items in the repository are automatically linked to the original publications, and some publishers, such as Taylor & Francis, link from their journal articles back to the QsarDB repository (*e.g.*, ref. 49, 51 and 52). The QsarDB repository makes mathematical representations of models available in PMML format. So far, the repository users have successfully applied this format to encode mathematical representations for the following classification and regression model types: decision trees, ensemble models (better known as consensus models in the (Q)SAR community), k-nearest neighbours, neural networks, random forests, linear regression models, support vector machines and Bayesian models. As of October 2016, the QsarDB repository includes 430 unique (Q)SAR models.

6.5.2 Integrated Modelling Environment Approaches

6.5.2.1 Standalone Software

(Q)SAR models for predictive toxicology can be found in a number of software packages that are available either freely or commercially. The following list is not complete but gives a brief overview of different efforts for solving the problem of making predictive toxicology QSAR models usable and accessible. Only specific examples of freely available software packages are listed.

ECOSAR[53] was developed by the United States Environmental Protection Agency (EPA) for estimating the short- and long-term aquatic toxicity of chemicals on different organisms. Currently, ECOSAR version 1.11 includes a classification scheme for grouping chemicals into 111 chemical classes, which are then connected with class-specific QSAR models (709 in total). These class-specific models are simple linear relationships between the toxicity and octanol–water partition coefficient, and they are documented in the help section of the software. The ECOSAR software has also been integrated with the EPI Suite™[54] and OECD QSAR Toolbox[55] software packages for predicting toxicological endpoints. The US EPA has also developed another software package called the Toxicity Estimation Software Tool (T.E.S.T.)[56] that includes different types of models for seven toxicological endpoints and nine physical properties. Data sets that were used for model development are included, and all the necessary descriptors for the prediction are calculated inside the software.

Toxtree[57] is a regulatory institution guided effort in the EU implementing 16 different decision trees from various authors and making them easily usable. Each decision tree can be used and visualised; however, the training data that was used for the development of these models is not provided.

Moreover, academic research groups often make their developed models available for general use. For example, QSARINS was developed at the University of Insubria, Italy.[58] Originally, it was designed for creating (Q)SAR models, but the latest version includes a QSARINS-Chem[59] module with 26 models for predicting various toxicological endpoints and physicochemical properties. Models from QSARINS-Chem are also available in the QsarDB repository,[60] along with many other models from the same group.[61] All these models were distributed with all the chemical structures, property data and descriptors that were used to develop the models. Another example is the VEGA platform,[62] which contains 31 different types of models developed in the course of multiple European Commission funded research projects. Data sets with experimental and predicted values are available; however, complete mathematical representations of the models are not available, and they are distributed as closed source software modules. ChemProp[63] was developed at the Helmholtz Centre for Environmental Research – UFZ in Leipzig, Germany, and includes models for the prediction of physicochemical properties, human health, environmental fate, degradation and ecotoxicology endpoints. They are available to registered users.

There are more software solutions for the prediction of toxicity that are distributed as commercial software, and such systems usually do not provide source data sets and mathematical representations of the models. Comprehensive overviews about the capabilities of such systems have been published by J.C. Dearden[64] and G. Patlewicz and J. M. Fitzpatrick.[65]

6.5.2.2 Web Solutions

In addition to standalone software, there are many web sites that provide the possibility of predicting several toxicological endpoints. Such web services provide a very attractive and accessible method for distributing predictive models because the end users do not need to install any additional software to use the models.

For example, the PBT Profiler[66] uses computerised methods, such as SARs, to estimate the persistence, bioaccumulation and aquatic toxicity of organic chemicals. The web site includes a description of the methodology together with literature references but the data sets are not included on the web site.

GUSAR@NIH[67] and GUSAR@Way2Drug[68] have similar functionalities to predict acute rat toxicity, ecotoxicity, and anti-target activity and also implement some models provided in the T.E.S.T. software. Absorption, distribution, metabolism, excretion and toxicity profiles can be predicted with admetSAR.[69]

The COSMOS KNIME Webportal[70] contains a collection of KNIME workflows from the COSMOS Project[26] related to the prediction of the human repeated dose toxicity of cosmetics-related substances.

Chembench[71] and OCHEM[72] are web-integrated modelling environments for model creation that both contain large collections of QSAR models that can be used for predictions. The Danish (Q)SAR Database[73] provides predictions for more than 600 000 chemical substances with a mixture of more than 200 commercial and open (Q)SARs but does not directly provide any QSAR models for IPR reasons. A recent review summarises (Q)SAR-s on the Web.[74]

6.6 Conclusions

The amount of chemical, toxicological, and biological data is increasing every day. However, the variety and complexity of the available data is an even more important issue, because small changes in chemical structures may cause significant changes in the behaviour of chemicals. Predictive models are abstractions of the available data and represent snapshots of the chemical space at different points in time. This leads to a requirement that originates from practical considerations – when using predictive models in decision support, the models must be kept alive for a substantially long time. When a model has been used in decision support for predictive purposes, it must be possible to backtrack the model, see how it was derived and what the underlying data was. On the other hand, the increasing amount of toxicological and chemical information determines that models will undergo constant re-evaluation that may eventually lead to new and improved models. Thus, models that hold today may not hold with the advent of new information about the chemical space, but the historically developed models are still required for comparison and evaluation purposes. Therefore, one needs file and/or data formats for predictive models and good documentation of related information for archiving and reusing models to keep past, present and future predictive models available.

The current book chapter provided an overview of modern solutions for organising qualitative and quantitative structure–activity relationships and how these solutions are used in the present day. One can see that predictive model use is still centred on standalone software solutions, but publicly available web-based solutions are emerging. At the same time, one should also appreciate that more important than the models are the problems that they solve or new knowledge that they provide about the physical, chemical, toxicological, biochemical or biomedical behaviour in concrete cases. Considering this, systematic access to the models and their transparency are essential and require a standardised and machine-readable organisation of the model information.

Today's state of the art in organising (Q)SAR models and data in a way that they can be documented transparently, understood and used in a systematic manner includes four publicly available efforts, namely, the QMRF, QSAR-ML,

PMML and QsarDB archive formats. Each of them has strengths and limitations, but all of them have made progress towards making new knowledge available together with models and underlying data, using dedicated repositories. With the advent of Big Data in toxicology and chemistry in general, it is becoming more and more important that derivatives of these data are more accessible and usable for scientific research and decision support and consequently also for independent evaluation.

References

1. Chemical Abstracts Service Home Page, https://www.cas.org/, accessed October 31, 2016.
2. T. Sterling and J. J. Irwin, ZINC 15 – Ligand Discovery for Everyone, *J. Chem. Inf. Model.*, 2015, **55**, 2324–2337.
3. S. Kim, P. A. Thiessen, E. E. Bolton, J. Chen, G. Fu, A. Gindulyte, L. Han, J. He, S. He, B. A. Shoemaker, J. Wang, B. Yu, J. Zhang and S. H. Bryant, PubChem substance and compound databases, *Nucleic Acids Res.*, 2016, **44**, D1202–D1213.
4. A. P. Bento, A. Gaulton, A. Hersey, L. J. Bellis, J. Chambers, M. Davies, F. A. Kruger, Y. Light, L. Mak, S. McGlinchey, M. Nowotka, G. Papadatos, R. Santos and J. P. Overington, The ChEMBL bioactivity database: an update, *Nucleic Acids Res.*, 2014, **42**, 1083–1090.
5. NCBI Resource Coordinators, Database resources of the National Center for Biotechnology Information, *Nucleic Acids Res.*, 2016, **44**, D7–D19.
6. ChEMBLdb FTP Directory, ftp://ftp.ebi.ac.uk/pub/databases/chembl/ChEMBLdb/releases/, accessed October 31, 2016.
7. ECOTOX User Guide: ECOTOXicology Database System, Version 4.0, https://cfpub.epa.gov/ecotox/, accessed October 31, 2016.
8. Toxicity Forecasting, http://www.epa.gov/chemical-research/toxicity-forecasting, accessed November 2016.
9. Toxicology in the 21st Century, http://www.ncats.nih.gov/tox21/, accessed November 2016.
10. iCSS chemstry dashboard, https://comptox.epa.gov/dashboard, accessed November 2016.
11. J.-L. Reymond, The chemical space project, *Acc. Chem. Res.*, 2015, **48**, 722–730.
12. C. Hansch and T. Fujita, p-σ-π Analysis. A Method for the Correlation of Biological Activity and Chemical Structure, *J. Am. Chem. Soc.*, 1964, **86**, 1616–1626.
13. Web of Science [v.5.22.3] – All Databases Home, http://apps.webofknowledge.com/, accessed November 2016.
14. V. Ruusmann, S. Sild and U. Maran, QSAR DataBank - an approach for the digital organization and archiving of QSAR model information, *J. Cheminf.*, 2014, **6**, 25.
15. X. Fu, A. Wojak, D. Neagu, M. Ridley and K. Travis, Data governance in predictive toxicology: A review, *J. Cheminf.*, 2011, **3**, 24.

16. V. Ruusmann, S. Sild and U. Maran, QSAR DataBank repository: open and linked qualitative and quantitative structure–activity relationship models, *J. Cheminf.*, 2015, **7**, 32.
17. A. Palczewska, X. Fu, P. Trundle, L. Yang., D. Neagu, K. Ridley and K. Travis, Towards model governance in predictive toxicology, *Int. J. Inf. Manage.*, 2013, **33**, 567–582.
18. A. Tropsha, Best practices for QSAR model development, validation, and exploitation, *Mol. Inf.*, 2010, **29**, 476–488.
19. J. C. Dearden, M. T. D. Cronin and K. L. E. Kaiser, How not to develop a quantitative structure–activity or structure–property relationship (QSAR/QSPR), *SAR QSAR Environ. Res.*, 2009, **20**, 241–266.
20. 5-star Open Data, http://5stardata.info/, accessed November 2016.
21. J. Hastings, L. Chepelev, E. Willighagen, N. Adams, C. Steinbeck and M. Dumontier, The chemical information ontology: provenance and disambiguation for chemical data on the biological semantic web, *PLoS One*, 2011, **6**, e25513.
22. V. Khatri and C. V. Brown, Designing data governance, *Commun. ACM*, 2010, **53**, 148–152.
23. NRC Committee on Models in the Regulatory Decision Process. National Academies Press, 2007, http://www.nap.edu/openbook.php?record id=11972, accessed November 2016.
24. M. N. Kooper, R. Maes and E. E. O. R. Lindgreen, On the governance of information: Introducing a new concept of governance to support the management of information, *Int. J. Inf. Manage.*, 2011, **31**, 195–200.
25. OECD, OECD principles for the validation, for regulatory purposes, of (quantitative) structure-activity relationship models, 2004, http://www.oecd.org/chemicalsafety/risk-assessment/37849783.pdf, accessed November 2016.
26. COSMOS, 2015, URL http://www.cosmostox.eu/, accessed November 2016.
27. C. Yang, D. Hristozov, A. Tarkhov, T. Kleinöder, I. Boyer, M. Cronin, E. Fioravanzo, H. Kim, B. Heldreth, A. Mostrag-Szylchtying, J. Rathman, A. Richarz, C. Schwab, V. Vitcheva and A. Worth, COSMOS DB as an international share point for exchanging regulatory and toxicity data of cosmetics ingredients and related substances, *Toxicol. Lett.*, 2015, **238**(2), S382.
28. A.-N. Richarz, P. Alov, S. J. Enoch, S. Kovarich, Y. Lan, T. Meinl, C. Mellor, D. Neagu, A. Paini, A. Palczewska, J. V. Sala Benito, F. Steinmetz and M. T. Cronin, In silico chemistry-based workflows to facilitate ADMET prediction for cosmetics-related substances, *Toxicol. Lett.*, 2015, **238**(2), S170.
29. M. Hall, E. Frank, G. Holmes, B. Pfahringer, P. Reutemann and I. H. Witten, The WEKA data mining software: an update, *SIGKDD Explor. Newsl.*, 2009, **11**, 10–18.
30. R. Core Team, *R: A Language and Environment for Statistical Computing*, R Foundation for Statistical Computing, Vienna, Austria, 2015.
31. M. R. Berthold, N. Cebron, F. Dill, T. R. Gabriel, T. Kötter, T. Meinl, P. Ohl, C. Sieb, K. Thiel and B. Wiswedel, in *Data Analysis, Machine Learning and Applications*, ed. C. Preisach, H. Burkhardt, L. Schmidt-Thieme and R. Decker, Springer, Heidelberg, 2008, ch. 38, pp. 319–326.

32. C. M. Ellison, J. C. Madden, M. T. D. Cronin and S. J. Enoch, Investigation of the Verhaar scheme for predicting acute aquatic toxicity: Improving predictions obtained from Toxtree ver. 2.6, *Chemosphere*, 2015, **139**, 146–154.
33. JRC QSAR Model Database User Manual, 2014, AnnexIII: QSAR Model Reporting Format (Version 2.0.0), https://eurl-ecvam.jrc.ec.europa.eu/laboratories-research/predictive_toxicology/doc/New User Manual_JRC QSAR Model Database 01–2014.pdf, accessed November 2016.
34. QMRF Editor 2.0.0, Ideaconsult Ltd., Sofia, Bulgaria, 2013.
35. O. Spjuth, E. L. Willighagen, R. Guha, M. Eklund and J. E. S. Wikberg, Towards interoperable and reproducible QSAR analyses: Exchange of datasets, *J. Cheminf.*, 2010, **2**, 5.
36. QSAR-ML Home Page, http://pele.farmbio.uu.se/qsar-ml, accessed November 2016.
37. O. Spjuth, J. Alvarsson, A. Berg, M. Eklund, S. Kuhn, C. Mäsak, G. Torrance, J. Wagener, E. L. Willighagen, C. Steinbeck and J. E. S. Wikberg, Bioclipse 2: A scriptable integration platform for the life sciences, *BMC Bioinf.*, 2009, **10**, 397.
38. QSAR.sf.net Descriptor Dictionary, http://qsar.sourceforge.net/dicts/qsar-descriptors/index.xhtml, accessed November 2016.
39. Data Mining Group, http://dmg.org/, accessed November 2016.
40. BibTex, http://www.bibtex.org, accessed November 2016.
41. The Unified Code for Units of Measure, http://unitsofmeasure.org, accessed November 2016.
42. Resources for QsarDB, https://qsardb.org/downloads, accessed November 2016.
43. (Q)SAR Model Reporting Format (QMRF) Inventory, http://qsardb.jrc.ec.europa.eu/qmrf/, accessed November 2016.
44. QSAR DataBank repository, http://qsardb.org, accessed November 2016.
45. A. Avdeef, Suggested improvements for measurement of equilibrium solubility-pH of ionizable drugs, *ADMET & DMPK*, 2015, **3**, 84–109.
46. A. Avdeef, Solubility Temperature Dependence Predicted from 2D Structure, *ADMET & DMPK*, 2015, **3**, 298–344.
47. V. Ruusmann and U. Maran, From data point timelines to a well curated data set, data mining of experimental data and chemical structure data from scientific articles, problems and possible solutions, *J. Comput.-Aided Mol. Des.*, 2013, **27**, 583–603.
48. V. Aruoja, M. Moosus, A. Kahru, M. Sihtmäe and U. Maran, Measurement of baseline toxicity and QSAR analysis of 50 non-polar and 58 polar narcotic chemicals for the alga Pseudokirchneriella subcapitata, *Chemosphere*, 2014, **96**, 23–32.
49. G. Piir, S. Sild and U. Maran, Classifying bio-concentration factor with random forest algorithm, influence of the bio-accumulative *vs.* non-bio-accumulative compound ratio to modelling result, and applicability domain for random forest model, *SAR QSAR Environ. Res.*, 2014, **25**, 967–981.

50. M. Oja and U. Maran, The Permeability of an Artificial Membrane for Wide Range of pH in Human Gastrointestinal Tract: Experimental Measurements and Quantitative Structure–Activity Relationship, *Mol. Inf.*, 2015, **34**, 493–506.
51. M. Oja and U. Maran, Quantitative structure–permeability relationships at various pH values for acidic and basic drugs and drug-like compounds, *SAR QSAR Environ. Res.*, 2015, **26**, 701–719.
52. M. Oja and U. Maran, Quantitative structure–permeability relationships at various pH values for neutral and amphoteric drugs and drug-like compounds, *SAR QSAR Environ. Res.*, 2016, **27**, 813–832.
53. US EPA, *The ECOSAR (ECOlogical Structure Activity Relationship) Class Program for Microsoft Windows*, United States Environmental Protection Agency, Washington, DC, USA, 2012.
54. US EPA, *Estimation Programs Interface Suite™ for Microsoft® Windows, v4.11*, United States Environmental Protection Agency, Washington, DC, USA, 2012.
55. OECD QSAR Toolbox 3.4, Laboratory of Mathematical Chemistry, Bulgaria, 2016.
56. US EPA, *T.E.S.T (Toxicity Estimation Software Tool) v4.2.1*, United States Environmental Protection Agency, Cincinatti, OH, USA, 2012.
57. G. Patlewicz, N. Jeliazkova, R. J. Safford, A. P. Worth and B. Aleksiev, An evaluation of the implementation of the Cramer classification scheme in the Toxtree software, *SAR QSAR Environ. Res.*, 2008, **19**, 495–524.
58. P. Gramatica, N. Chirico, E. Papa, S. Cassani and S. Kovarich, *J. Comput. Chem.*, 2013, **34**, 2121–2132.
59. P. Gramatica, S. Cassani and N. Chirico, QSARINS: A new software for the development, analysis, and validation of QSAR MLR models, *J. Comput. Chem.*, 2014, **35**, 1036–1044.
60. G. Piir, QDB archive #177, QsarDB repository, 2016, https://dx.doi.org/10.15152/QDB.177, accessed November 2016.
61. University of Insubria (Italy), QSAR Research Unit in Environmental Chemistry and Ecotoxicology Community and Collections at QsarDB: http://hdl.handle.net/10967/155, accessed November 2016.
62. VEGA 1.1.3, *Istituto di Ricerche Farmacologiche Mario Negri (Laboratory of Environmental Chemistry and Toxicology) and Kode srl*, Italy, 2015. http://www.vega-qsar.eu/, accessed November 2016.
63. UFZ Department of Ecological Chemistry, *ChemProp 6.4*, 2016.
64. J. C. Dearden, Expert systems for toxicity prediction, in *In Silico Toxicology: Principles and Applications*, ed. M. T. D. Cronin and J. C. Madden, Royal Society of Chemistry, London, 2010, pp. 478–507.
65. G. Patlewicz and J. M. Fitzpatrick, Current and Future Perspectives on the Development, Evaluation, and Application of *in Silico* Approaches for Predicting Toxicity, *Chem. Res. Toxicol.*, 2016, **29**, 438–451.
66. PBT Profiler, http://www.pbtprofiler.net, accessed November 2016.
67. Chemical Activity Predictor—GUSAR, http://cactus.nci.nih.gov/chemical/apps/cap, accessed November 2016.

68. Gusar – create QSAR/QSPR models on the basis of the appropriate training sets, http://www.way2drug.com/gusar/, accessed November 2016.
69. F. Cheng, W. Li, Y. Zhou, J. Shen, Z. Wu, G. Liu, P. W. Lee and Y. Tang, admetSAR: a comprehensive source and free tool for assessment of chemical ADMET properties, *J. Chem. Inf. Model.*, 2012, **52**, 3099–3105.
70. COSMOS KNIME Webportal, http://knimewebportal.cosmostox.eu, accessed November 2016.
71. CHEMBENCH, https://chembench.mml.unc.edu/, (accessed November 2016).
72. I. Sushko, S. Novotarskyi, R. Körner, A. K. Pandey, M. Rupp, W. Teetz, S. Brandmaier, A. Abdelaziz, V. V. Prokopenko, V. Y. Tanchuk, R. Todeschini, A. Varnek, G. Marcou, P. Ertl, V. Potemkin, M. Grishina, J. Gasteiger, C. Schwab, I. I. Baskin, V. A. Palyulin, E. V. Radchenko, W. J. Welsh, V. Kholodovych, D. Chekmarev, A. Cherkasov, J. Aires-de-Sousa, Q.-Y. Zhang, A. Bender, F. Nigsch, L. Patiny, A. Williams, V. Tkachenko and I. V. Tetko, Online chemical modeling environment (OCHEM): web platform for data storage, model development and publishing of chemical information, *J. Comput.-Aided Mol. Des.*, 2011, **25**, 533–554.
73. Danish (Q)SAR Database, *Division of Diet, Disease Prevention and Toxicology, National Food Institute*, Technical University of Denmark, http://qsar.food.dtu.dk/, accessed November 2016.
74. I. V. Tetko, U. Maran and A. Tropsha, Public (Q)SAR Services, Integrated Modeling Environments, and Model Repositories on the Web: State of the Art and Perspectives for Future Development, *Mol. Inf.*, 2017, **36**, 1600082.

CHAPTER 7

Toxicogenomics and Toxicoinformatics: Supporting Systems Biology in the Big Data Era

TEREZINHA M. SOUZA, JOS C. S. KLEINJANS AND DANYEL G. J. JENNEN*

Maastricht University, Department of Toxicogenomics, Universiteitssingel 40, Maastricht, The Netherlands
*E-mail: danyel.jennen@maastrichtuniversity.nl

7.1 Introduction to Toxicogenomics and Toxicoinformatics

In the last decades, the field of Toxicogenomics consolidated its position within Toxicology by improving toxicity evaluation and aiding risk assessment in drug development, chemical exposure and biomarker discovery. The underlying assumption is that molecular alterations arising from (external) perturbations precede and reflect adverse effects observed at cellular and organism levels.[1] Biomolecules investigated by Toxicogenomics include genes (sequence and structure), gene products (transcripts), as well as proteins and metabolites that take part in a myriad of biological reactions. Together, they

Issues in Toxicology No. 41
Big Data in Predictive Toxicology
Edited by Daniel Neagu and Andrea-Nicole Richarz
© The Royal Society of Chemistry 2020
Published by the Royal Society of Chemistry, www.rsc.org

comprise the "-omes" and are the object of study of the "-omics" fields *in vivo*, *in vitro* and *in silico*: genomics, epigenomics, transcriptomics, proteomics and metabolomics.

Repeated dose toxicity testing in rodent and non-rodent species has been the core of traditional toxicological screening. Such investigations are carried out for a minimum of 28 days and may last up to two years if the substance is being assessed for carcinogenicity.[2] In addition to the laborious and time-consuming aspect of these approaches, extrapolation to humans is non-trivial and the predictive value may be as low as 33%.[3]

Alternatively, Toxicogenomics offers a human-based approach to overcome the main limitations of animal-based toxicological studies and optimise all stages throughout risk assessment (hazard identification, dose–response and exposure assessments and risk characterisation). First, while aforementioned models are exposed to high doses associated with a specific phenotype (*e.g.*, organ toxicity, cancer, death), measurements at the molecular level provide higher resolution portrayals of adverse effects at more realistic dose range and scheduling in a timely and cost-effective fashion. Second, the use of cells and tissues originating from different individuals portrays the heterogeneity of the population in contrast with the customary inbred animal strains, which show very low genetic variability. Finally, it prevents uncertainties inherent to interspecies comparisons, which may interfere with biological interpretations and subsequent extrapolation to human physiology and biochemistry.[3]

Toxicogenomics improves prediction of toxicity and related modes-of-action by employing simplified *in vitro* biological systems. Common models include cultured cells (immortalised or of primary origin), microtissues, co-culture systems and precision-cut organ slices, which, in spite of the variable levels of functional similarity to the tissue of origin, allow a detailed investigation of mechanisms of action and adverse effects.[4] Toxicogenomics may also assist the discovery of biomarkers for the safety assessment of drug exposure by screening components of blood (circulating cells or serum) and other bodily fluids (saliva, urine) as surrogates to organ toxicity. Thorough and unbiased analyses of such data may yield promising biomarkers – molecules or biological processes that correlate to an increased risk of a pathogenic process or pharmacological response, such as drug-induced liver injury (DILI), muscle toxicity and kidney and cardiac injury.[5]

Successful application of such collective approaches instigated the introduction of innovative techniques to increase data throughput, which in turn demanded improvements in computational power, data storage, handling and analyses as well as standardisation of information in order to facilitate data sharing and knowledge exchange. Toxicoinformatics comprises *in silico* tools to collect, analyse and interpret Toxicogenomics data. The Environmental Bioinformatics Knowledge Base (ebKB; http://ebkb.org) places Toxicoinformatics, along with Virtual cells, tissues and organs and Toxicokinetic/Toxicodynamic modelling as a branch of Computational Toxicology. Although Cheminformatics was previously addressed as a problem for

Toxicoinformatics, this subdivision was upgraded to a whole field of knowledge, comprising chemical structure elucidation, prediction of activity and modelling of interaction; here, Toxicoinformatics will generally refer to the tasks (and challenges) involved in storage, analyses and handling of 'omics Big Data.

This chapter will therefore provide an overview of the main 'omics technologies and the steps involved in processing, analysing and ensuring data availability, and finally will discuss examples of successful applications and future developments within the scope of predictive toxicology and Big Data.

7.2 Current 'Omics Technologies

7.2.1 Transcriptomics

Transcriptomics, also known as gene expression profiling, is aimed at the identification and quantification of transcripts. The first technique used for quantification dates back to the mid-80s and is still the gold standard for accurate measurement of transcript levels: the polymerase chain reaction (PCR). Since then, some technical upgrades have been made (as real-time quantitative PCR), but this approach remains a low-throughput technique that requires pre-defined targets (one to a few genes of interest at a time). This scenario changed after the advent and popularisation of the first high-throughput, hybridization-based technology – the microarray – which allowed the assessment of hundreds of transcripts simultaneously and led to the expansion of several areas of knowledge, ultimately leading to the emergence of Toxicogenomics as a research field.

In an initial stage, PCRs and even cDNA microarrays were intended for quantifying the expression of genes of toxicological interest. One example is Phase-1 Human 600 array, which contained 597 probes related to genes of several categories, including xenobiotic metabolism, inflammation, DNA damage, fibrosis and mitochondrial function.[6] Even then, investigations showed the validity of transcriptomics as predictors of organ toxicity with high accuracy.[7] Advances in genomics and functional annotation gradually increased the number of genes in a single array, and nowadays many commercial vendors (*e.g.*, Agilent, Affymetrix, Illumina and GE Healthcare) offer a wide range of platforms to measure the expression of almost ~50 000 probes comprising approximately 20 000 genes in a single experiment. Furthermore, platforms for unicellular eukaryotes, plants, insects and non-rodents (fish, primates and birds) are also available.

Initially applied to sequencing genomes, next-generation sequencing (NGS) emerged as a new set of technologies to improve throughput, sensitivity and running time in transcriptomics. Pyrosequencing (Roche/454), ion torrent sequencing, sequencing by synthesis (Illumina/Solexa) and sequencing by ligation (SOLiD) are among the most common NGS methods that may also be applied to RNA-sequencing (RNA-seq), enabling higher resolution and eliminating oligonucleotide chips from the process. Instead, transcripts are directly sequenced, which offers a series of advantages over microarrays. First, background signals or probe saturation do not affect overall quantification, benefitting genes at the extremities of the abundance scale. Second, the use of existing genome/transcriptome sequences is optional. This allows not only the discovery of new, unannotated variants of a gene transcript and alternative splicing events but also enables the use of non-model organisms. Investigations of more discrete changes of distinct types of RNA (microRNA, circular RNA, long non-coding RNA) as well as gene fusions, single nucleotide variants, indels (small insertions and deletions) are also possible.

Within Toxicogenomics, gene expression profiling prevails as the main technique, being the core of toxicity prediction studies targeting the liver, kidneys, lungs, skin and heart.[8] MicroRNAs (22-nucleotide endogenous RNAs) are particularly sought as safety markers due to their tissue-specific expression and early detection in surrogate tissues.[5] An example is miRNA-122, targeted as a biomarker of drug-induced liver injury (DILI). Traditional biomarkers of DILI often lack specificity and sensitivity, and miRNA-122 showed superior performance over alanine aminotransferase (ALT) in two human cohorts.[9] Such results may help early diagnostics with increased specificity for the prevention of organ-related morbidities.

7.2.2 (Epi)Genomics

Understanding genomic information – gene sequence, location and chromosomic context – was among the earliest concerns of the molecular era. From 1990 until 2003, a joint effort from the scientific community worldwide resulted in the complete sequencing of the human genome. In 2001, preliminary assessment of the first draft showed that human complexion could not be explained solely by the number of gene-coding proteins (estimated 30 000–40 000 at that time), and pointed at the influential effects of genome landscape, single nucleotide polymorphisms (SNPs) and alternative splicing.[10] As a result, new consortia, such as the Haplotype Map (HapMap, 2002, ftp://ftp.ncbi.nlm.nih.gov/hapmap/), the Human Epigenome Project (HEP, 2000, http://www.epigenome.org), Human Variome (2006, http://www.humanvariomeproject.org) and others, were launched to address sources of genetic variability and their impacts on health, disease susceptibility and response to pharmacological and environmental agents.[11,12]

Early technologies (first-generation sequencing) relied on automatisation of the Sanger methodology, which was costly and time consuming with outputs ranging from hundreds to thousands of nucleotide bases. Popularisation of NGS technologies allowed the generation of up to billions of base reads in just a few days and since then large research centers and academic institutions have embraced them. More recently, a third generation of technologies encompasses single-molecule and real-time sequencing, in which DNA amplification is not required.[13]

Applications of NGS technologies have so far yielded many high-resolution investigations. The '1000 genomes' project, completed in 2015, reconstructed the genomes of 2504 individuals from 26 populations[14] and generated data has since been widely targeted in genome-wide association studies (GWAS) to evaluate relationships between single nucleotide polymorphisms (SNPs) and diseases. Although most SNPs have a negligible impact on the outcome in toxic responses, extremely polymorphic genes involved in xenobiotic metabolism such as CYPs[15] or genes from DNA repair machinery may prove useful indicators of individual susceptibility to environmental, chemical, and drug toxicity.

Epigenomics, or the investigation of epigenetic mechanisms, has gained some notoriety by investigating reversible (and yet heritable) genetic changes regardless of modifications in nucleotide sequence. Covalent modifications on DNA and associated proteins – the histones – have a substantial impact on gene expression. Post-translational modifications of histone residues (*e.g.*, acetylation, phosphorylation and ubiquitination) play critical roles in chromatin structure and methylation of genomic regions, especially promoters, holding an inverse relation with transcriptional activity. ChIP-seq combines chromatin immunoprecipitation (ChIP) and DNA sequencing to identify histone modifications and DNA-binding proteins (*e.g.*, transcription factors), driving differential gene expression. Using ChIP-on-chip and ChIP-seq, researchers identified targets of the farnesoid X receptor (FXR), a transcription factor that may exert hepatoprotective effects in acetaminophen-induced injury by repressing the activity of many genes from xenobiotic metabolism.[16] Furthermore, MeDIP-chip and its NGS counterpart MeDIP-seq have been used to specifically quantify methylation patterns in genomic regions after chemical exposure.[17] For instance, arsenic, nickel, cadmium, hormones and reactive oxygen species (ROS) are some agents that may change methylation patterns,[18] and a study has shown that differential prenatal methylation induced by cigarette smoke last from birth throughout childhood until adolescence.[19] Such long-lasting alterations may be used in epigenome-wide association studies (EWAS) to establish a link to lung cancer, for example.

7.2.3 Proteomics

Proteomics is the study of the proteome (protein set) of an organism, tissue or cell in a given moment. As an extension of the transcriptome, it is dynamic and constantly adapting to accommodate changes in the surrounding environment. Despite their direct role in defining the phenotype, a vast set of structural variations arising from intracellular signalling (*e.g.*, phosphorylation, acetylation) or external damage (*e.g.*, adduction by intermediary metabolites or reactive oxygen species) still defies high-throughput assessment of the proteome.

Current approaches include gel-based and shotgun proteomics. Gel-based proteomics uses bidimensional electrophoresis (2D-SDS-PAGE) to segregate proteins based on isoelectric point and molecular mass.

Differential gel electrophoresis (DIGE) allows fluorescent labelling of different samples, which are analysed simultaneously. Each spot, presumably containing a single protein type, is subsequently identified by mass spectrometry. On the other hand, shotgun proteomics takes digested protein mixtures as direct input for liquid-chromatography-coupled mass spectrometry (LC/MS). Both approaches have a series of disadvantages, including limited resolution and effects of protein modifications on reliable protein identification.[20] Although protein, antibody-based arrays have been developed in order to increase throughput, such a method lacks robustness since unspecific binding (either to individual proteins or to protein complexes) may occur. The creation of a "proteome library" that can be immobilised onto a solid surface as in the case of cDNA libraries has also proven impractical.

Despite the technical challenges, toxicoproteomics is an interesting field and several investigations have been conducted to this date, targeting the liver as the main subject of study.[21] An interesting toxicoproteomics (LC/MS) approach evaluated 20 compounds classified as toxic and non-toxic drugs, leading to the discovery of two potential protein markers for drug-induced idiosyncratic hepatotoxicity.[22] Since idiosyncrasy by default is not detected in controlled, small sample size pre-clinical trials, identification of toxicity predictors is a valuable finding in drug development. HepG2, a common *in vitro* model for the study of liver toxicants, has been used to identify markers of toxic exposure to ethanol[23] and drug hepatotoxicants causing different pathogenesis (steatosis, cholestasis, necrosis).[24]

7.2.4 Metabolomics

Metabolomics is aimed at the analyses of molecules that act as intermediary metabolites or final products in biological reactions. Unlike transcripts and proteins, these molecules are not affected by post-transcriptional/-translational modifications or epigenetic changes, directly reflecting biochemical activity and therefore the phenotype.[25]

Within metabolomics, one can choose between targeted and untargeted approaches. The first scenario is appropriate for pathway-driven investigations in which several analytical tools (mass spectrometry, nuclear magnetic resonance, ultraviolet-visible spectroscopy and gas chromatography *etc.*) determine the relative abundance of a predefined set of metabolites. The latter approach, also known as "global metabolome analysis" is aimed at simultaneously measuring as many metabolites as possible from a single sample. Targeted metabolomics suits pharmacokinetic studies, quantification of enzymatic activity or pathway responsiveness following chemical or therapeutic exposure. Although untargeted metabolomics may be a powerful tool in mechanistic studies and biomarker discovery, the amount of data generated may be difficult to interpret when entities are not components of known, canonical metabolic pathways.

Blood, serum and urine are common targets of *in vivo* metabolomics assessments, but *in vitro* models are also employed.[26] A model using human embryonic stem cells was able to predict the (non-)teratogenic potential of several common marketed drugs[27] and metabolomics profiling of lung cells identified relevant metabolites related to exposure to cigarette smoke.[28] Metabolomics is experiencing a steady expansion and is often considered a valuable addition to transcriptomic and proteomic data in toxicity studies with an integrative focus.[26]

7.3 Data Handling

7.3.1 Toxicoinformatics: Standardisation, Storage and Availability of 'Omics Data

Within a decade, outputs from 'omics experimental approaches escalated from dozens of megabytes (10^6 bytes) to terabytes (10^{12} bytes), demanding several improvements in storage capability, computational power and the development of analyses tools. Furthermore, guidelines and regulatory aspects were created and adjusted as needed to ensure sharing and integration of high-quality, standardised data of multiple origin.

Compared to other 'omics technologies, these criteria were met particularly quickly for the field of transcriptomics. In 2001, researchers from the Microarray Gene Expression Data Society conceived the Minimum Information About a Microarray Experiment (MIAME), which includes essential information from the experimental approach. These guidelines are currently the gold standard for publishing and sharing microarray-based data. Standard guidelines for other 'omics technologies include Minimum Information about a Genome Sequence (MIGS), Minimum Information About a Proteomics Experiment (MIAPE) and the Core Information for Metabolomics Reporting (CIMR).[29]

The MicroArray Quality Control Consortium (MAQC, 2006),[30] a multistakeholder association, evaluates technical aspects and best practices for publishing microarray data. The first phase (MAQC-I) assessed methodologies focusing on cross-platform, inter-laboratory reproducibility, while MAQC-II aimed at reaching a consensus on the best practices for the development and validation of predictive models in Toxicogenomics.[31] As of 2014, the MAQC project entered its third phase – the Sequencing Quality Control (SEQC) – which is aimed at establishing the utility and best practices underlying sequencing-based toxicological assessments.[32] Preliminary findings have shown that RNA-seq outperforms microarrays in detecting differentially expressed genes (93% *versus* 75%), consistent with PCR profiles in the livers of rats exposed to chemicals with different modes of action.[33]

In terms of storage of experimental outputs, **ArrayExpress** (EBI), **GEO** (NCBI) and **CIBEX** (DDBJ) are three large repositories containing high-throughput studies originating from several 'omics. Until April 2016, ArrayExpress storage comprised 40.98 terabytes of raw and processed data from

65 003 experiments; from these, almost 82% encompassed RNA assays (53 267 experiments, 7954 from RNA-seq profilings), followed by DNA (10 752 experiments) and protein assays (175 experiments). In addition, EBI-based **PRIDE** (PRoteomics IDEntifications), **ENA** (European Nucleotide Archive) and **MetaboLights** have additional datasets for proteomics, genomics/nucleotide sequencing and metabolomics investigations. Due to the miscellaneous nature of these repositories, however, an initial mining step is necessary to find and retrieve toxicity studies.

Databases exclusive for Toxicogenomics investigations are also available. The Comparative Toxicogenomics Database (**CTD**) includes almost 12 000 chemicals, describing more than one million curated chemical–gene interactions and over 200 000 curated chemical–disease associations. CTD also provides gene ontologies, pathways and tools for visualisation and enrichment of chemicals or gene sets. The Library of Integrated Network Based Cellular Signature (**LINCS**) consortium offers network-based alterations induced by almost 400 compounds in multiple cell lines. **DrugMatrix** is one of the largest databases for multi-organ, rodent-based Toxicogenomics studies, comprising datasets for over 600 compounds at toxic and non-toxic doses and multiple time points; comparative analyses to user-submitted data are supported through the **ToxFX** or **DrugMatrix Toolbox** platforms. **Open TG-GATEs** offers *in vivo* (primary rat hepatocytes) and *in vitro* (primary rat hepatocytes and primary human hepatocytes) exposure datasets for over 170 compounds at multiple doses and time points, freely available for download and local analyses. Hematology, biochemistry and histopathology imaging from the liver and kidneys of exposed rats are also accessible.

A list of relevant databases, their content and URLs is summarised in Table 7.1. For a more comprehensive list, **ebKB** (http://ebkb.org) provides links to an extensive set of tools used in Toxicoinformatics/Computational Toxicology (in addition to the fields of Cheminformatics, Enviroinformatics, Bioinformatics, Exposure/Risk Analysis and Pattern Recognition).

All abovementioned databases are intended as information resources and do not accept data submissions. **CEBS** (Chemical Effects in Biological Systems) and **diXa** (Data Infrastructure for Chemical Safety) are constantly updated, curated databases. **CEBS** contains over 9000 measurements collected from several experimental approaches generated in universities, industry and governmental institutes, allowing extensive integration across datasets.

diXa is a unique multi-omics database and web interface that centralises toxicity studies from public databases (*e.g.*, TG-GATEs ArrayExpress, PRIDE), the Netherlands Toxicogenomics Centre and several EU Framework Programme (FP) research projects assessing carcinogenicity, pre-clinical safety, immunotoxicity as well as multi-organ toxicity (renal, hepatic, cardiac *etc.*).[34] The EU FP6 carcinoGENOMICS project[35,36] assessed the use of 'omics technologies for screening chemical-induced carcinogenicity; extensive datasets from exposures targeting *in vitro* models of lung, liver and kidney tissues are available on the website. PredTox (EU FP6)

Table 7.1 Summary of relevant data repositories in 'omics technologies.

Institution	Database	URL	Content
Miscellaneous databases			
European Bioinformatics Institute (EBI)	ArrayExpress	https://www.ebi.ac.uk/arrayexpress	Transcriptomic data
	PRIDE	https://www.ebi.ac.uk/pride	Proteomic data
	ENA	http://www.ebi.ac.uk/ena	Transcriptomic data (next-generation sequencing) and DNA-based approaches (epigenome)
	MetaboLights	http://www.ebi.ac.uk/metabolights	Metabolomic data
National Center for Biotechnology Information (NCBI)	GEO	http://www.ncbi.nlm.nih.gov/geo	Transcriptomic data
DNA Data Bank of Japan (DDBJ)	CIBEX	http://cibex.nig.ac.jp/data/	Transcriptomic data
Toxicogenomics			
MDI Biological Laboratory & NC State University	CTD	http://ctdbase.org	Chemical–gene interactions
National Institutes of Health (NIH)	LINCS	http://www.lincsproject.org/	Network-based alterations induced by chemical exposure
National Institutes of Health (NIH)/ National Institute of Environmental Health Sciences (NIEHS)	DrugMatrix, ToxFX and DrugMatrix Toolbox	https://ntp.niehs.nih.gov/drugmatrix https://ntp.niehs.nih.gov/toxfx	Multi-organ, rodent-based microarrays
	CEBS	http://tools.niehs.nih.gov/cebs3/ui/	Transcriptomic and proteomic data (human, rat, mouse)
National Institutes of Biomedical Innovation, Health and Nutrition (NIBIOHN)	Open TG-GATEs	https://toxico.nibiohn.go.jp	*In vivo* (rat) and *in vitro* (rat and human) transcriptomic data from 170 chemicals
Netherlands Toxicogenomics Centre (NTC)	diXa	http://wwwdev.ebi.ac.uk/fg/dixa/index.html	mRNA and microRNA arrays from 110 compounds

(https://www.genedata.com/lp/innomed-predtox.html) and Predict-IV (EU FP7) (http://www.predict-iv.toxi.uni-wuerzburg.de) were projects focused on the discovery of early markers of toxicity using 'omics technologies, offering rat (PredTox and Predict-IV), mouse and human (Predict-IV) transcriptomic data. As of May 2016, the **diXa** data warehouse comprised 29 609 measurements, including transcriptomics, metabolomics and proteomics data, 469 compounds accessible through **ChEMBL** (https://www.ebi.ac.uk/chembl) and the **JRC ChemAgora portal** (https://chemagora.jrc.ec.europa.eu) and 188 experimental disease datasets.

7.3.2 From Bits to Annotation

Although 'omics technologies are highly variable in terms of technical aspects, running time, output files and analyses tools, four basic steps mediate the process of turning raw data files into biologically relevant information. The first step is quality control (QC), a series of standard metrics and concepts that helps the identification (and removal) of possible outliers resulting from technical issues. Normalisation follows as a strategy to remove technical artefacts and sources of variability other than exposure status (a well-known example is the dye bias from two-color cDNA microarrays, causing lower signals for Cy3-labeled samples in comparison to Cy5). Data transformation, *i.e.*, conversion of intensities/readings into a logarithmic scale – usually $\log_2$ – is an optional step (but required for microarray-based experiments) that may help to reduce heteroscedasticity and level the data. Finally, annotation is a crucial step to convert unknown entities into recognisable molecules in a biological and functional background. In view of the massive amount of data generated from transcriptomics measurements within Toxicogenomics, in this section we will describe in detail the aforementioned steps for processing microarrays and sequencing datasets.

A vast set of tools is currently available for pre-processing of one-color cDNA microarrays and a deep understanding of programming languages commonly used in bioinformatics (R, MATLAB, Python) is no longer a requirement; the most relevant ones are detailed in Table 7.2. For instance, **ArrayAnalysis**[37] provides a user-friendly platform to analyse, pre-process and perform statistical analysis from data generated in Affymetrix and Illumina platforms. Following quality control, it generates a file containing normalised data with the preferred annotation type (Entrez, Ensembl, RefSeq *etc.*). One- and two-color microarrays require further steps during processing (probe flagging, filtering and merging), but these services are also available in a web-based suite, **Babelomics.**[38] **GenePattern**[39] offers modules for pre-processing and analyses of several types of 'omics data (proteomics, genomics and transcriptomics). Optionally, customised approaches or pipelines for analyses of specific types of molecules – *e.g.*, microRNAs – may be designed for scratch and code repositories such as Github (*e.g.*, ArrayQC module)[40] or Bitbucket (*e.g.*, MagiCmicroRNA)[41] provide several scripts that can be adapted and executed locally.

Data generated from RNA-seq experiments, on the other hand, offer more flexibility and may be used for several purposes other than gene mining depending on the underlying hypothesis. A standard pipeline for analyses of sequencing data consists of quality control, alignment

Table 7.2 List of computational tools commonly used in transcriptomic analysis (microarray and next-generation sequencing).

Tool	URL	Function
Microarray		
ArrayAnalysis[a]	www.arrayanalysis.org	Pre-processing, statistical and pathway analysis of one-color microarray data (*e.g.*, Affymetrix and Illumina)
Babelomics[a]	http://www.babelomics.org/	Pre-processing of two-color microarrays (*e.g.*, Agilent)
GenePattern[a]	http://www.genepattern.org	Wide range of modules and pipelines for analysing transcriptomics, proteomics and metabolomics data
ArrayQC[b]	https://github.com/BiGCAT-UM/arrayQC_Module	Pre-processing, statistical and pathway analysis of one-color microarray data (*e.g.*, Affymetrix and Illumina)
MagiCmicroRNA[b]	https://bitbucket.org/mutgx/magicmicrorna.git	Pre-processing and analysis of microRNA data
Next generation sequencing		
TopHat	http://ccb.jhu.edu/software/tophat/index.shtml	Splice-aware aligner
STAR	https://github.com/alexdobin/STAR	Splice-aware aligner
Bowtie	http://bowtie-bio.sourceforge.net	Reference-based aligner
BWA	http://bio-bwa.sourceforge.net	Reference-based aligner
SOAP2	http://soap.genomics.org.cn/soapaligner.html	Reference-based aligner
RSEM	http://deweylab.github.io/RSEM	Reference-based software to estimate abundance of aligned transcripts
Tuxedo pipeline	http://cole-trapnell-lab.github.io/cufflinks	Pipeline to assemble, estimate transcript abundance and test for differential expression. Allows discovery of novel transcript variants
Trinity	https://github.com/trinityrnaseq/trinityrnaseq	Performs *de novo* assembly
Trans-AbySS	http://www.bcgsc.ca/platform/bioinfo/software/trans-abyss	Performs *de novo* assembly
Velvet/OASES	https://www.ebi.ac.uk/~zerbino/velvet http://www.ebi.ac.uk/~zerbino/oases	Performs *de novo* assembly

[a]Web-based suites.
[b]developed for R users.

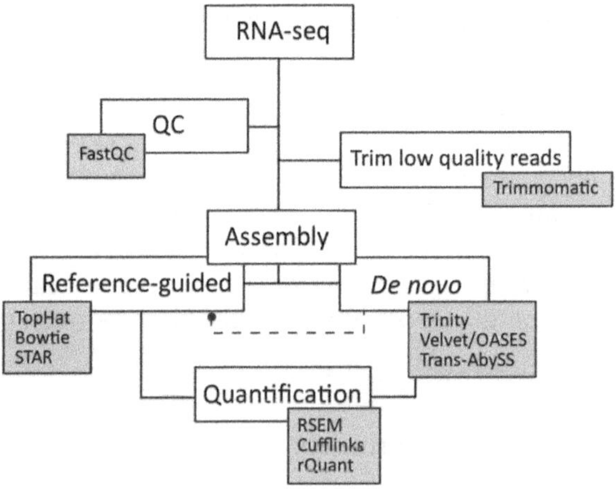

Figure 7.1 Standard RNA-seq pipeline. Dashed path indicates that genomes/transcriptomes assembled *de novo* may serve as a reference to guide mappings.

or sequence assembly, quantification and differential analysis (Figure 7.1). Following QC, high-quality reads are either mapped to a reference genome/transcriptome sequence (reference-based approach) or assembled without a reference (*de novo* assembly). When using reference-based alignment, splice aware aligners such as TopHat[42] and STAR[43] allow the discovery of new, potentially exposure-dependent alternative splicing (AS) events. However, it should be noted that long, paired-end (~80–100 bp) are preferred over single-end reads to increase confidence. The Tuxedo (Cufflinks, Cuffdiff) pipeline[44] is widely used and allows identification of novel transcript variants of a gene and the putative involvement of differential splicing. Alternatively, Bowtie[45] and RSEM[46] may be used to map and quantify genes or (already annotated) transcript variants. Bowtie, BWA[47] and SOAP2[48] are alternative, non-splice aware aligners, while Trinity,[49] Trans-AbySS[50] and Velvet/OASES[51] are popular tools to assemble a transcriptome when working with a non-model organism. Although most NGS utilities are command-line-driven, the Galaxy Tool Shed[52] provides a web-based repository for 3830 valid tools that can be used locally. Several algorithms are available for quantification and differential expression analysis of transcripts mapped to genomic regions and one may consider the best trade-off between output (quantitative and qualitative) and computational requirements.

It is important to stress that there are no 'gold standard' pipelines; initial assumptions, available resources and even programming skills are critical factors ultimately influencing the choice of the most appropriate methodology for each case. As evidenced by a MAQC-II survey conducted with a heterogeneous group of data analysts, predictive models generated by simple approaches may perform as well as complex ones, provided that good modelling practices are employed.[31]

7.4 Use of 'Omics Data in Predictive Toxicology

Once 'omics data are generated, processed and annotated, they may serve different purposes; a single dataset may explain or refute different hypotheses depending on the approach and the levels of data integration. A notable example is the widespread use of TG-GATEs datasets: at least 23 research papers have been published between 2009 and 2016[53–56] and almost half of those (43%) describe predictive models and/or biomarkers for hepatotoxicity.[57,58] In the next section, we will describe the main approaches used in transcriptomics – but generally applicable to other 'omics – and some practical examples.

7.4.1 Comparative Assessments

Class comparison (such as treated *versus* control groups) constitutes the core of many Toxicogenomics studies, aimed at the identification of differentially expressed features among phenotypes tested. Differential expression analysis depends on statistical testing and methods vary among technologies.

Common statistics for microarrays include the *t*-test (comparing two groups) and analysis of variance (three or more groups), which return *p*-values based on the significance to distinguish between groups. Confidence levels at 95 or 99% ($p \leq 0.05$ or $p \leq 0.01$) are common thresholds to select significant features. In addition, multiple testing (*e.g.*, Bonferroni correction) is strongly advisable in high-throughput experiments (>10 000 measurements) to reduce the number of features deemed significant due to chance. Fold change cutoffs (±1.2, ±1.5) may be further combined to filter the most biologically relevant features and create expression matrices that can be used in (un)supervised approaches and mechanistic studies.

7.4.2 Unsupervised and Supervised Pattern Recognition

As constantly emphasised throughout this chapter, 'omics technologies, in particular gene expression profiling, generate a large amount of measurements that cannot be analysed individually. In a first moment, it is important to detect possible trends in data profiles using unsupervised methods. This set of approaches does not take into account the labels of the samples, *i.e.*, the groups to which they belong (toxic, non-toxic), but groups similar profiles together based on experimental measurements. "Are there common patterns of expression related to exposure to this range of toxicants?" exemplifies a general question for unsupervised methods, leading to equally broad answers that are nonetheless essential to guide further assessments. One- and two-dimensional hierarchical clustering, principal component analysis (PCA), self-organising maps and k-means clustering are notable tools for pattern recognition and identification of clusters with similar expression across samples.

Unbiased integration of several types of data is possible through Bayesian inference, which generates probabilistic graphs for conditional interactions between genes, microRNAs, protein, Gene Ontology (GO; http://www.geneontology.org) terms and even phenotypic measurements. A study successfully related transcriptomic responses in smoking individuals to cotinine levels (a marker for exposure) and DNA adducts (a marker for carcinogenic risk) through Bayesian network inference and presented novel biomarker candidates linked to carcinogenesis and autoimmune diseases arising from exposure.[59]

On the other hand, the use of supervised methods is suitable for classification and prediction of toxicity relying on the assumption that chemicals with similar activity elicit similar responses at the molecular level. A set of well-characterised chemicals (training set) is used to retrieve the most informative features distinguishing between the phenotypes (classes) of interest and *train* an algorithm to recognise these features as classifiers. The trained algorithm is then employed to classify a set of unknown compounds (test set). A cross-validation step determines the robustness of the model by testing its ability to classify properly an additional set of well-characterised chemicals. Support vector machines (SVMs), k-nearest neighbours, artificial neural networks (ANN), prediction analysis for microarrays (PAM) and random forests are some algorithms that can be trained to solve categorical classification problems.

For instance, a SVM module trained with histopathological and gene expression data from exposed rats correctly classified nephrotoxicants in 82% of the cases and identified potential biomarkers.[60] Supervised methods may also discriminate chemicals with different mechanisms of toxicity, as shown by a discriminating model capable of identifying genotoxic (*i.e.*, toxicity is mediated by DNA-damaging properties) and non-genotoxic (*i.e.*, toxicity is not related to DNA damage) compounds based on specific gene expression patterns *in vivo*[61] and *in vitro*.[62,63]

Such sensitive approaches, however, often require high-quality data and well-defined groups and endpoints to yield useful information for predictive toxicology. To overcome these limitations, integration of several 'omics technologies or other fields of knowledge may help to improve sensitivity and specificity.[64]

7.4.3 Connectivity Mapping (CMap)

Connectivity mapping establishes associations (positive, negative or null) between expression profiles by comparing a reference gene signature of a biological state of interest (or a group of reference compounds with well-defined biological activity) to a query expression set.[65] CMaps have been considered valid predictors for drug discovery, toxicity and carcinogenicity.

For instance, datasets from the PredTox consortium were used to build a signature set for cholestasis (impairment of bile flow) based on liver transcriptomics of rats exposed to griseofulvin, a cholestatic drug. Sixteen compounds were queried against the griseofulvin signature and from those, five were found to be positively connected to the reference. Four had their effects substantiated by histopathological findings showing liver hypertrophy and bile duct damage.[66]

In a more recent, human-based approach, the carcinogenic potential of 139 compounds from exposed primary human hepatocytes (2320 microarrays, TG-GATEs) was assessed by using a reference signature of patients suffering from hepatocellular carcinoma. Known (aflatoxin B1) and potentially human (carbon tetrachloride, cyclosporine A) liver carcinogens were among the top scoring compounds. Furthermore, the lowest dose and latest exposure period (24 h) showed the best classification as potential carcinogens, indicating the dose factor as an important variable for *in vitro* predictions of carcinogenicity.[55]

A web version of CMap allows comparative assessments of a user-submitted gene expression set to over 7000 expression profiles from 1309 compounds (broadinstitute.org/cmap).

7.4.4 Mechanistic Analysis

Identifying the mechanism of action and the biological mechanisms underlying chemical exposure is, if not the most important, a strong asset in Toxicogenomics investigations. During this stage, differential expression of genes, proteins and metabolites is interpreted by means of pathway analysis and gene ontology enrichment, which describe cellular localisation, molecular function and biological processes. Processes are deemed significantly enriched/overrepresented when the number of features belonging to the same pathway or process is higher than expected. Several tools are available for functional analyses (*e.g.*, MetaCore™, Ingenuity Pathway Analysis, PathVisio); most interesting is **ConsensusPathDB** (CPDB; http://cpdb.molgen.mpg.de), which centralises biological interactions from 32 public resources and integrates protein–protein, genetic, metabolic, gene regulatory and even drug–target interactions.[67]

In the last few years, integration of classical toxicology, quantitative analysis and multi-omic functional changes measured at several levels of biological organisation has been targeted as an attempt to attain a systematic portrait, or Systems Toxicology. It is expected that Systems Toxicology will create profound knowledge about molecular events arising from external perturbations that can be ultimately summarised into a series of events leading to an outcome. This concept is the core of adverse outcome pathways (AOPs), a multi-step, sequential chain that results in an adverse effect – which can be specified at many levels, including organ, tissue or even population.[68,69] An AOP is comprised of three main blocks: a molecular initiating event (MIE) that triggers the cascade, a series of intermediary, possibly multi-level, key events, and one adverse outcome (see also Chapter 9). 'Omics technologies and integrative approaches are therefore important assets for detecting MIEs or key intermediary steps that can be applied to predictive toxicology in AOPs.

A practical example is the AOP designed for chemically-induced liver fibrosis, a reversible regenerative process that follows extensive damage to hepatic cells. Exhaustive injury–healing cycles associated with repeated-dose toxicity strongly correlates with development of primary liver cancer. By investigating two hepatotoxicants, allyl alcohol and carbon tetrachloride, the Joint Research Centre of the European Commission has suggested that protein alkylation may act as a common molecular initiating event leading to hepatocyte injury, oxidative stress, cell death and chronic inflammation (alterations at the cellular level). Progressive collagen accumulation and changes in extracellular matrix composition ultimately build a fibrotic liver.[70,71]

7.5 Big ('Omics) Data

Despite the flood of tools, storage, visualisation, querying and global interpretation of 'omics data is difficult. Therefore, most, if not all, studies conduct a previous selection of relevant features to reduce complexity. As a result, valuable information may be lost in the Big Data of high-throughput screenings. Besides computational power, the availability of tools capable of grasping the subtleties of biological interactions as a whole are important factors that may advance or limit researches. In this section, we discuss the limitations and possible solutions for calculation and integration of massive datasets.

For comparison purposes, we will exemplify the calculation issue in Toxicogenomics using datasets generated from time-series analyses in two platforms, Agilent Human 1A Microarray (V2) (microarray) and Illumina HiSeq 2000 (RNA-seq) (detailed in Table 7.3). Both experiments were conducted with HepG2 – an immortalised cell line originated from hepatocytes – and the environmental carcinogen benzo(*a*)pyrene, BaP. The first experiment consists of measurements of 11 time points over a time span of 60 h.[72] The sequencing approach comprised paired-end reads from ribo-depleted libraries in six time points over 48 h, with an average of 93 million reads per sample.[73] In the first dataset, raw readings from 48 samples measuring 22 575 probes each amounted to 300 megabytes of data. The second experiment, with only half the number of samples (24), yielded raw files requiring at least 400 000 megabytes (0.4 terabytes) of storage, a thousand times more than in the first scenario. A dual-core computer, as routinely used in everyday life, could easily perform locally all the tasks for the first dataset using a standard microarray pipeline (QC, processing, normalisation) within one day or less (<24 Central Processing Unit (CPU) hours). For the second experiment, however, almost 400 CPU hours would be necessary to perform a reference-based pipeline. Excluding all other variables and assuming operating system stability, these analyses would not be completed in less than 100 days using the same machine. It is also worth noting that intermediary files may be 5–10 times larger than the raw data, creating a storage challenge. This is, however, only the "tip of the iceberg", and to fully profit from NGS results, one must investigate deeper. A reference-based approach using the results mentioned

Table 7.3 Comparison between high-throughput transcriptomics data generated from microarray and RNA-sequencing technologies.

Feature	Microarray	RNA-sequencing
Platform	G4110B Agilent 22K	Illumina HiSeq 2000
Study	E-TOXM-22 (ArrayExpress)[72]	ERP009514 (ArrayExpress/ENA)[73]
RNA fraction	mRNA	ribo-depleted
Number of reads per sample	22 575 probes	93 million reads (average)
Number of samples	48 (11 time points)	24 (6 time points)
File sizes (raw data, megabytes)	300	>400 000
CPU hours (estimated)	<24	~400
Output	22 575 probes	>46 000 genes ~200 000 gene transcripts microRNA, circRNA

returns expression values for approximately 85 000 entities (coding and non-coding RNAs, pseudogenes) using the latest Ensembl annotation (v. 84, March 2016). It also yields expression values for almost 200 000 gene transcripts (transcript variants), and depending on the methodology used to build the expression libraries (*e.g.*, polyA-enriched, ribo-depleted *etc.*), it is possible to investigate the behaviour of RNA classes other than mRNAs (microRNAs, circRNAs) – which in turn increases calculation time.

It is clear, therefore, that large-scale, 'omics Big Data is heavily dependent on high-performance computing (HPC) and sophisticated storage infrastructures.[74] Multi-threaded servers (several CPUs) in a stable environment that can remain active for long periods without rebooting is a common resolution; on the downside, it requires a team of bioinformaticians or IT specialists to aid inexperienced biologists and toxicologists. An alternative is the creation of multi-institutional clusters for parallel computing and data storage. Biowulf, implemented by the National Institutes of Health (NIH), is a HPC comprising 24 600 computer cores that accept data submission from registered users. Within 24 h, this massive computational grid may process almost 50 000 jobs, the equivalent to almost 690 000 CPU hours (6 billion CPU hours in a year). Nationwide HPC projects have also been successfully implemented in several countries. The "UPPMAX cluster and storage for next-generation sequencing" (UPPNEX), as the name suggests, is a cluster based in Sweden for the sole purpose of analysing NGS data. Operating at a rate of 1 million CPU hours per month, UPPNEX has pre-installed software intended for alignment, *de novo* assembly and other general tools. Reference genomes are also available and are constantly updated by supporting staff.[75] The Life Sciences Grid (LSG), designed by the HPC SURFSara, is dedicated for research groups in Life Sciences across the Netherlands. The clusters are located in several academic institutions and large-scale calculation jobs may be submitted and processed on-site or scaled to other centralised clusters. Table 7.4 contains a list of some HPCs currently used in biosciences.

Table 7.4 List of some currently active high-performance computing (HPCs) clusters.

HPC	Institution	URL	Number of cores
Biowulf	National Institutes of Health (NIH)	https://hpc.nih.gov/systems	24 600
UPPNEX	Multiple Swedish institutions	www.scilifelab.se/facilities/uppnex	2748
SurfSara/Life Science Grid	Multiple Dutch institutions	https://serinfo.surfsara.nl/systems/grid	1152
HPCBio	University of Illinois (US)	https://hpcbio.illinois.edu	NA[a]
HiperGator 2.0	University of Florida (US)	https://www.rc.ufl.edu/services/computation/hipergator	30 000

[a]NA: information not available.

Calculation requirements surpassed, data mining remains an important (and challenging) step to retrieve additional information that can be coupled to observed patterns. In addition to the Toxicoinformatics databases mentioned in the previous subsection (CTD, CEBS, DrugMatrix), several other resources for chemical structure and biological effects comprising results from various cell-free and cell-based high-throughput screenings are available (see also Chapter 4). **PubChem** (https://pubchem.ncbi.nlm.nih.gov) is a NIH database that offers results from more than one million different assays performed with almost 90 million compounds, including gene and protein targets (~30 000).[76] **ChEMBL** contains binding, functional and ADMET (absorption, distribution, metabolism, excretion – toxicity) information[77] for over 1.5 million compounds.

Also, the integration of multi-level, multi-source data ('omics, target sites, phenotype or trait), although essential for extrapolating results from *in vitro* assays to a whole organism,[78] is still a challenge and a key issue for 'omics Big Data. Molecular diversity and data standardisation are complex, and statistical calculation for all possible combinations among entities of each 'omics input is impractical, even with HPCs (Figure 7.2). Meta-analysis, which compiles results from several studies, usually uses one type of data (*e.g.*, gene expression): an investigation of 6567 *in vivo* rat samples (collected from GEO, ArrayExpress, TG-GATEs and CEBS) and 453 distinct compounds found at least 18 genes related to toxicity and identified cathepsin D as a gene commonly upregulated after exposure to toxic concentrations of chemicals.[57] Alternatively, two- and three-level 'omics integration (*e.g.*, transcriptomics – proteomics or transcriptomics – metabolomics) stands out as the most common practice being successfully used to identify specific signatures for toxic exposures.[79]

To achieve that, conceptual- or statistical-based integration may be employed to retrieve biological information.[80] Conceptual-based integration

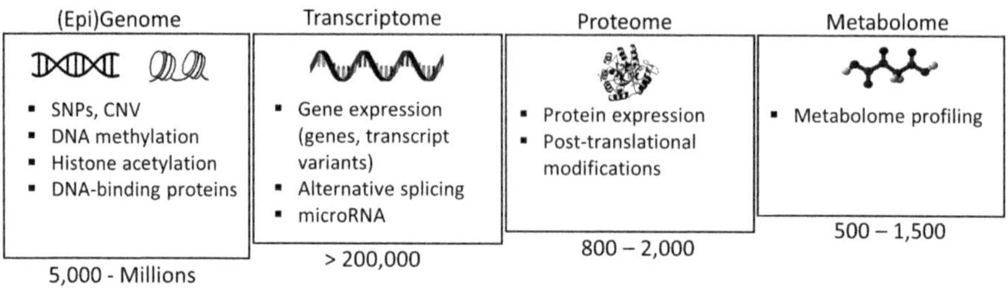

Figure 7.2 'Omes and estimated number of measurements per experiment. A combinatorial analysis yields 4×10^{14} possible interactions to be evaluated.

draws conclusions from merging data types analysed independently. Cyslosporine A (CsA) is an immunosuppressant drug that may induce cholestasis *in vivo*. A study integrating protein-mRNA-microRNA data obtained from cultured rat hepatocytes identified for the first time the involvement of endoplasmic reticulum (ER) stress and unfolding protein response (UPR) in toxic responses to CsA.[79] However, in spite of valuable insights, important information may be lost. Statistical-base integration solves this problem by analysing all datasets at once, employing network-free or network-based methods. These methods may integrate multi-omics data for several purposes, including discovery of molecular mechanisms, sample clustering and outcome prediction.[81] iClusterPlus, a tool for sample clustering, integrates binary, categorical and continuous values.[82] An investigation comprising gene scores arising from previous meta-analysis, structural similarities (3D Tanimoto score) and protein target predictions led to improved clustering of hepatotoxic compounds compared to hierarchical cluster analysis.[83] Camelot (CAusal Modelling with Expression Linkage for cOmplex Traits) integrates genotype, phenotype and gene expression data and may predict the molecular mechanisms culminating in the phenotype under investigation.[81] In a study with two strains of yeast, Camelot was able to predict the response of 87 drugs (out of 94) and infer 25 (out of 27) gene–drug interactions.[84] Other tools are also available, but types of data accepted as input may vary considerably across methods.[81]

7.6 Toxicogenomics in the Big Data Era: Challenges, Perspectives and Opportunities

After years of explosive growth, Toxicogenomics matured into a reliable field of knowledge; well-established techniques are now routinely used to answer questions from basic and applied science. In order to keep moving forward, we should look beyond current practices and focus on innovative ways to tackle current challenges, which arise from two main sources: biology and bioinformatics.

As animal-based Toxicogenomics was gradually (but not entirely) replaced by *in vitro*, human-based models, questions were raised with respect to the biological relevance for such findings in the context of a whole organism. An overwhelming amount of models has its roots in neoplastic tissues, which may lack functional genes or enzymes essential for toxic responses.[85,86] On the other hand, advantages offered by cells of primary origin (obtained from healthy tissues) are often counterbalanced by their limited life span (5–10 days) and inability to proceed with cell cycles when plated.[4,87] Three-dimensional cultures are sound alternatives for mimicking tissue arrangements, which may help to improve or even rescue enzymatic activity.[88] Organotypic cultures (microtissues) resembling the cellular composition of the original tissue restore cell–cell communication, regarded as essential for maintenance of normal tissue physiology.[89] Microtissues from the heart,[90] liver[91] and kidney,[92] for example, may remain viable and functional for several weeks, making them suitable for repeated dose, long-term toxicity studies[93] – often pillars of chemical-induced carcinogenicity. More recently, in order to address systemic effects and multi-organ toxicity, much effort has been directed towards the development of a "body-on-a-chip" (BOC).[94–96] For example, a recent study assessing the pharmacological relevance of a four-organ module measuring cardiac, muscle, neuronal and hepatic response to five drugs for 14 days presented novel and known effects of the drugs tested.[97] Despite that, market availability may not be a reality for at least three years, and in the meantime, non-mammalian models such as zebrafish may be suitable replacements to study whole-organism effects. Their small size enables large-scale screenings for predicting toxicity[98] and transcriptomics investigations have highlighted the sensitivity and specificity of the model to toxicant exposure: gene expression changes are detected prior to morphological alterations and pathways, usually conserved, are related to known human effects.[99,100]

Moreover, molecular responses show strong time- and dose-dependency. Prediction of drug carcinogenicity in rodents was shown to diverge by almost 30% depending on dose and exposure times used.[101] Translating organism-level toxicity to *in vitro* models (and *vice versa*) and identifying how and when crucial molecular events leading to an observed adverse outcome occur are unrealistic tasks for wet bench research – and therefore assigned to Bioinformatics (dry bench). In the first scenario, *in silico* toxico-kinetic/dynamic (*e.g.*, physiologically-based pharmacokinetic/dynamic (PBPK/PD)) models may provide an estimate of physiologically relevant doses, which may be associated to a molecular signature. The biology underlying the "how and when", on the other hand, poses more challenges to bioinformaticians. First, integration of multi-level 'omics data often yields only a few logical (or at least known) interactions since the complexity of biological systems cannot (yet?) be addressed by current models. Furthermore, networks arising from these complex interactions are dynamic by nature, dictating cell fate under physiological and perturbed conditions. Therefore, (improved) tools for analysis of network structure and its evolution over time would help to identify turning points arising from exposure.

In conclusion, Toxicogenomics is a fertile research ground. The amount of easily accessible data generated to date is massive and diverse, and asking the right questions may lead to revolutionary discoveries.

References

1. S. Farr and R. T. Dunn, Concise review: gene expression applied to toxicology, *Toxicol. Sci.*, 1999, **50**(1), 1–9.
2. S. Parasuraman, Toxicological screening, *J. Pharmacol. Pharmacother.*, 2011, **2**(2), 74.
3. H. Olson, G. Betton, D. Robinson, K. Thomas, A. Monro and G. Kolaja, *et al.*, Concordance of the toxicity of pharmaceuticals in humans and in animals, *Regul. Toxicol. Pharmacol.*, 2000, **32**(1), 56–67.
4. J. Jiang, J. E. J. Wolters, S. G. van Breda, J. C. Kleinjans and T. M. de Kok, Development of novel tools for the in vitro investigation of drug-induced liver injury, *Expert Opin. Drug Metab. Toxicol.*, 2015, **11**(10), 1523–1537.
5. S. Campion, J. Aubrecht, K. Boekelheide, D. W. Brewster, V. S. Vaidya and L. Anderson, *et al.*, The current status of biomarkers for predicting toxicity, *Expert Opin. Drug Metab. Toxicol.*, 2013, **9**(11), 1391–1408.
6. Y. C. M. Staal, M. H. M. van Herwijnen, F. J. van Schooten and J. H. M. van Delft, Application of four dyes in gene expression analyses by microarrays, *BMC Genomics*, 2005, **6**(1), 101.
7. L. D. Kier, R. Neft, L. Tang, R. Suizu, T. Cooki and K. Onsurez, *et al.*, Applications of microarrays with toxicologically relevant genes (tox genes) for the evaluation of chemical toxicants in sprague dawley rats in vivo and human hepatocytes in vitro, *Mutat. Res.*, 2004, **549**(1–2), 101–113.
8. *Toxicogenomics-based Cellular Models*, ed. J. C. Kleinjans, Elsevier, 2014.
9. P. Thulin, G. Nordahl, M. Gry, G. Yimer, E. Aklillu and E. Makonnen, *et al.*, Keratin-18 and microRNA-122 complement alanine aminotransferase as novel safety biomarkers for drug-induced liver injury in two human cohorts, *Liver Int.*, 2014, **34**(3), 367–378.
10. E. S. Lander, L. M. Linton, B. Birren, C. Nusbaum, M. C. Zody and J. Baldwin, *et al.*, Initial sequencing and analysis of the human genome, *Nature*, 2001, **409**(6822), 860–921.
11. D. M. Altshuler, R. A. Gibbs, L. Peltonen, E. Dermitzakis, S. F. Schaffner and F. Yu, *et al.*, Integrating common and rare genetic variation in diverse human populations, *Nature*, 2010, **467**(7311), 52–58.
12. M. Esteller, The necessity of a human epigenome project, *Carcinogenesis*, 2006, **27**(6), 1121–1125.
13. J. M. Heather and B. Chain, The sequence of sequencers: The history of sequencing DNA, *Genomics*, 2015, **107**(1), 1–8.

14. A. Auton, G. R. Abecasis, D. M. Altshuler, R. M. Durbin, D. R. Bentley and A. Chakravarti, *et al.*, A global reference for human genetic variation, *Nature*, 2015, **526**(7571), 68–74.
15. I. Johansson and M. Ingelman-Sundberg, Genetic polymorphism and toxicology–with emphasis on cytochrome p450, *Toxicol. Sci.*, 2011, **120**(1), 1–13.
16. F. Y. Lee, T. Q. de Aguiar Vallim, H. K. Chong, Y. Zhang, Y. Liu and S. A. Jones, *et al.*, Activation of the farnesoid x receptor provides protection against acetaminophen-induced hepatic toxicity, *Mol. Endocrinol.*, 2010, **24**(8), 1626–1636.
17. R. E. Watson and J. I. Goodman, Epigenetics and dna methylation come of age in toxicology, *Toxicol. Sci.*, 2002, **67**(1), 11–16.
18. M. Verma, Epigenome-wide association studies (ewas) in cancer, *Curr. Genomics*, 2012, **13**(4), 308–313.
19. K. W. K. Lee, R. Richmond, P. Hu, L. French, J. Shin and C. Bourdon, *et al.*, Prenatal exposure to maternal cigarette smoking and dna methylation: epigenome-wide association in a discovery sample of adolescents and replication in an independent cohort at birth through 17 years of age, *Environ. Health Perspect.*, 2015, **123**(2), 193–199.
20. National Research Council (US) Committee on Applications of Toxicogenomic Technologies to Predictive Toxicology, *Applications of Toxicogenomic Technologies to Predictive Toxicology and Risk Assessment*, National Academies Press, Washington, D.C, 2007.
21. C. M. McHale, L. Zhang, A. E. Hubbard and M. T. Smith, Toxicogenomic profiling of chemically exposed humans in risk assessment, *Mutat. Res.*, 2010, **705**(3), 172–183.
22. J. Gao, L. Ann Garulacan, S. M. Storm, S. A. Hefta, G. J. Opiteck and J.-H. Lin, *et al.*, Identification of in vitro protein biomarkers of idiosyncratic liver toxicity, *Toxicol. In Vitro*, 2004, **18**(4), 533–541.
23. J. A. Lewis, W. E. Dennis, J. Hadix and D. A. Jackson, Analysis of secreted proteins as an in vitro model for discovery of liver toxicity markers, *J. Proteome Res.*, 2010, **9**(11), 5794–5802.
24. A. Van Summeren, J. Renes, F. G. Bouwman, J.-P. Noben, J. H. M. van Delft and J. C. S. Kleinjans, *et al.*, Proteomics investigations of drug-induced hepatotoxicity in HepG2 cells, *Toxicol. Sci.*, 2011, **120**(1), 109–122.
25. G. J. Patti, O. Yanes and G. Siuzdak, Innovation: metabolomics: the apogee of the omics trilogy, *Nat. Rev. Mol. Cell Biol.*, 2012, **13**(4), 263–269.
26. M. Bouhifd, T. Hartung, H. T. Hogberg, A. Kleensang and L. Zhao, Review: toxicometabolomics, *J. Appl. Toxicol.*, 2013, **33**(12), 1365–1383.
27. P. R. West, A. M. Weir, A. M. Smith, E. L. R. Donley and G. G. Cezar, Predicting human developmental toxicity of pharmaceuticals using human embryonic stem cells and metabolomics, *Toxicol. Appl. Pharmacol.*, 2010, **247**(1), 18–27.

28. S. V. Vulimiri, M. Misra, J. T. Hamm, M. Mitchell and A. Berger, Effects of mainstream cigarette smoke on the global metabolome of human lung epithelial cells, *Chem. Res. Toxicol.*, 2009, **22**(3), 492–503.
29. C. F. Taylor, D. Field, S.-A. Sansone, J. Aerts, R. Apweiler and M. Ashburner Michael, *et al.*, Promoting coherent minimum reporting guidelines for biological and biomedical investigations: the mibbi project, *Nat. Biotechnol.*, 2008, **26**(8), 889–896.
30. L. Shi, L. H. Reid, W. D. Jones, R. Shippy, J. A. Warrington and S. C. Baker, *et al.*, The microarray quality control (maqc) project shows inter- and intraplatform reproducibility of gene expression measurements, *Nat. Biotechnol.*, 2006, **24**(9), 1151–1161.
31. L. Shi, G. Campbell, W. D. Jones, F. Campagne, Z. Wen and S. J. Walker, *et al.*, The MicroArray Quality Control (MAQC)-II study of common practices for the development and validation of microarray-based predictive models, *Nat. Biotechnol.*, 2010, **28**(8), 827–838.
32. SEQC/MAQC-III Consortium, A comprehensive assessment of RNA-seq accuracy, reproducibility and information content by the Sequencing Quality Control Consortium, *Nat. Biotechnol.*, 2014, **32**(9), 903–914.
33. C. Wang, B. Gong, P. R. Bushel, J. Thierry-Mieg, D. Thierry-Mieg and J. Xu, *et al.*, The concordance between rna-seq and microarray data depends on chemical treatment and transcript abundance, *Nat. Biotechnol.*, 2014, **32**(9), 926–932.
34. D. M. Hendrickx, H. J. W. L. Aerts, F. Caiment, D. Clark, T. M. D. Ebbels and C. T. Evelo, *et al.*, Dixa: a data infrastructure for chemical safety assessment, *Bioinformatics*, 2015, **31**(9), 1505–1507.
35. M. Vinken, T. Doktorova, H. Ellinger-Ziegelbauer, H.-J. Ahr, E. Lock and P. Carmichael, *et al.*, The carcinoGENOMICS project: critical selection of model compounds for the development of omics-based in vitro carcinogenicity screening assays, *Mutat. Res.*, 2008, **659**(3), 202–210.
36. R. Herwig, H. Gmuender, R. Corvi, K. M. Bloch, A. Brandenburg and J. Castell, *et al.*, Inter-laboratory study of human in vitro toxicogenomics-based tests as alternative methods for evaluating chemical carcinogenicity: a bioinformatics perspective, *Arch. Toxicol.*, 2016, **90**(9), 2215–2229.
37. L. M. T. Eijssen, M. Jaillard, M. E. Adriaens, S. Gaj, P. J. de Groot and M. Müller, *et al.*, User-friendly solutions for microarray quality control and pre-processing on ArrayAnalysis.org, *Nucleic Acids Res.*, 2013, **41**(Web Server issue), W71–W76.
38. F. Al-Shahrour, P. Minguez, J. M. Vaquerizas, L. Conde and J. Dopazo Joaquín, Babelomics: a suite of web tools for functional annotation and analysis of groups of genes in high-throughput experiments, *Nucleic Acids Res.*, 2005, **33**(Web Server issue), W460–W464.
39. M. Reich, T. Liefeld, J. Gould, J. Lerner, P. Tamayo and J. P. Mesirov, Genepattern 2.0, *Nat. Genet.*, 2006, **38**(5), 500–501.

40. L. Rieswijk, K. J. J. Brauers, M. L. J. Coonen, S. G. J. van Breda, D. G. J. Jennen and J. C. S. Kleinjans, Evaluating microRNA profiles reveals discriminative responses following genotoxic or non-genotoxic carcinogen exposure in primary mouse hepatocytes, *Mutagenesis*, 2015, **30**(6), 771–784.
41. M. L. Coonen, D. H. Theunissen, J. C. Kleinjans and D. G. Jennen, Magicmicrorna: a web implementation of agimicrorna using shiny, *Source Code Biol. Med.*, 2015, **10**(1), 4.
42. C. Trapnell, L. Pachter and S. L. Salzberg, Tophat: discovering splice junctions with rna-seq, *Bioinformatics*, 2009, **25**(9), 1105–1111.
43. A. Dobin, C. A. Davis, F. Schlesinger, J. Drenkow, C. Zaleski and S. Jha, *et al.*, Star: ultrafast universal rna-seq aligner, *Bioinformatics*, 2013, **29**(1), 15–21.
44. C. Trapnell, D. G. Hendrickson, M. Sauvageau, L. Goff, J. L. Rinn and L. Pachter, Differential analysis of gene regulation at transcript resolution with rna-seq, *Nat. Biotechnol.*, 2013, **31**(1), 46–53.
45. B. Langmead, C. Trapnell, M. Pop and S. L. Salzberg, Ultrafast and memory-efficient alignment of short dna sequences to the human genome, *Genome Biol.*, 2009, **10**(3), R25.
46. B. Li and C. N. Dewey, Rsem: accurate transcript quantification from rna-seq data with or without a reference genome, *BMC Bioinf.*, 2011, **12**(1), 323.
47. H. Li and R. Durbin, Fast and accurate short read alignment with burrows-wheeler transform, *Bioinformatics*, 2009, **25**(14), 1754–1760.
48. R. Li, C. Yu, Y. Li, T.-W. Lam, S.-M. Yiu and K. Kristiansen, *et al.*, Soap2: an improved ultrafast tool for short read alignment, *Bioinformatics*, 2009, **25**(15), 1966–1967.
49. B. J. Haas, A. Papanicolaou, M. Yassour, M. Grabherr, P. D. Blood and J. Bowden, *et al.*, De novo transcript sequence reconstruction from rna-seq using the trinity platform for reference generation and analysis, *Nat. Protoc.*, 2013, **8**(8), 1494–1512.
50. G. Robertson, J. Schein, R. Chiu, R. Corbett, M. Field and S. D. Jackman, *et al.*, De novo assembly and analysis of rna-seq data, *Nat. Methods*, 2010, **7**(11), 909–912.
51. M. H. Schulz, D. R. Zerbino, M. Vingron and E. Birney, Oases: robust de novo rna-seq assembly across the dynamic range of expression levels, *Bioinformatics*, 2012, **28**(8), 1086–1092.
52. E. Afgan, D. Baker, M. van den Beek, D. Blankenberg, D. Bouvier and M. Čech, *et al.*, The Galaxy platform for accessible, reproducible and collaborative biomedical analyses: 2016 update, *Nucleic Acids Res.*, 2016, **44**, W3–W10.
53. J. D. Zhang, N. Berntenis, A. Roth and M. Ebeling, Data mining reveals a network of early-response genes as a consensus signature of drug-induced in vitro and in vivo toxicity, *Pharmacogenomics J.*, 2014, **14**(3), 208–216.

54. Y. Hizukuri, R. Sawada, Y. Yamanishi, D. C. Swinney, J. Anthony and N. Nagamine, *et al.*, Predicting target proteins for drug candidate compounds based on drug-induced gene expression data in a chemical structure-independent manner, *BMC Med. Genomics*, 2015, **8**(1), 82.
55. F. Caiment, M. Tsamou, D. Jennen and J. Kleinjans, Assessing compound carcinogenicity in vitro using connectivity mapping, *Carcinogenesis*, 2014, **35**, 201–207.
56. S. M. Bell, M. M. Angrish, C. E. Wood and S. W. Edwards, Integrating publicly available data to generate computationally predicted adverse outcome pathways for fatty liver, *Toxicol. Sci.*, 2016, **150**(2), 510–520.
57. H. Kim, J.-H. Kim, S. Y. Kim, D. Jo, H. J. Park and J. Kim, *et al.*, Meta-analysis of large-scale toxicogenomic data finds neuronal regeneration related protein and cathepsin d to be novel biomarkers of drug-induced toxicity, *PLoS One*, 2015, **10**(9), e0136698.
58. Y. R. An, J. Y. Kim and Y. S. Kim, Construction of a predictive model for evaluating multiple organ toxicity, *Mol. Cell. Toxicol.*, 2016, **12**(1), 1–6.
59. D. G. J. Jennen, D. M. van Leeuwen, D. M. Hendrickx, R. W. H. Gottschalk, J. H. M. van Delft and J. C. S. Kleinjans, Bayesian network inference enables unbiased phenotypic anchoring of transcriptomic responses to cigarette smoke in humans, *Chem. Res. Toxicol.*, 2015, **28**(10), 1936–1948.
60. S. K. Thukral, P. J. Nordone, R. Hu, L. Sullivan, E. Galambos and V. D. Fitzpatrick, *et al.*, Prediction of nephrotoxicant action and identification of candidate toxicity-related biomarkers, *Toxicol. Pathol.*, 2005, **33**(3), 343–355.
61. H. Ellinger-Ziegelbauer, H. Gmuender, A. Bandenburg and H. J. Ahr, Prediction of a carcinogenic potential of rat hepatocarcinogens using toxicogenomics analysis of short-term in vivo studies, *Mutat. Res.*, 2008, **637**(1–2), 23–39.
62. K. Mathijs, K. J. J. Brauers, D. G. J. Jennen, A. Boorsma, M. H. M. van Herwijnen and R. W. H. Gottschalk, *et al.*, Discrimination for genotoxic and nongenotoxic carcinogens by gene expression profiling in primary mouse hepatocytes improves with exposure time, *Toxicol. Sci.*, 2009, **112**(2), 374–384.
63. C. Magkoufopoulou, S. M. H. Claessen, M. Tsamou, D. G. J. Jennen, J. C. S. Kleinjans and J. H. M. van Delft, A transcriptomics-based in vitro assay for predicting chemical genotoxicity in vivo, *Carcinogenesis*, 2012, **33**(7), 1421–1429.
64. L. Rieswijk, K. J. J. Brauers, M. L. J. Coonen, D. G. J. Jennen, S. G. J. van Breda and J. C. S. Kleinjans, Exploiting microRNA and mRNA profiles generated in vitro from carcinogen-exposed primary mouse hepatocytes for predicting in vivo genotoxicity and carcinogenicity, *Mutagenesis*, 2016, **31**(5), 603–615.
65. J. Lamb, E. D. Crawford, D. Peck, J. W. Modell, I. C. Blat and M. J. Wrobel, *et al.*, The connectivity map: using gene-expression signatures to connect small molecules, genes, and disease, *Science*, 2006, **313**(5795), 1929–1935.

66. T. W. Gant, P. R. Baus, B. Clothier, J. Riley, R. Davies and D. J. Judah, *et al.*, Gene expression profiles associated with inflammation, fibrosis, and cholestasis in mouse liver after griseofulvin, *EHP Toxicogenomics*, 2003, **111**(1T), 37–43.
67. A. Kamburov, U. Stelzl, H. Lehrach and R. Herwig, The consensuspathdb interaction database: 2013 update, *Nucleic Acids Res.*, 2013, **41**, D793–D800.
68. M. Vinken, The adverse outcome pathway concept: a pragmatic tool in toxicology, *Toxicology*, 2013, **312**, 158–165.
69. T. Horvat, B. Landesmann, A. Lostia, M. Vinken, S. Munn and M. Whelan, Adverse outcome pathway development from protein alkylation to liver fibrosis, *Arch. Toxicol.*, 2017, **91**(4), 1523–1543.
70. B. Landesmann, M. Gomenou, S. Munn and M. Whelan, Description of Prototype Modes-of-Action Related to Repeated Dose Toxicity, Publications Office of the European Union, EUR 25631 EN, 2012.
71. G. T. Ankley, R. S. Bennett, R. J. Erickson, D. J. Hoff, M. W. Hornung, R. D. Johnson, D. R. Mount, J. W. Nichols, C. L. Russom, P. K. Schmieder, J. A. Serrrano, J. E. Tietge and D. L. Villeneuve, Adverse outcome pathways: a conceptual framework to support ecotoxicology research and risk assessment, *Environ. Toxicol. Chem.*, 2010, **29**(3), 730–741.
72. J. H. M. van Delft, K. Mathijs, Y. C. M. Staal, M. H. M. van Herwijnen, K. J. J. Brauers and A. Boorsma, *et al.*, Time series analysis of benzo[a]pyrene-induced transcriptome changes suggests that a network of transcription factors regulates the effects on functional gene sets, *Toxicol. Sci.*, 2010, **117**, 381–392.
73. F. Caiment, S. Gaj, S. Claessen and J. Kleinjans, High-throughput data integration of rna-mirna-circrna reveals novel insights into mechanisms of benzo[a]pyrene-induced carcinogenicity, *Nucleic Acids Res.*, 2015, **43**(5), 2525–2534.
74. D. C. Koboldt, L. Ding, E. R. Mardis and R. K. Wilson, Challenges of sequencing human genomes, *Briefings Bioinf.*, 2010, **11**(5), 484–498.
75. S. Lampa, M. Dahlö, P. I. Olason, J. Hagberg and O. Spjuth, Lessons learned from implementing a national infrastructure in sweden for storage and analysis of next-generation sequencing data, *GigaScience*, 2013, **2**(1), 9.
76. S. Kim, P. A. Thiessen, E. E. Bolton, J. Chen, G. Fu and A. Gindulyte, *et al.*, Pubchem substance and compound databases, *Nucleic Acids Res.*, 2015, **44**(D1), D1202–D1213.
77. A. Gaulton, L. J. Bellis, A. P. Bento, J. Chambers, M. Davies and A. Hersey, *et al.*, ChEMBL: a large-scale bioactivity database for drug discovery, *Nucleic Acids Res.*, 2012, **40**(Database issue), D1100–D1107.
78. H. Zhu, J. Zhang, M. T. Kim, A. Boison, A. Sedykh and K. Moran, Big data in chemical toxicity research: the use of high-throughput screening assays to identify potential toxicants, *Chem. Res. Toxicol.*, 2014, **27**(10), 1643–1651.

79. W. F. P. M. Van den Hof, A. Ruiz-Aracama, A. Van Summeren, D. G. J. Jennen, S. Gaj and M. L. J. Coonen, *et al.*, Integrating multiple omics to unravel mechanisms of cyclosporin a induced hepatotoxicity in vitro, *Toxicol. In Vitro*, 2015, **29**(3), 489–501.
80. R. Cavill, D. Jennen, J. Kleinjans and J. J. Briedé, Transcriptomic and metabolomic data integration, *Briefings Bioinf.*, 2016, **17**(5), 891–901.
81. M. Bersanelli, E. Mosca, D. Remondini, E. Giampieri, C. Sala and G. Castellani, *et al.*, Methods for the integration of multi-omics data: mathematical aspects, *BMC Bioinf.*, 2016, **17**(S2), 15.
82. Q. Mo, S. Wang, V. E. Seshan, A. B. Olshen, N. Schultz and C. Sander, *et al.*, Pattern discovery and cancer gene identification in integrated cancer genomic data, *Proc. Natl. Acad. Sci. U. S. A.*, 2013, **110**(11), 4245–4250.
83. D. G. A. J. Hebels, A. Rasche, R. Herwig, G. J. P. van Westen, D. G. J. Jennen and J. C. S. Kleinjans, A systems biology approach for identifying hepatotoxicant groups based on similarity in mechanisms of action and chemical structure, in *In Silico Methods for Predicting Drug Toxicity*, 2016.
84. B.-J. Chen, H. C. Causton, D. Mancenido, N. L. Goddard, E. O. Perlstein and D. Pe'er, Harnessing gene expression to identify the genetic basis of drug resistance, *Mol. Syst. Biol.*, 2009, **5**, 310.
85. S. N. Hart, Y. Li, K. Nakamoto, E. Subileau, D. Steen and X. Zhong, A comparison of whole genome gene expression profiles of HepaRG cells and HepG2 cells to primary human hepatocytes and human liver tissues, *Drug Metab. Dispos.*, 2010, **38**, 988–994.
86. Y. Dai and A. I. Cederbaum, Cytotoxicity of acetaminophen in human cytochrome p4502e1-transfected HepG2 cells, *J. Pharmacol. Exp. Ther.*, 1995, **273**(3), 1497–1505.
87. M. J. A. Jetten, J. C. S. Kleinjans, S. M. Claessen, C. Chesné and J. H. M. van Delft, Baseline and genotoxic compound induced gene expression profiles in HepG2 and HepaRG compared to primary human hepatocytes, *Toxicol. In Vitro*, 2013, **27**, 2031–2040.
88. K. Takayama, K. Kawabata, Y. Nagamoto, K. Kishimoto, K. Tashiro and F. Sakurai, *et al.*, *Biomaterials*, 2013, **34**(7), 1781–1789.
89. S. Heydarkhan-Hagvall, G. Helenius, B. R. Johansson, J. Y. Li, E. Mattsson and B. Risberg, Co-culture of endothelial cells and smooth muscle cells affects gene expression of angiogenic factors, *J. Cell. Biochem.*, 2003, **89**(6), 1250–1259.
90. P. Beauchamp, W. Moritz, J. M. Kelm, N. D. Ullrich, I. Agarkova and B. D. Anson, *et al.*, Development and characterization of a scaffold-free 3d spheroid model of induced pluripotent stem cell-derived human cardiomyocytes, *Tissue Eng., Part C*, 2015, **21**(8), 852–861.
91. R. Bruderer, O. M. Bernhardt, T. Gandhi, S. M. Miladinović, L.-Y. Cheng and S. Messner, *et al.*, Extending the limits of quantitative proteome profiling with data-independent acquisition and application to acetaminophen-treated three-dimensional liver microtissues, *Mol. Cell. Proteomics*, 2015, **14**(5), 1400–1410.

92. J. A. Prange, M. Bieri, S. Segerer, C. Burger, A. Kaech and W. Moritz, *et al.*, Human proximal tubule cells form functional microtissues, *Pflügers Arch.*, 2016, **468**(4), 739–750.
93. S. Letzsch, K. Boettcher, J. M. Kelm and S. Messner, Quantifying efflux activity in 3d liver spheroids, *Genet. Eng. Biotechnol. News*, 2015, **35**(7), 14–15.
94. O. Frey, P. M. Misun, D. A. Fluri, J. G. Hengstler and A. Hierlemann, Reconfigurable microfluidic hanging drop network for multi-tissue interaction and analysis, *Nat. Commun.*, 2014, **5**, 4250.
95. J.-Y. Kim, D. A. Fluri, R. Marchan, K. Boonen, S. Mohanty and P. Singh, *et al.*, 3d spherical microtissues and microfluidic technology for multi-tissue experiments and analysis, *J. Biotechnol.*, 2015, **205**, 24–35.
96. J.-Y. Kim, D. A. Fluri, J. M. Kelm, A. Hierlemann and O. Frey, 96-well format-based microfluidic platform for parallel interconnection of multiple multicellular spheroids, *J. Lab. Autom.*, 2015, **20**(3), 274–282.
97. C. Oleaga, C. Bernabini, A. S. T. Smith, B. Srinivasan, M. Jackson and W. McLamb, *et al.*, Multi-organ toxicity demonstration in a functional human in vitro system composed of four organs, *Sci. Rep.*, 2016, **6**, 20030.
98. S. Ali, H. G. J. van Mil and M. K. Richardson, Large-scale assessment of the zebrafish embryo as a possible predictive model in toxicity testing, *PLoS One*, 2011, **6**(6), e21076.
99. S. H. Lam, M. M. Hlaing, X. Zhang, C. Yan, Z. Duan and L. Zhu, *et al.*, Toxicogenomic and phenotypic analyses of bisphenol-a early-life exposure toxicity in zebrafish, *PLoS One*, 2011, **6**(12), e28273.
100. L. Yang, J. R. Kemadjou, C. Zinsmeister, M. Bauer, J. Legradi and F. Müller, *et al.*, Transcriptional profiling reveals barcode-like toxicogenomic responses in the zebrafish embryo, *Genome Biol.*, 2007, **8**(10), R227.
101. D. Gusenleitner, S. S. Auerbach, T. Melia, H. F. Gómez, D. H. Sherr and S. Monti, Genomic models of short-term exposure accurately predict long-term chemical carcinogenicity and identify putative mechanisms of action, *PLoS One*, 2014, **9**(7), e102579.

CHAPTER 8

Profiling the Tox21 Chemical Library for Environmental Hazards: Applications in Prioritisation, Predictive Modelling, and Mechanism of Toxicity Characterisation

S. SAKAMURU, H. ZHU, M. XIA, A. SIMEONOV AND R. HUANG*

National Center for Advancing Translational Sciences, NIH, 9800 Medical Center Drive, Rockville, MD, 20850 USA
*E-mail: huangru@mail.nih.gov

8.1 Introduction

More than 80 000 chemicals are registered for use in the United States; for 95 percent of them, there are no data on human exposure and/or hazard to inform society about their effects on human health.[1] In addition, about 2000 new compounds are introduced into our environment every year.[1] However, traditional toxicity testing methods heavily rely on animal studies, which are low-throughput, expensive and time-consuming. It is impossible to

Issues in Toxicology No. 41
Big Data in Predictive Toxicology
Edited by Daniel Neagu and Andrea-Nicole Richarz
© The Royal Society of Chemistry 2020
Published by the Royal Society of Chemistry, www.rsc.org

evaluate the *in vivo* toxicity of the fast-growing number of compounds using traditional toxicity testing methods. To overcome these limitations, the U.S. Tox21 (Toxicology in the 21st Century) program was established in 2008 with an emphasis on developing new methodology to evaluate the potential risk of thousands of environmental chemicals on human health.[2–5] The Tox21 program is a collaboration involving the National Toxicology Program (NTP) at the National Institute of Environmental Health Sciences (NIEHS), the National Center for Computational Toxicology (NCCT) at the US Environmental Protection Agency (EPA), the intramural research division of the National Center for Advancing Translational Sciences (NCATS), and the US Food and Drug Administration (FDA).

The Tox21 program aims to:

1. analyse the patterns of compound-induced biological responses in order to identify toxicity pathways and compound mechanism of toxicity;
2. prioritise compounds for further extensive toxicological evaluation;
3. develop predictive models for biological response in human beings.

The ultimate goal of the Tox21 program is to identify *in vitro* chemical signatures that could act as predictive surrogates for *in vivo* toxicity.

This inter-federal agency partnership integrates the expertise from different areas to profile a library of ~10 000 environmental chemicals (Tox21 10K library) for their potential effect on human health. The Tox21 10K compound collection is screened against a large panel of cell-based assays in a quantitative high-throughput screening (qHTS) format as a 15 pt titration series.[6] qHTS tests compounds in titration series and thus generates a concentration–response curve for each compound, which greatly reduces the frequency of false positives and negatives. During phase II (2011–2016), the production phase, the Tox21 10K library has been screened against a panel of 47 nuclear receptor (NR)[7–10] and stress response (SR) pathway assays (see Table 8.1)[11,12] in qHTS format in triplicate producing over 70 million data points.[13,14] Here, we describe the workflow (Figure 8.1) employed by the Tox21 program to generate this large, robust data set, and a few examples on how to use this data set to build *in silico* models to predict *in vivo* toxicity, generate hypotheses on mechanism of compound toxicity, and prioritise chemicals for further advanced toxicological studies.

8.2 The Tox21 Process

8.2.1 Tox21 Compound Collection

The Tox21 library consists of over 10 000 environmental chemicals (approximately 8599 unique chemical entities) representing a wide range of structural diversity. The main criteria for selection of the Tox21 compounds included, but were not limited to, known or perceived environmental hazards or exposure

Table 8.1 Tox21 phase II assays.

Assay	Assay name	PubChem AID	Reproducibility score
Stress Response Pathway Assays			
ATAD5	ATAD5	651632, 720516, 651634	99.51
DT40 Rad54/Ku70	DNA repair Rad54/Ku70 deficient isogenic chicken DT40 cell viability	70743015	90.61
DT40 Rev3	DNA repair Rev3 deficient isogenic chicken DT40 cell viability	743014	83.02
DT40 WT	Wild type chicken DT40 cell viability	743015	79.58
H2AX	H2AX	1224845, 1224847	93.71
P53	P53	651631, 720552, 651633	103.02
AP-1-BLA	Activator protein-1	1159526, 1159525, 1159528	91.65
ARE	Antioxidant responsive element	743202, 743219, 743203	81.85
HRE-BLA agonist	Hypoxia-inducible factor-1	1224846, 1224844	96.96
HSE-BLA	Heat shock factor response element	743210, 743228, 743209	95.46
NFκB-BLA agonist	Nuclear factor-kappa B	1159509, 1159515, 1159518	100.31
ER stress	Endoplasmic reticulum stress	1159516, 1159517, 1159519	92.97
Mitochondria toxicity	Mitochondrial toxicity	720635, 720637, 720634	87.20
Nuclear Receptor Assays			
Aromatase	Aromatase	743083, 743139, 743084	93.66
AhR	Aryl hydrocarbon receptor	743085, 743122, 743086	84.05
AR-BLA agonist	Androgen receptor, ligand binding domain, agonist mode	743036, 743053	89.55
AR-BLA antagonist	Androgen receptor, ligand binding domain, antagonist mode	743035, 743063, 743033	96.44
AR-MDA agonist	Androgen receptor, full length, agonist mode	743040	99.74
AR-MDA antagonist	Androgen receptor, full length, antagonist mode	743042, 743054, 743041	92.28
CAR agonist	Constitutive androstane receptor, agonist mode	1224839, 1224836	99.93
CAR antagonist	Constitutive androstane receptor, antagonist mode	1224838, 1224837	89.24
ER-BG1 agonist	Estrogen receptor alpha, full length, agonist mode	743079	91.46
ER-BG1 antagonist	Estrogen receptor alpha, full length, antagonist mode	743080, 743091, 743081	95.25
ER-BLA agonist	Estrogen receptor alpha, ligand binding domain, agonist mode	743075, 743077	95.25
ER-BLA antagonist	Estrogen receptor alpha, ligand binding domain, antagonist mode	743069, 743078, 743074	84.90
ERR	Estrogen related receptor	1224834, 1224849, 1224848	95.66

ERR-PGC	Estrogen related receptor with the pleiotropic PPAR-gamma coactivator (PGC)	1224842, 1224841, 1224840	93.35
FXR-BLA agonist	Farnesoid X receptor, agonist mode	743220, 743239, 743218	95.09
FXR-BLA antagonist	Farnesoid X receptor, antagonist mode	743217, 743240, 743221	88.48
GR-BLA agonist	Glucocorticoid receptor, agonist mode	720691, 720719	94.89
GR-BLA antagonist	Glucocorticoid receptor, antagonist mode	720692, 720725, 720693	77.40
PPAR-delta-BLA agonist	Peroxisome proliferator-activated receptor delta, agonist mode	743212, 743227, 743211	91.91
PPAR-delta-BLA antagonist	Peroxisome proliferator-activated receptor delta, antagonist mode	743215, 743226, 743213	90.37
PPAR-gamma agonist	Peroxisome proliferator-activated receptor gamma, agonist mode	743094, 743140	93.81
PPAR-gamma antagonist	Peroxisome proliferator-activated receptor gamma, antagonist mode	743191, 743199, 743194	85.15
RAR agonist	Retinoic acid receptor, agonist mode	1159553	89.15
RAR antagonist	Retinoic acid receptor, antagonist mode	1159552	71.50
ROR antagonist	Retinoid-related orphan receptor gamma, antagonist mode	1159521	72.95
RXR-BLA agonist	Retinoid X receptor, agonist mode	1159527, 1159529, 1159531	79.20
TR-beta agonist	Thyroid hormone receptor, agonist mode	743066	87.84
TR-beta antagonist	Thyroid hormone receptor, antagonist mode	743065, 743067, 743064	87.82
VDR-BLA agonist	Vitamin D receptor, agonist mode	743222, 743241, 743224	91.79
VDR-BLA antagonist	Vitamin D receptor, antagonist mode	743223, 743242, 743225	88.97
Other Assays			
TSHR agonist	Thyroid stimulating hormone receptor, agonist mode	1224843	90.49
Real time viability (HepG2)	Real time cell viability in HepG2 cells	1224882, 1224876, 1224878, 1224883, 1224879, 1224890, 1224889, 1224885, 1224867, 1224877, 1224870, 1224873	N/A
Real time viability (HEK293)	Real time cell viability in HEK293 cells	1224869, 1224888, 1224875, 1224881, 1224871, 1224884, 1224880, 1224872, 1224886, 1224868, 1224874, 1224887	N/A
Luciferase biochemical	Firefly luciferase, biochemical	1224835	99.97

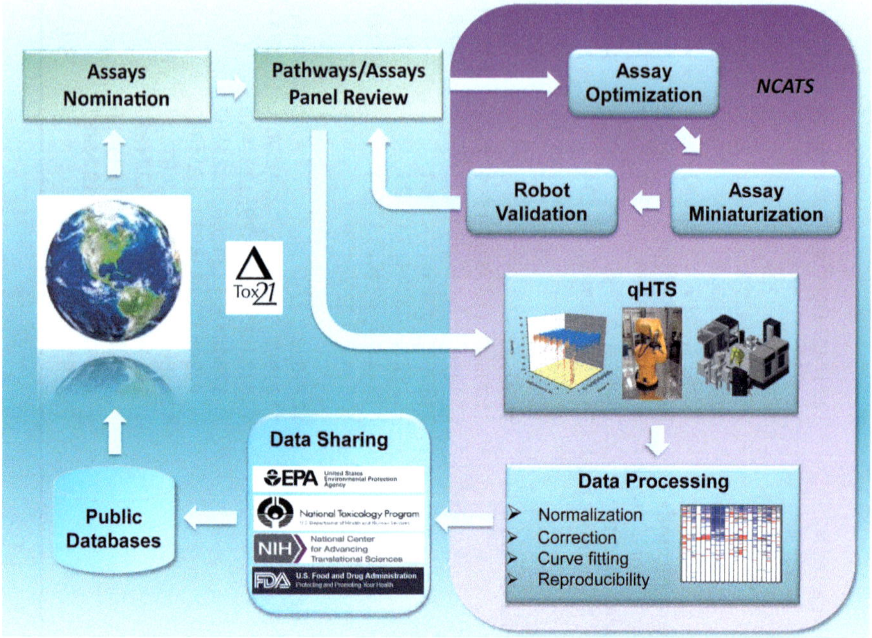

Figure 8.1 The Tox21 process. Assays nominated by researchers are reviewed and approved by the Tox21 Pathways/Assays group and the Tox21 leadership. The selected assays are first optimised, validated and miniaturised into a 1536-well plate format. After robotic validation, the Tox21 10K library is run in a qHTS format on the robotic platform. The qHTS data are then processed at NCATS and the screening results are first shared among the Tox21 partners, and then made publicly accessible.

concerns, physicochemical properties (molecular weight, volatility, solubility, logP) suitability for qHTS, commercial availability, and cost. The types of compounds selected for constructing the library include pesticides, food additives, flame retardants, industrial chemicals, drinking water disinfection by-products, preservatives, household cleaning agents, and drugs (Figure 8.2). In addition, a set of 88 diverse compounds was included as duplicates in every screening plate and designated as internal controls for assay reproducibility.[15]

Three Tox21 partners – EPA, NTP, and NCATS – each contributed over 3000 physical compounds to the collection primarily procured from commercial sources. All of the compounds in the Tox21 10K collection are currently going through analytical quality control (QC) that tests for chemical purity and identity. The structures and annotations of the Tox21 10K library as well as the QC results have been made publicly available.[16,17] The compound library was plated in 1536-well plates as 15 concentration series with √5-fold dilution in dimethylsulfoxide (DMSO), covering a concentration range up to four orders of magnitude. To assess the reproducibility of data, three physical copies of the library were prepared in three different formats, in which the same compounds were plated in a different well location in each copy.

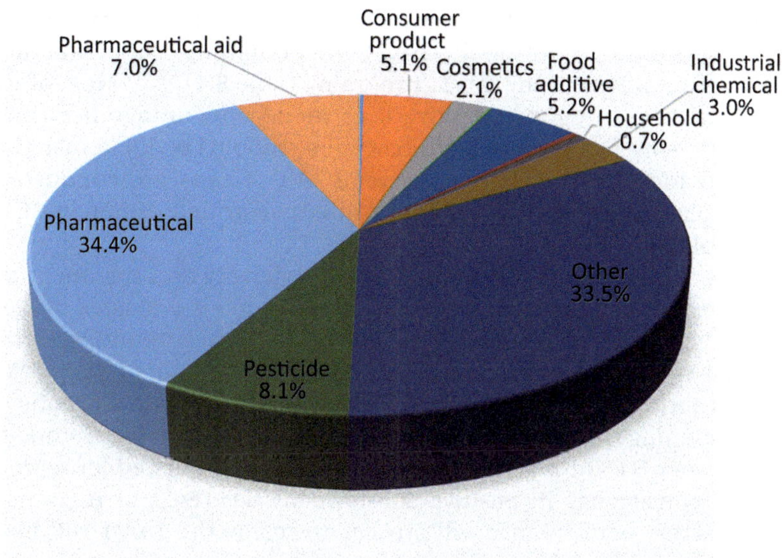

Figure 8.2 Composition of the Tox21 10K compound library.

8.2.2 Tox21 Robotic Platform

The Tox21 screens were performed on a dedicated Tox21 robotic system consisting of a high-precision robotic arm surrounded by assay plate incubators, compound storage units, liquid dispensers, Pintool work station and microplate readers.[15] The assay plates are stored in the incubators that are equipped with temperature, CO_2, and humidity controls. Compound addition to assay plates is performed as a 23 nL transfer from 1536-well compound plates to 1536-well assay plates by a Pintool workstation (Wako Automation, CA, USA). Two types of liquid dispensers are integrated on the system: one is a Bioraptr Flying Reagent Dispenser (FRD; Beckman Coulter, Inc. CA, USA), which allows for a liquid transfer of 0.2–10 μL to 1536-well assay plates, and it is equipped with four different tips to dispense four different reagents simultaneously, and the second one is a Multidrop Combi (Thermo Scientific, MA, USA), which is capable of dispensing one reagent at high speed using eight tips. For the final detection of signals, the system uses two types of detectors, ViewLux and Envision plate readers (PerkinElmer, MA, USA), which are capable of measuring absorbance, fluorescence, and luminescence.

8.2.3 Tox21 Screening Process

A battery of cell-based assays (https://tripod.nih.gov/tox21/assays/),[18] including phenotypic assays (*e.g.*, cell viability, DNA damage, mitochondrial toxicity), target-specific assays (*e.g.*, nuclear receptors) and pathway-based assays (*e.g.*, Nrf2/ARE, HIF-1α, p53), has been developed and validated in 1536-well format at NCATS.[5,19] Nearly 50 of these assays have been selected, with a primary focus on nuclear receptors and stress response pathways, to profile the Tox21 10K compound collection during the production phase of the Tox21 program (Table 8.1).[7,9–11] Most of these assays were run in agonist (activation) mode and antagonist (inhibition) mode using β-lactamase (fluorescence readout) or luciferase (luminescence readout) reporter gene systems. All screens are conducted in 1536-well plates, with initial optimisations performed either in 384- or 1536-well plates.

The selected assay is first optimised for cell density and treatment duration using appropriate positive controls. Optimisation is evaluated by performance statistics including high (>two-fold) signal to background ratios (S/B), low (<10%) coefficient of variations (CV) and high *Z*-factors (>0.5).[20] When all these criteria are met, the assay will then be validated on a robotic platform by screening the Library of 1280 Pharmacologically Active Compounds ($LOPAC^{1280}$) in qHTS format. The validation data are evaluated for reproducibility and consistency in positive control activities (EC_{50} or IC_{50}). Assays with satisfactory performance will proceed to screen the Tox21 10K library, where all the compounds are tested in triplicate in three separate runs with each run conducted on a different day. All assays are multiplexed with cell viability or cytotoxicity counter screens in order to distinguish true positive inhibitors from cytotoxicity artefacts, especially for antagonist mode assays. Usually, β-lactamase assays are multiplexed with a luminescence-based viability assay (CellTiter-Glo Viability assay, Promega) and luciferase assays are multiplexed with a fluorescence-based viability assay (CellTiter-Fluor Viability assay, Promega).

8.2.4 Tox21 Data Analysis Process Including Data Quality Control

Screening the Tox21 10K library during phase II of the Tox21 program has generated a large amount of data. Counter screens are also employed to minimise interferences from off-target, assay-specific artefacts, such as compound auto fluorescence and cytotoxicity. New data analysis approaches are needed to integrate these data and characterise the activities observed from these assays. A standardised qHTS data analysis process[21] (Figure 8.3) has been developed at NCATS:

(1) raw data processing;
(2) concentration response curve fitting and classification;
(3) data reproducibility evaluation;
(4) assignment of activity to compounds.

During the initial evaluation process, quality metrics, such as CV, S/B, and *Z*-factor, from each plate[20] are calculated using raw fluorescence or luminescence reads to evaluate gross assay performance. "Failed plates," identified by abnormally poor values, are further inspected manually and, if necessary, excluded from further data analysis. After the initial evaluation process, raw

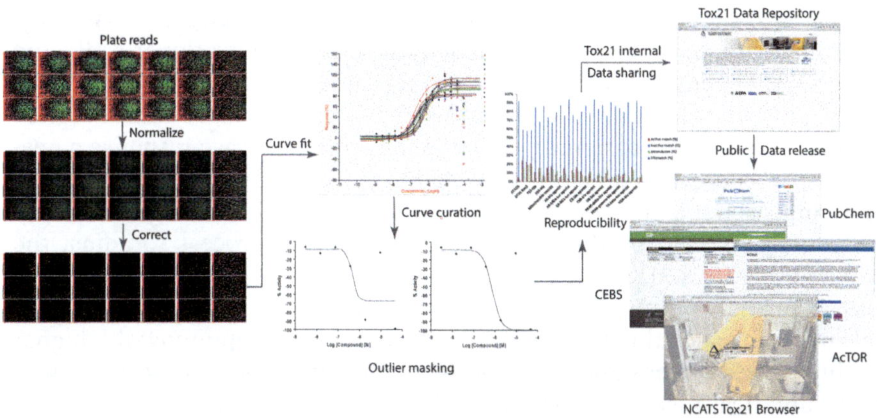

Figure 8.3 Tox21 data pipeline. The raw qHTS plate reads are first normalised, corrected and pivoted to form concentration–response curves (CRCs). The CRCs are then fit to the Hill equation and classified into different curve classes based on potency, efficacy and quality of curve fit. The curve fits are further inspected manually to correct any misfit curves. The "clean" curve fitting results are then assessed for activity reproducibility to determine the final assay performance. After the initial data parsing and assessment at the NCATS are complete, the qHTS data are shared with the Tox21 partners where the data are further scrutinised for quality and utility. The data will then be released to the public domain in a number of public databases including PubChem, CEBS and ACToR.

plate reads are first normalised to the positive control (100% for agonist mode and −100% for antagonist mode) and negative control (DMSO) (0% for both agonist and antagonist modes) as follows:

$$\% \text{ Activity} = ((V_{\text{cmpd}} - V_{\text{neg}})/(V_{\text{pos}} - V_{\text{neg}})) \times 100 \quad (8.1)$$

where V_{cmpd} denotes the compound well values, V_{pos} denotes the median value of the positive control wells, and V_{neg} denotes the median values of the negative control, and then corrected using compound-free control plates (*i.e.*, DMSO-only plates) at the beginning and end of the compound plate stack to remove background patterns and subtle abnormalities such as tip effects or blotting from cell dispenses.[22]

Corrected plate data are pivoted to form concentration–response series, which are subsequently fit to a four-parameter Hill equation,[23] yielding concentrations of half-maximal activity (AC_{50}) and maximal response (efficacy) values.[24] Concentration–response curves are designated as Class 1–4 based on efficacy, the number of data points observed above background activity, and the quality of fit.[6] Curve classes are heuristic measures of data confidence. The qHTS curve classification scheme has been recently amended to better suit the needs of toxicology research:[8] The most problematic concentration responses are automatically assigned curve class 5 based on considerations like the direction of activity (observing alternately both increases and decreases in signal over a short concentration range) and unusually large signal at low sample concentrations (activity at zero concentration is estimated to be >3SD of control).[8] Class 5 curves and other cases in which an inconsistency between the highest compound activity and the curve class assigned is identified, such as assigning a compound with a positive response a negative curve class, are manually inspected to correct the curve class, if necessary. Adjustment is normally done by masking or by unmasking data points improperly masked by the automated curve fitting process to adjust the curve fit (Figure 8.3). To facilitate analysis and activity profiling, each curve class is further combined with an efficacy cutoff and converted to a numerical curve rank such that more potent and efficacious compounds with higher quality curves are assigned a higher rank.[8] Curve ranks should be viewed as a numerical measure of compound activity. After manual curation, the "clean" curve fitting results from the replicate assay runs are assessed for activity reproducibility to determine the final assay performance. Each assay is assigned a score based on the reproducibility of the data (Table 8.1).[8]

In fluorescence-based agonist mode assays, autofluorescent compounds can produce the same phenotype as agonists.[25] Several approaches are used to identify potential autofluorescent artefacts to distinguish them from true agonists, such as using the autofluorescence detection counter screen or by examining the activity of each compound in all the assays screened having the same reporter. The compounds identified by either approach are flagged as potential false positive due to autofluorescence. The same methods could be applied to luciferase reporter gene assays to identify compounds that are promiscuous luciferase stabilisers. Luciferase counter screens are also currently underway as an auxiliary approach to identify such compounds in the Tox21 10K library. In antagonist mode assays, cytotoxic compounds can show the same inhibitory phenotype as those of true antagonists. For this reason, each antagonist mode assay screened for Tox21 is accompanied with a cell viability readout that serves as the counter screen to distinguish true antagonists from cytotoxicity artefacts. Compounds that show no significant difference in activities between the antagonist mode readout and the cell viability readout are flagged as potential cytotoxicity-related false positive responses. The final activity outcome of a compound is determined based on its reproducibility and multi-channel activity. For antagonist mode assays, the cell viability counter screen data are used to flag potential cytotoxic artefacts. For agonist mode assays, potential artefacts produced by compounds that autofluoresce in the signal channel (*e.g.*, 460 nm readout in the ER-bla assay[7]) are flagged using both the compound autofluorescence profiling data and the promiscuous compound activity shown in the signal readout of all the assays screened in Tox21 that have the same reporter (*e.g.*, β-lactamase).

As soon as the initial data parsing and assessment at the NCATS are complete, the concentration response data, curve fitting results, the raw plate reads, the assay conditions, and the sample mapping information are shared with the Tox21 partners through a suite of databases and software tools custom built by the NCATS for the Tox21 program (https://tripod.nih.gov/tox/).

Within the first six months of data generation, the assay data are only made available to the Tox21 partners through the aforementioned controlled-access site in which the data are further scrutinised for quality and utility. The data are then released to the public domain in a number of public databases (Figure 8.2) including.

- PubChem (http://pubchem.ncbi.nlm.nih.gov/),
- The NCATS Tox21 Browser (https://tripod.nih.gov/tox21/),
- The NIEHS Chemical Effects in Biological Systems (CEBS) database (http://tools.niehs.nih.gov/cebs3/ui/) and
- EPA's Aggregated Computational Toxicology Online Resource (ACToR) (http://actor.epa.gov).

In phase II, by 2016, over 70 million data points generated from nearly 50 assays on the Tox21 10K collection have been deposited in PubChem (see Table 8.1 for a list of PubChem assay IDs).[13]

8.3 Using the Tox21 Big Data Collection for Predictive Modelling

8.3.1 Profiling for Compound Mechanism of Toxicity

One of the Tox21 goals is to identify patterns of compound-induced biological response to better understand the compounds' mechanisms of toxicity. The Tox21 10K collection screened against around 50 assays provides a rich and robust set of compound activity profiles that could serve this purpose. Previous studies with the data generated from screening the Tox21 Phase I collection[26] of approximately 3000 compounds against a panel of cell viability and caspase-3/7 assays have shown that these compound activity profiles or signatures are useful for hypotheses generation on compound mode of action (MoA).[27] To evaluate the utility of the Tox21 10K data toward this goal, the 10K compounds were clustered by their activity profile[13] similarity using the self-organising map (SOM) algorithm.[28] Each cluster was evaluated for enrichment of Medical Subject Headings (MeSH)[29] pharmacological action (PA) terms using the Fisher's exact test. This result indicates that compounds with similar activity profiles tend to share similar MoAs. Examples include clusters of compounds that are cardiotonic agents, including deslanoside, digitoxin, digoxin, ouabain and proscillaridin, which are cardiac glycosides that act by inhibiting Na/K channels;[30] statins, including atorvastatin, fluvastatin and cerivastatin, which are hydroxymethylglutaryl-coenzyme A reductase inhibitors;[31] and antineoplastic antimetabolites.[32] Estrogenic compounds belonging to different structural classes were found in neighbouring clusters segregated by structural class. One cluster is enriched with steroid hormones, such as 17-alpha-ethinylestradiol, and non-steroidal estrogens such as diethylstilbestrol and zearalenone. Bisphenol-type compounds, which are also known estrogenic compounds,[33] such as bisphenols A, B, Z and AF, were found in a neighbouring cluster with similar activity profiles.

Many clusters were found to contain both compounds with known MeSH PA and compounds with no MeSH PA annotation available.[13] If compounds in the same cluster share similar MoAs, then we can use this information to hypothesise on the MoA of the unknown compounds. The pesticide fludioxonil does not have a MeSH PA assigned but co-clustered with the estrogenic compounds, consistent with recent reports on the endocrine disrupting activity of fludioxonil.[34] The MeSH term "anti-inflammatory agents" is significantly enriched in a cluster of glucocorticoids. The other compounds in this cluster do not have MeSH PA annotations, including metformin, a diabetes drug that has a completely distinct structure. Metformin is known to decrease hyperglycemia primarily by suppressing glucose production by the liver (hepatic gluconeogenesis)[35] with its exact molecular mechanism not clearly understood. The co-clustering of metformin with the anti-inflammatory glucocorticoids indicates that metformin may act in a fashion similar to the glucocorticoids. Ampiroxicam is another non-steroidal, anti-inflammatory drug found in the same cluster without a MeSH PA annotation,[36] supporting the utility of clustering by activity profile as an indicator for compound MoA. These results show that the Tox21 10K activity profiles are useful for compound MoA hypotheses generation. Compounds with unknown MoA that share similar profiles with compounds with known MoA could be prioritised and tested for that hypothesised MoA.

8.3.2 Modelling Tox21 Data for *In Vivo* Toxicity Prediction

The Tox21 10K data were also evaluated for their utility in realising another Tox21 goal – serving as *in vitro* signatures that are predictive of *in vivo* toxicity.[13] For this purpose, 72 *in vivo* toxicity endpoints (Table 8.2) were derived from the Registry of Toxic Effects of Chemical Substances (RTECS) database compiled by Leadscope® (Leadscope, Inc. Columbus, Ohio, USA). Models were built for the 72 *in vivo* toxicity endpoints using either the structure (structure-based models) or assay activity (activity-based models) in self-organising map (SOM) clusters, or both (combined models). Model performance was measured by the area under the receiver operating characteristic (ROC) curve (AUC-ROC).[37] The premise for these models is that compounds that share similar *in vitro* signatures and/or structure features are likely to show similar *in vivo* effects. There are five human toxicity endpoints, including standard Draize test for human skin irritation, multiple dose toxicity data (TDLo) through oral exposure from human females, human males, and humans (gender not specified), and reproductive toxicity data (TDLo) through oral exposure from human females. The activity-based models built for the human endpoints performed significantly better than the models of the animal (mouse/rat/rabbit) toxicity endpoints. The

Table 8.2 Performance of *in vivo* toxicity prediction models measured by AUC-ROC values.

In vivo endpoint	Activity model	Structure model	Combined model
Acute toxicity LC50 Mouse Inhalation	0.55 ± 0.05	0.74 ± 0.04	0.84 ± 0.14
Acute toxicity LC50 Rat Inhalation	0.55 ± 0.03	0.68 ± 0.04	0.7 ± 0.12
Acute toxicity LC50 Rat Inhalation	0.56 ± 0.07	0.63 ± 0.06	0.69 ± 0.21
Acute toxicity LD50 Bird Oral	0.56 ± 0.1	0.68 ± 0.07	N/A
Acute toxicity LD50 Cat Oral	0.7 ± 0.13	0.86 ± 0.07	N/A
Acute toxicity LD50 Chicken Oral	0.74 ± 0.11	0.83 ± 0.05	0.9 ± 0.14
Acute toxicity LD50 Dog Injection intravenous	0.63 ± 0.07	0.93 ± 0.04	0.93 ± 0.05
Acute toxicity LD50 Dog Oral	0.59 ± 0.05	0.82 ± 0.03	0.95 ± 0.02
Acute toxicity LD50 Duck Oral	0.65 ± 0.11	0.86 ± 0.05	0.9 ± 0.15
Acute toxicity LD50 Guinea pig Intraperitoneal	0.67 ± 0.13	0.89 ± 0.04	0.94 ± 0.07
Acute toxicity LD50 Guinea pig Oral	0.64 ± 0.03	0.8 ± 0.03	0.87 ± 0.06
Acute toxicity LD50 Mammal	0.59 ± 0.09	0.92 ± 0.03	N/A
Acute toxicity LD50 Mammal Oral	0.58 ± 0.08	0.84 ± 0.07	N/A
Acute toxicity LD50 Mouse Injection intravenous	0.64 ± 0.02	0.85 ± 0.01	0.87 ± 0.03
Acute toxicity LD50 Mouse Injection subcutaneous	0.62 ± 0.02	0.75 ± 0.01	0.82 ± 0.03
Acute toxicity LD50 Mouse Intramuscular	0.82 ± 0.06	0.86 ± 0.06	0.84 ± 0.1
Acute toxicity LD50 Mouse Intraperitoneal	0.61 ± 0.01	0.73 ± 0.01	0.8 ± 0.02
Acute toxicity LD50 Mouse	0.67 ± 0.05	0.81 ± 0.03	0.89 ± 0.07
Acute toxicity LD50 Mouse Oral	0.59 ± 0.01	0.71 ± 0.01	0.8 ± 0.02
Acute toxicity LD50 Quail Oral	0.55 ± 0.08	0.82 ± 0.05	0.67 ± 0.22
Acute toxicity LD50 Rabbit Injection intravenous	0.61 ± 0.06	0.89 ± 0.03	0.91 ± 0.06
Acute toxicity LD50 Rabbit Oral	0.61 ± 0.03	0.77 ± 0.03	0.89 ± 0.06
Acute toxicity LD50 Rabbit Skin	0.55 ± 0.03	0.72 ± 0.03	0.77 ± 0.15
Acute toxicity LD50 Rat Injection intravenous	0.68 ± 0.02	0.88 ± 0.01	0.9 ± 0.03
Acute toxicity LD50 Rat Injection subcutaneous	0.61 ± 0.02	0.76 ± 0.02	0.81 ± 0.03

(*continued*)

Table 8.2 (*continued*)

In vivo endpoint	Activity model	Structure model	Combined model
Acute toxicity LD50 Rat Intramuscular	0.72 ± 0.08	0.86 ± 0.04	0.97 ± 0.04
Acute toxicity LD50 Rat Intraperitoneal	0.65 ± 0.02	0.74 ± 0.01	0.82 ± 0.03
Acute toxicity LD50 Rat	0.58 ± 0.05	0.78 ± 0.04	0.84 ± 0.09
Acute toxicity LD50 Rat Oral	0.58 ± 0.01	0.72 ± 0.01	0.76 ± 0.02
Acute toxicity LD50 Rat Skin	0.6 ± 0.04	0.72 ± 0.05	0.73 ± 0.11
Hepatotoxicity	0.55 ± 0.02	0.63 ± 0.02	0.69 ± 0.04
Irritation standard Draize test Human Skin	0.89 ± 0.05	0.84 ± 0.04	0.8 ± 0.09
Irritation standard Draize test Rabbit Eyes	0.51 ± 0.03	0.71 ± 0.03	0.74 ± 0.07
Irritation standard Draize test Rabbit Skin	0.51 ± 0.03	0.72 ± 0.02	0.76 ± 0.07
Multiple dose TCLo Rat Inhalation micro	0.55 ± 0.04	0.65 ± 0.04	0.78 ± 0.11
Multiple dose TCLo Rat Inhalation	0.5 ± 0.05	0.59 ± 0.06	N/A
Multiple dose TDLo Dog Injection intravenous	0.9 ± 0.06	0.89 ± 0.05	0.97 ± 0.04
Multiple dose TDLo Dog Oral	0.63 ± 0.03	0.72 ± 0.03	0.83 ± 0.04
Multiple dose TDLo Human Female Oral	0.64 ± 0.06	0.85 ± 0.06	0.97 ± 0.03
Multiple dose TDLo Human Male Oral	0.76 ± 0.06	0.8 ± 0.05	0.93 ± 0.06
Multiple dose TDLo Human Oral	0.72 ± 0.05	0.75 ± 0.04	0.95 ± 0.03
Multiple dose TDLo Monkey Oral	0.59 ± 0.13	0.93 ± 0.04	0.99 ± 0.08
Multiple dose TDLo Mouse Injection subcutaneous	0.74 ± 0.08	0.78 ± 0.05	0.96 ± 0.05
Multiple dose TDLo Mouse Intraperitoneal	0.73 ± 0.05	0.83 ± 0.03	0.91 ± 0.03
Multiple dose TDLo Mouse Oral	0.63 ± 0.02	0.67 ± 0.02	0.78 ± 0.04
Multiple dose TDLo Rabbit Oral	0.64 ± 0.07	0.74 ± 0.06	0.65 ± 0.14
Multiple dose TDLo Rat Injection intravenous	0.84 ± 0.06	0.9 ± 0.04	0.94 ± 0.04
Multiple dose TDLo Rat Injection subcutaneous	0.7 ± 0.04	0.78 ± 0.03	0.79 ± 0.05
Multiple dose TDLo Rat Intramuscular	0.9 ± 0.08	0.89 ± 0.05	0.97 ± 0.04
Multiple dose TDLo Rat Intraperitoneal	0.63 ± 0.03	0.71 ± 0.03	0.81 ± 0.05
Multiple dose TDLo Rat	0.62 ± 0.09	0.75 ± 0.05	0.89 ± 0.1
Multiple dose TDLo Rat Oral	0.56 ± 0.01	0.64 ± 0.01	0.71 ± 0.03
Multiple dose TDLo Rat Skin	0.71 ± 0.08	0.86 ± 0.06	0.89 ± 0.14

Reproductive TDLo Human Female Oral	0.74 ± 0.09	0.8 ± 0.06	0.62 ± 0.15
Reproductive TDLo Mouse Injection subcutaneous	0.67 ± 0.05	0.87 ± 0.03	0.91 ± 0.04
Reproductive TDLo Mouse Intraperitoneal	0.62 ± 0.06	0.81 ± 0.04	0.87 ± 0.07
Reproductive TDLo Mouse Oral	0.73 ± 0.03	0.77 ± 0.03	0.91 ± 0.04
Reproductive TDLo Rabbit Oral	0.68 ± 0.05	0.79 ± 0.04	0.92 ± 0.04
Reproductive TDLo Rat Injection intravenous	0.67 ± 0.1	0.87 ± 0.05	0.89 ± 0.09
Reproductive TDLo Rat Injection subcutaneous	0.66 ± 0.05	0.81 ± 0.03	0.8 ± 0.05
Reproductive TDLo Rat Intraperitoneal	0.66 ± 0.05	0.8 ± 0.04	0.9 ± 0.04
Reproductive TDLo Rat	0.85 ± 0.07	0.86 ± 0.06	0.83 ± 0.09
Reproductive TDLo Rat Oral	0.62 ± 0.02	0.7 ± 0.02	0.82 ± 0.03
Tumorigenic TD Mouse Oral	0.61 ± 0.09	0.92 ± 0.04	0.95 ± 0.07
Tumorigenic TD Rat Oral	0.52 ± 0.09	0.77 ± 0.05	0.59 ± 0.21
Tumorigenic TDLo Mouse Injection subcutaneous	0.7 ± 0.1	0.72 ± 0.07	0.69 ± 0.12
Tumorigenic TDLo Mouse Oral	0.58 ± 0.04	0.75 ± 0.04	0.92 ± 0.05
Tumorigenic TDLo Mouse Skin	0.69 ± 0.1	0.86 ± 0.04	0.83 ± 0.19
Tumorigenic TDLo Rat Oral	0.55 ± 0.05	0.67 ± 0.03	0.93 ± 0.04
Acute toxicity	0.57 ± 0.01	0.73 ± 0.01	0.72 ± 0.01
Reproductive	0.62 ± 0.01	0.7 ± 0.01	0.78 ± 0.02
Tumorigenic	0.56 ± 0.03	0.69 ± 0.02	0.83 ± 0.04

compound structure-based models showed overall better performance than the activity-based models with an average AUC-ROC of 0.78, underlying the ongoing need to further expand the battery of *in vitro* assays. However, the performance of the structure-based models did not show any species difference. Compared to the activity-based models, the performance improved significantly for the mouse/rat/rabbit toxicity models, but not as significantly for the human toxicity models. Combining the compound activity and structure data resulted in significantly better models, with an average AUC-ROC of 0.84, than the models built with activity or structure alone. Similar to the structure-based models, the species difference between the model performances disappeared.

The models built with the current set of Tox21 assay activity profiles showed reasonable but less than ideal performance for most *in vivo* toxicity endpoints, and appeared to be less predictive than the models built with chemical structures. As the current Tox21 assays focused primarily on nuclear receptor signalling and stress response pathways, all aspects of biology involved in toxic response are not covered sufficiently, indicating the need to expand the coverage of the biological space by including assays that target additional pathways relevant for toxicity in the continuation of the Tox21 program. In the meantime, Tox21 is transitioning into Phase III, where challenges such as assay pathway coverage and metabolic capacity of cells will be addressed. On the other hand, the assay data-based models performed markedly better in predicting human toxicity, though just a few endpoints, than animal toxicity, whereas the structure-based models did not show this species selectivity. As all of the Tox21 assays screened so far are cell-based assays constructed from human cells or cell lines, the species difference may be a significant contributing factor to the less-than-ideal performance of the activity-based models on predicting animal toxicity endpoints. New drugs that passed animal testing are often known to fail in human clinical trials because of lack of effect or unexpected toxicity.[38] Other studies also show that animal data only predicted human outcomes approximately half of the time.[39] This exemplifies the need to have *in vivo* human toxicity data, *i.e.* clinical toxicity data presently not readily available to the public, in order to better assess the predictive value of the human *in vitro* assay data. Nevertheless, combining activity data with structure information significantly improved the model performance for most of the *in vivo* endpoints (Table 8.2), which serves as a validation of the value of the *in vitro* assay data when applied to *in vivo* toxicity prediction. The model performance in addition highlights the importance of data quality. The Tox21 *in vitro* qHTS data have shown good reproducibility (Table 8.1).[13] The *in vivo* data used in modelling were also evaluated for reproducibility, and a significant positive correlation was found between the reproducibility of the *in vivo* endpoint and the performance of the model built for that endpoint, suggesting that the performance of *in vivo* toxicity prediction models could be further improved if better quality *in vivo* data were available.

8.4 The Tox21 Data Challenge – New Methods for Data Modelling in the Big Data Era

The high-quality concentration response data generated by the Tox21 program on a wide spectrum of pathways and phenotypic toxicity endpoints provide a valuable resource for predictive toxicity modelling. These data can not only serve as *in vitro* signatures that could be used to predict *in vivo* toxicity endpoints[13,40,41] and to prioritise chemicals for more in depth toxicity testing[42] that help to fulfil the Tox21 goals, but also serve as a knowledge base to correlate chemical structures to their biological activities for the QSAR (quantitative structure–activity relationship) modelling community to build more robust models.[43,44] There is no human exposure and/or hazard data for 95 percent of the >80 000 chemicals registered for use in the U.S. to inform society about their potential toxic effects.[1] *In silico* approaches, such as QSAR models that infer biological activity from chemical structure similarity, provide a viable alternative to fill in the experimental data gap.[45,46]

To encourage the mining and usage of the Tox21 data, NCATS launched the Tox21 Data Challenge 2014[47,48] to "crowdsource" data analysis by independent researchers to obtain new models and methods that can predict the potential toxicity of compounds by disrupting cellular and biochemical pathways using chemical structure data. The Tox21 10K qHTS data from 12 assays, seven nuclear receptor and five stress response pathway assays, were selected based on data quality and hit rate for the challenge. The competition attracted 125 participants representing 18 different countries, with 378 model submissions from 40 teams received for final evaluation.

The winning models all achieved >80% accuracy. Several models exceeded 90% accuracy. The high-quality winning models serve as a confirmation of the ability of computational approaches to provide meaningful predictions of toxicity responses in terms of pathway disruption upon environmental compound exposure, and also as a validation of the quality of datasets produced from the Tox21 qHTS assays. Consensus models constructed by combining the individual models from all participating teams resulted in improved predictive performance, with some outperforming the winning models, showing the wisdom of the crowd. The winning models (Table 8.3) as well as methods used by some other challenge participants have been published in a special issue of the journal *Frontiers in Environmental Science* as Research Topic "Tox21 Challenge to Build Predictive Models of Nuclear Receptor and Stress Response Pathways as Mediated by Exposure to Environmental Toxicants and Drugs" (http://journal.frontiersin.org/researchtopic/2954/tox21-challenge-to-build-predictive-models-of-nuclear-receptor-and-stress-response-pathways-as-media).[49] A number of new methods, such as associative neural networks (ASNN) with stratified bagging[50] and multi-tree ensembles (*e.g.*, Random Forest, Extra Trees) with assorted feature selection,[51] were employed by the winning teams to achieve high performance models (Table 8.3). In addition to the traditional machine learning methods, the Grand Challenge winning team applied novel Deep Learning[52]

Table 8.3 Tox21 Challenge winning models.

Subchallenge assay	Team	Model performance (AUC-ROC)	Method
NR-AhR	Bioinf@JKU	0.93	Deep Learning
NR-AR	dmlab	0.83	Multi-tree ensemble (*e.g.*, Random Forest, Extra Trees) with assorted feature selection
NR-AR-LBD	Bioinf@JKU-ensemble3	0.88	Deep Learning
NR-Aromatase	dmlab	0.84	Multi-tree ensemble (*e.g.*, Random Forest, Extra Trees) with assorted feature selection
NR-ER	Bioinf@JKU-ensemble1	0.81	Deep Learning
NR-ER-LBD	Microsomes	0.83	Random Forest
NR-PPAR-gamma	Bioinf@JKU-ensemble4	0.86	Deep Learning
SR-ARE	Bioinf@JKU	0.84	Deep Learning
SR-ATAD5	AMAZIZ	0.83	Associative neural networks (ASNN) with stratified bagging
SR-HSE	Bioinf@JKU-ensemble1	0.86	Deep Learning
SR-MMP	AMAZIZ	0.95	Associative neural networks (ASNN) with stratified bagging
SR-p53	dmlab	0.88	Multi-tree ensemble (*e.g.*, Random Forest, Extra Trees) with assorted feature selection

techniques to their winning models.[53] Deep Learning, as a field of machine learning, has gained popularity in recent years, having been widely applied in fields such as signal and information processing, speech recognition, as well as physics and life sciences. Deep Learning studies artificial neural networks and related machine learning algorithms that utilise a cascade of many layers of non-linear processing units for feature extraction and transformation. Deep Learning has also been applied to predict the outcome of biological assays. The Grand Challenge winners were the first to apply Deep Learning to computational toxicity.

All winning models, or better performing consensus models, can be applied in parallel to establish activity/toxicity profiles for data-poor environmental chemicals and obtain an estimate of their toxicity potential in a matter of hours of computational time. These computational models could become decision-making tools for government agencies in determining which environmental chemicals and drugs are of the greatest potential concern to human health. Chemicals estimated to have a high potential for toxicity, which would be a much smaller number, could be prioritised for experimental evaluation and validation. Combining these computational models with existing experimental data[17,54] will make chemical prioritisation more time and cost efficient.

8.5 Conclusion

During phase II, the production phase, of the Tox21 program, a multiagency federal collaboration to advance *in vitro* toxicological testing in the 21st century, a battery of *in vitro* assays with target-specific and mechanism-based readouts has been developed, validated and adapted to a quantitative high-throughput screening platform. A standardised operation process has been established from assay optimisation, offline and online validation, robotic screening to performance evaluation, data processing and sharing to ensure the production of high-quality data. In phase II, the Tox21 10K compound collection has been successfully screened against a panel of 47 nuclear receptor and stress response pathway assays in 15 concentration series and in triplicate, generating over 70 million publicly available data points as a big data resource for toxicology.

The high-quality data have been shown to be instrumental in identifying mechanisms of compound toxicity, prioritising substances for further toxicological evaluation and developing models for predicting *in vivo* toxicity response. All results provide rich datasets to researchers for further data mining, generation of new hypotheses and developing new methods for activity modelling. The predictive computational models generated from these high-quality datasets can help shed light on the potential of using *in vitro* assays as an alternative approach for assessing chemical toxicity. In phase III of its research program, Tox21 aims to increase the biological diversity of human cell lines used in the HTS assays, as well as add medium- to high-throughput transcriptomics assays to test for effects of chemicals on gene expression, providing new and more diverse data resources to better identify chemicals that may cause adverse human health effects.

Acknowledgements

This work was supported by the Intramural Research Programs of the National Toxicology Program (Interagency agreement #Y2-ES-7020-01), National Institute of Environmental Health Sciences, the US Environmental Protection Agency (Interagency Agreement #Y3-HG-7026-03) and the National Center for Advancing Translational Sciences, National Institutes of Health. We also thank Samuel Michael and Carleen Klumpp-Thomas for assisting with the screens, Paul Shinn, Misha Itkin and Danielle VanLeer for compound management and William Leister for the Tox21 10K library quality control.

References

1. R. Judson, *et al.*, The toxicity data landscape for environmental chemicals, *Environ. Health Perspect.*, 2009, **117**, 685–695.
2. NRC, *Toxicity Testing in the 21st Century: A Vision and a Strategy*, The National Academies Press, 2007.
3. F. S. Collins, G. M. Gray and J. R. Bucher, Toxicology. Transforming environmental health protection, *Science*, 2008, **319**, 906–907.
4. R. J. Kavlock, C. P. Austin and R. R. Tice, Toxicity testing in the 21st century: implications for human health risk assessment, *Risk Anal.*, 2009, **29**, 485–487.
5. R. R. Tice, C. P. Austin, R. J. Kavlock and J. R. Bucher, Improving the human hazard characterization of chemicals: a Tox21 update, *Environ. Health Perspect.*, 2013, **121**, 756–765.
6. J. Inglese, *et al.*, Quantitative high-throughput screening: a titration-based approach that efficiently identifies biological activities in large chemical libraries, *Proc. Natl. Acad. Sci. U. S. A.*, 2006, **103**, 11473–11478.
7. R. Huang, *et al.*, Profiling of the Tox21 10K compound library for agonists and antagonists of the estrogen receptor alpha signaling pathway, *Sci. Rep.*, 2014, **4**, 5664.
8. R. Huang, *et al.*, Chemical genomics profiling of environmental chemical modulation of human nuclear receptors, *Environ. Health Perspect.*, 2011, **119**, 1142–1148.
9. C. W. Hsu, *et al.*, Quantitative high-throughput profiling of environmental chemicals and drugs that modulate farnesoid X receptor, *Sci. Rep.*, 2014, **4**, 6437.
10. S. Chen, *et al.*, Cell-Based High-Throughput Screening for Aromatase Inhibitors in the Tox21 10K Library, *Toxicol. Sci.*, 2015, **147**, 446–457.
11. M. S. Attene-Ramos, *et al.*, Profiling of the Tox21 chemical collection for mitochondrial function to identify compounds that acutely decrease mitochondrial membrane potential, *Environ. Health Perspect.*, 2015, **123**, 49–56.
12. K. Nishihara, *et al.*, Identification of genotoxic compounds using isogenic DNA repair deficient DT40 cell lines on a quantitative high throughput screening platform, *Mutagenesis*, 2015, **31**, 69–81.
13. R. Huang, *et al.*, Modelling the Tox21 10 K chemical profiles for *in vivo* toxicity prediction and mechanism characterization, *Nat. Commun.*, 2016, **7**, 10425.
14. PubChem, *Tox21 phase II data*, http://www.ncbi.nlm.nih.gov/pcassay?term=tox21, 2016.
15. M. S. Attene-Ramos, *et al.*, The Tox21 robotic platform for the assessment of environmental chemicals – from vision to reality, *Drug Discovery Today*, 2013, **18**, 716–723.
16. PubChem., *Tox21 Phase II compound collection*, http://www.ncbi.nlm.nih.gov/pcsubstance/?term=tox21, 2013.

17. NCATS, *Tox21 Data Browser*, https://tripod.nih.gov/tox21/, 2016.
18. Tox21, *Tox21 assays*, https://tripod.nih.gov/tox21/assays, 2017.
19. S. J. Shukla, R. Huang, C. P. Austin and M. Xia, The Future of Toxicity Testing: A Focus on *In Vitro* Methods Using a Quantitative High Throughput Screening Platform, *Drug Discovery Today*, 2010, **15**, 997–1007.
20. J. H. Zhang, T. D. Chung and K. R. A. Oldenburg, Simple Statistical Parameter for Use in Evaluation and Validation of High Throughput Screening Assays, *J. Biomol. Screening*, 1999, **4**, 67–73.
21. R. Huang, in *High-Throughput Screening Assays in Toxicology*, ed. H. Zhu and M. Xia, *Methods in Molecular Biology*, Humana Press, 2016, vol. 1473, ch. 12.
22. N. T. Southall, A. Jadhav, R. Huang, T. Nguyen and Y. Wang, in *Handbook of Drug Screening*, ed. R. Seethala and L. Zhang, Taylor and Francis, 2009, pp. 442–463.
23. A. V. Hill, The possible effects of the aggregation of the molecules of haemoglobin on its dissociation curves, *J. Physiol. (London)*, 1910, **40**, 4–7.
24. Y. Wang, A. Jadhav, N. Southal, R. Huang and D. T. Nguyen, A grid algorithm for high throughput fitting of dose-response curve data, *Curr. Chem. Genomics*, 2011, **4**, 57–66.
25. A. Simeonov, *et al.*, Fluorescence spectroscopic profiling of compound libraries, *J. Med. Chem.*, 2008, **51**, 2363–2371.
26. PubChem, *Tox21 Phase I compound collection*, http://www.ncbi.nlm.nih.gov/sites/entrez?db=pcsubstance&term=EPA_NCGC_Tox21_Plate0, 2009.
27. R. Huang, *et al.*, Characterization of diversity in toxicity mechanism using *in vitro* cytotoxicity assays in quantitative high throughput screening, *Chem. Res. Toxicol.*, 2008, **21**, 659–667.
28. T. Kohonen, Self-organizing neural projections, *Neural Networks*, 2006, **19**, 723–733.
29. NCBI, *MeSH, Medical Subject Headings*, http://www.ncbi.nlm.nih.gov/mesh, 2013.
30. P. Babula, M. Masarik, V. Adam, I. Provaznik and R. Kizek, From Na+/K+-ATPase and cardiac glycosides to cytotoxicity and cancer treatment, *Anticancer Agents Med. Chem.*, 2013, **13**, 1069–1087.
31. T. T. Abd and T. A. Jacobson, Statin-induced myopathy: a review and update, *Expert Opin. Drug Saf.*, 2011, **10**, 373–387.
32. B. Vidal, *et al.*, The alkylating carcinogen N-methyl-N'-nitro-N-nitrosoguanidine activates the plasminogen activator inhibitor-1 gene through sequential phosphorylation of p53 by ATM and ATR kinases, *Thromb. Haemostasis*, 2005, **93**, 584–591.
33. J. A. Rogers, L. Metz and V. W. Yong, Review: Endocrine disrupting chemicals and immune responses: A focus on bisphenol-A and its potential mechanisms, *Mol. Immunol.*, 2012, **53**, 421–430.

34. S. Medjakovic, *et al.*, Effect of nonpersistent pesticides on estrogen receptor, androgen receptor, and aryl hydrocarbon receptor, *Environ. Toxicol.*, 2013, **29**, 1201–1216.
35. D. Kirpichnikov, S. I. McFarlane and J. R. Sowers, Metformin: an update, *Ann. Intern. Med.*, 2002, **137**, 25–33.
36. T. J. Carty, *et al.*, Ampiroxicam, an anti-inflammatory agent which is a prodrug of piroxicam, *Agents Actions*, 1993, **39**, 157–165.
37. M. H. Zweig and G. Campbell, Receiver-operating characteristic (ROC) plots: a fundamental evaluation tool in clinical medicine, *Clin. Chem.*, 1993, **39**, 561–577.
38. FDA, Innovation or Stagnation: Challenge and Opportunity on the Critical Path to New Medical Products, http://www.fda.gov/ScienceResearch/SpecialTopics/CriticalPathInitiative/CriticalPathOpportunitiesReports/ucm077262.htm (2004).
39. M. I. Martic-Kehl, R. Schibli and P. A. Schubiger, Can animal data predict human outcome? Problems and pitfalls of translational animal research, *Eur. J. Nucl. Med. Mol. Imaging*, 2012, **39**, 1492–1496.
40. N. S. Sipes, *et al.*, Predictive models of prenatal developmental toxicity from ToxCast high-throughput screening data, *Toxicol. Sci.*, 2011, **124**, 109–127.
41. M. T. Martin, *et al.*, Predictive model of rat reproductive toxicity from ToxCast high throughput screening, *Biol. Reprod.*, 2011, **85**, 327–339.
42. R. S. Judson, *et al.*, *In vitro* screening of environmental chemicals for targeted testing prioritization: the ToxCast project, *Environ. Health Perspect.*, 2010, **118**, 485–492.
43. R. Huang, *et al.*, Weighted feature significance: a simple, interpretable model of compound toxicity based on the statistical enrichment of structural features, *Toxicol. Sci.*, 2009, **112**, 385–393.
44. H. Sun, M. Xia, C. P. Austin and R. Huang, Paradigm shift in toxicity testing and modeling, *AAPS J.*, 2012, **14**, 473–480.
45. A. Vedani and M. Smiesko, In silico toxicology in drug discovery – concepts based on three-dimensional models, *Altern. Lab. Anim.*, 2009, **37**, 477–496.
46. W. Muster, *et al.*, Computational toxicology in drug development, *Drug Discovery Today*, 2008, **13**, 303–310.
47. R. Huang, *et al.*, Tox21 Challenge to Build Predictive Models of Nuclear Receptor and Stress Response Pathways as Mediated by Exposure to Environmental Chemicals and Drugs, *Front. Environ. Sci.*, 2016, **3**, 1–9.
48. NCATS, *Tox21 Data Challenge*, https://tripod.nih.gov/tox21/challenge/, 2014.
49. R. Huang and M. Xia, Editorial: Tox21 Challenge to Build Predictive Models of Nuclear Receptor and Stress Response Pathways as Mediated by Exposure to Environmental Toxicants and Drugs, *Front. Environ. Sci.*, 2017, **5**, 3.

50. A. Abdelaziz, H. Spahn-Langguth, K.-W. Schramm and I. V. Tetko, Consensus Modeling for HTS Assays Using *In silico* Descriptors Calculates the Best Balanced Accuracy in Tox21 Challenge, *Front. Environ. Sci.*, 2016, **4**, 1–12.
51. G. Barta, Identifying biological pathway interrupting toxins using multi-tree ensembles, *Front. Environ. Sci.*, 2016, **4**, 52.
52. Y. LeCun, Y. Bengio and G. Hinton, Deep learning, *Nature*, 2015, **521**, 436–444.
53. A. Mayr, G. Klambauer, T. Unterthiner and S. Hochreiter, DeepTox: Toxicity Prediction using Deep Learning, *Front. Environ. Sci.*, 2016, **3**, 1–15.
54. USEPA, *ToxCast Data*, http://www2.epa.gov/chemical-research/toxicity-forecaster-toxcasttm-data, 2017.

CHAPTER 9

Big Data Integration and Inference

KAREN H. WATANABE-SAILOR*[a], HRISTO ALADJOV[b], SHANNON M. BELL[c,e], LYLE BURGOON[d], WAN-YUN CHENG[c,e], RORY CONOLLY[e], STEPHEN W. EDWARDS[e], NÀTALIA GARCIA-REYERO[d], MICHAEL L. MAYO[d], ANTHONY SCHROEDER[f], CLEMENS WITTWEHR[g] AND EDWARD J. PERKINS[d]

[a]Arizona State University, School of Mathematical and Natural Sciences, West Campus, 4701 W. Thunderbird Rd., Glendale, AZ 85308-4908, USA; [b]Organisation for Economic Co-operation and Development, 2 rue André-Pascal, 75775 Paris Cedex 16, France; [c]Oak Ridge Institute for Science and Education, Oak Ridge, Tennessee, USA; [d]US Army Engineer Research and Development Center, Environmental Laboratory, Vicksburg, MS 39180, USA; [e]U.S. Environmental Protection Agency, Integrated Systems Toxicology Division, National Health and Environmental Effects Research Laboratory, Research Triangle Park, NC 27711, USA; [f]University of Minnesota – Twin Cities, Water Resources Center, 1985 Lower Buford Circle, St. Paul, MN 55108, USA; [g]European Commission, Joint Research Centre, Via E. Fermi 2749, 21027 Ispra, VA, Italy
*E-mail: Karen.Watanabe@asu.edu

Issues in Toxicology No. 41
Big Data in Predictive Toxicology
Edited by Daniel Neagu and Andrea-Nicole Richarz
© The Royal Society of Chemistry 2020
Published by the Royal Society of Chemistry, www.rsc.org

9.1 Introduction: New Toxicology Paradigm and Challenges of Big Data Integration and Application

There is increasing global interest in developing a better understanding of the health hazards of most, if not all, chemicals in use today to better support safety evaluation and risk-based decision-making. For example, the European Union has passed the Regulation on the Registration, Evaluation, Authorisation and Restriction of Chemicals (REACH), requiring extensive characterisation of chemicals in production,[1,100] while the United States Environmental Protection Agency (USEPA) has initiated the Toxicity Forecaster (ToxCast™) programme to characterise the toxicity of thousands of chemicals (see Chapter 8).[2] This, along with a need to reduce animal use due to exorbitant costs[3] and societal concerns,[4] has created a demand for large amounts of toxicological information that is more directly related to the species of concern (*e.g.*, human rather than rat).[5]

In 2007, the U. S. National Research Council (NRC) published an influential report on a vision and strategy for improving and modernising the practice of toxicology used to inform regulatory efforts.[5] The NRC vision incorporated advances in new technologies including *in vitro* screening, computational methods, bioinformatics, and systems biology to provide more accurate and rapid assessments of chemical hazards to humans. The vision was based on "a new toxicity-testing system that relies mainly on understanding toxicity pathways."[5] The report recommended development of

1. approaches to replace animal testing with *in vitro* assays that assess perturbation of toxicity pathways;
2. methods to assess chemical effects in an automated, high-throughput manner; and
3. the use of human cells or cell lines rather than animal cell lines so that testing results are more relevant to human populations.[5]

Computational toxicology, systems biology, and bioinformatics play key roles in interpreting and applying this approach, especially when extrapolating to humans and other species, population, and ecosystem level effects.[6–8]

In parallel to the demand to better understand the hazards of chemicals around us, technology trends over the last 20 years have exponentially increased the amount and variety of types of data that are collected from a limited number of biological measurements to tens of thousands of measurements.[102] As a result, the practice of toxicology has changed to one where traditional whole animal experiments and expert knowledge is being integrated with large datasets and computational approaches to enable a more in-depth understanding of the mechanistic basis of chemical effects to support toxicological and regulatory decisions. This chapter describes how big data (*e.g.*, from toxicogenomic studies) and small data (*e.g.*, traditional whole-animal toxicology studies, specific case studies) can be combined and used in the context of the 21st Century Toxicology vision of toxicity pathways to help with understanding and predicting chemical toxicity.

The mechanisms by which chemicals cause adverse effects *via* toxicity pathways can be summarised using an Adverse Outcome Pathway (AOP) framework.[9] AOPs incorporate many types of data to define:

1. how a chemical initiates a toxicity pathway *via* a molecular initiating event (MIE); and
2. a toxicity pathway as a sequence of key events (KE) required to cause an adverse outcome (AO) of regulatory concern.

An AOP covers events at the molecular, cellular, tissue, organ, individual, and population levels and reflects the many scales and types of information needed to describe how a chemical causes toxicity. Since events at many biological levels must be integrated in an AOP, data from many different sources must also be combined to support this understanding.

Big data can be used in several different ways to support regulatory toxicology. Generally, data integration is done to help bolster or complement existing data. In a traditional, small data experiment we may be asking targeted questions such as "at what dose do we see tumorigenesis", "how fast does this drug clear the liver", or "what fraction of the population has an allelic variant that increases their susceptibility to an environmental chemical". We may want to see if there are relationships between exposure rates to an environmental compound and increased likelihood of any adverse effect in order to prioritise research efforts. Or we may be interested in determining what processes are affected that lead from an exposure event to a known AO for designing targeted testing and monitoring schemes. Having access to multiple data sources spanning the chemical, molecular, and physiological space helps supply a richness and depth to the analysis that enables us to address these questions.

This chapter discusses several key issues related to the integration and application of big data in toxicology, especially in a toxicological or AOP context (see Figure 9.1). These issues are essential toward applying big data to toxicological problems. In Section 9.2, we first address how currently known information, chemical and toxicological data, and expert knowledge can be structured in the development of a knowledge base for AOPs so that it is possible to find information relevant to toxicological questions using data mining to support modelling and decision-making. Development of such a resource is essential in analysing relationships between data types and permitting communication and synchronisation between different analytical tools. In Section 9.3, we discuss how the large-scale production of high-throughput screening (HTS) assay data for cellular functions (*e.g.*, *in vitro* assays for receptor binding, enzymes, cell signaling, *etc.*) can be used in combination with gene expression and modelling to predict chemical effects *in vivo*. HTS datasets have brought new dimensions to regulatory toxicology by capturing a very wide range of defined biological responses that can be used

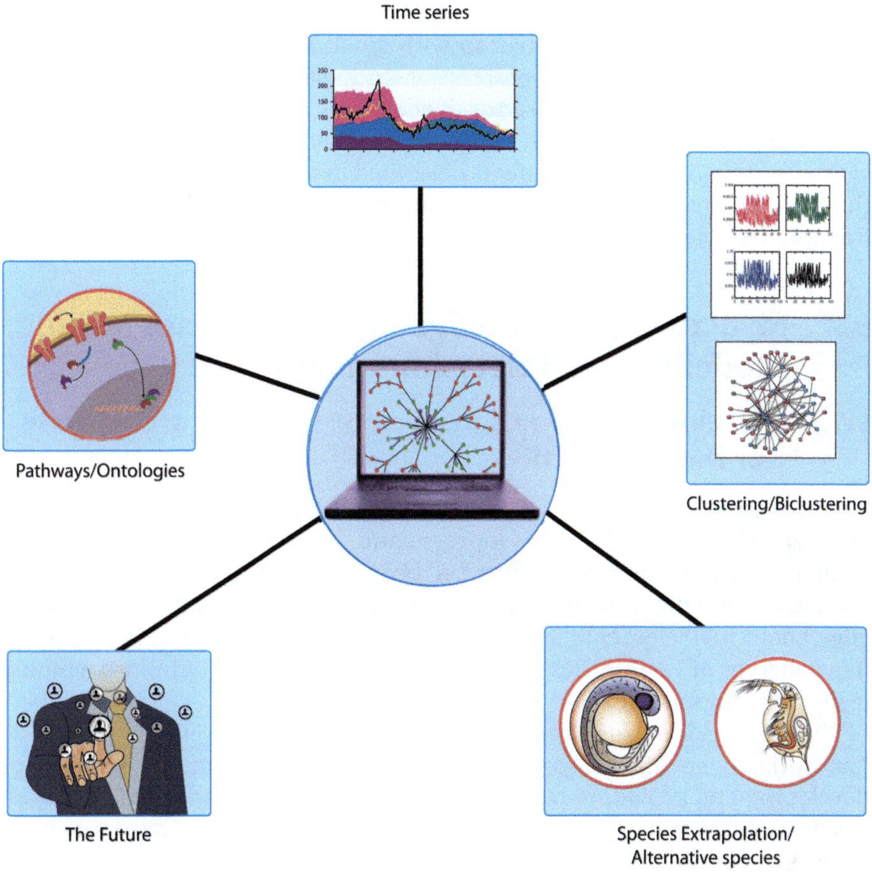

Figure 9.1 Integration of big and small data in network analysis and modelling for predictive toxicology.

to characterise the potential hazards of a chemical. In Section 9.4, we discuss how the availability of big biological datasets permits development of data-driven network models that facilitate understanding the complex relationships underlying chemical effects on animals. The ability to use experimental data to generate hypotheses about toxicological pathways affected by a chemical complements expert knowledge, which can reveal unkown relationships such as feedback control mechanisms. In Section 9.5, we discuss how big datasets, including literature, can be mined to produce a large library of computationally predicted AOPs (cpAOPs), and with the help of expert knowledge they can be refined into putative AOPs. The use of AOPs in risk assessment is covered in Section 9.6, and finally, in Section 9.7, we discuss how experimental data are translated into models for predicting toxicological effects using a combination of high-throughput datasets and expert knowledge to shape model development.

As in many fields, big data has changed the way toxicology is being done today. Big datasets in combination with small data and expert knowledge may provide the necessary answers for understanding and predicting the potential hazards of the thousands of chemicals for which no test data exist. The approaches for handling, analysing and modelling the biological processes leading to hazardous effects described in this chapter are just some of the tools that can be applied to big datasets to better support safety evaluation and risk-based decision-making.

9.2 Structuring Knowledge to Support Data Mining, Modelling, and Decision-making

9.2.1 Development of a Globally Accessible Knowledge Base for Data Integration

Large amounts of knowledge (here defined as including expert opinion, literature and data) needed for development and use of AOPs already exists. The challenge was that no global effort had been undertaken to collect this information (here defined as documentation, literature, and data) since knowledge about AOP elements is distributed among many parties, is not stored in a coherent or easily accessible way, and needs to be brought into a biologically meaningful order and perspective.

The Organisation for Economic Co-operation and Development (OECD) has initiated an international AOP Development Programme focused on developing and providing resources for sharing AOP knowledge. To provide a central location for this AOP-related knowledge, the OECD has launched a project to develop the "*Adverse Outcome Pathway Knowledge Base*" (AOP-KB; https://aopkb.oecd.org/) as depicted in Figure 9.2.

The AOP-KB allows all stakeholders to build AOPs by entering and then linking information about chemical initiators, MIEs, KEs, and AOs. Pathway elements are not necessarily unique to a single AOP, and AOPs are also chemical-agnostic, *i.e.*, they apply to all chemicals triggering the same pathway, thus facilitating the re-use of MIE, KE and AO information in multiple AOPs to prevent redundancy and make the collective knowledge about those entities available in all AOPs in which they appear. The AOP-KB is a combination of individually-developed platforms, synchronised and orchestrated in a way that gives users the possibility to capture, review, browse and comment on AOPs shared by the AOP stakeholder community. The AOP-KB project integrates four independently developed software platforms (https://aopkb.oecd.org/index.html). The **AOP-Wiki** (https://aopwiki.org) organises, *via* crowd-sourcing, available knowledge and published research into a textual description of individual pathways, with a user-friendly Wiki interface. Controlled-vocabulary drop-down lists allow users to select Methods, Actions, Biological Objects, Life stages, Species, *etc.* related to the AOP and simplify the entry of ontology-based

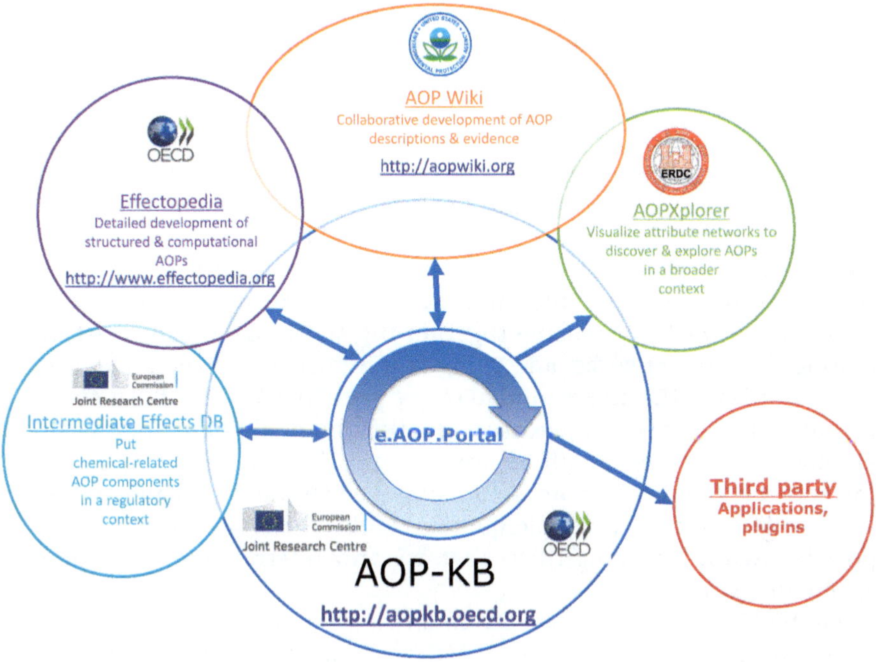

Figure 9.2 The Adverse Outcome Pathway Knowledge Base (AOP-KB) and its components.

information. **Effectopedia** (https://www.effectopedia.org) is a knowledge aggregation and quantitative modelling platform designed for collaborative development and utilisation of AOPs. It uses a visually-expressed modular structure, which captures semantically-annotated knowledge, computational models and algorithms along with supporting evidence/data. The **AOPXplorer** is a computational tool that enables automated graphical representation of AOPs in a network context. The **Intermediate Effects DB** will follow the OECD Harmonised Templates (OHT) format and host chemical-related data derived from non-classical methods to inform users how individual compounds trigger MIEs and/or KEs. Cross-referencing these empirical observations with AOP descriptions in the other tools will make it possible to predict the extent to which a chemical compound is involved in an AOP.

9.2.2 Use of Structured Knowledge

The assembly of structured knowledge for AOPs is an obvious complement to the methods for AOP development and use described later in this chapter. An ideal AOP-KB will capture AOP information from all AOP development efforts such that the findings can be evaluated by experts in the field and incorporated into formal AOP descriptions. As the expert evaluation proceeds, the results should be made readily available for developers of the methods to use when refining the methods. As focus shifts toward more quantitative analysis and modelling of AOPs, data need to shift.

At this point, knowledge must be organised in a way that is accessible by computers and quantitative values must be explicitly captured for easy retrieval. The advantage of using a formal ontology, such as the AOP Ontology,[101] to structure the knowledge base and the data within it is that first-order logic and inference methods can be used to automatically generate new AOPs by piecing together the AOPs that already exist, integrating them in networks, and using that information to create new connections from MIEs, to KEs and AOs (see Section 9.5). This will also allow risk assessors and investigators to identify potential points for cross-talk amongst AOPs. This can all be done automatically using computational tools. If the AOP-KB is structured using Web Ontology Language (OWL), like the AOP Ontology, simple queries can be performed that retrieve all possible AOPs, including those that are only inferred and not explicitly specified *a priori*.

For AOP information to be used in a risk assessment context, the provenance and evidence supporting an AOP is essential. This will not be a problem for cpAOPs, as each of the KEs and AOs already have their evidence and provenance captured in the AOP-KB.

9.2.2.1 Data Mining

The expert-based development of formal AOPs is a labour-intensive and time-consuming process that has greatly limited the number of AOPs developed to date. Data mining is a valuable tool for rapidly generating hypothetical AOPs to increase the coverage of biological space as will be described later in this chapter. The AOPXplorer is being designed to facilitate this process by merging information derived from high-throughput methods with expert-curated AOP information from the AOP-Wiki. Since AOPs normally exist within an interconnected network of biological processes, most data mining methods naturally create full AOP networks from which individual AOPs can be selected.

By gathering AOP networks from multiple input sources (the AOP-Wiki, Effectopedia, and the AOP-KB), the AOPXplorer provides much of the background information needed to assemble expert-curated AOP descriptions in the AOP-Wiki. Since the content of the AOP-Wiki is displayed along with information from data mining, users can readily determine which biological processes are included in existing AOPs and identify novel pathways that are candidates for AOP development. The AOPXplorer will also provide a graphical interface for users to perform analyses of how toxicological data from HTS assays, toxicogenomic studies, and other sources can be associated with KEs in an AOP network. This process will be described in more detail in Section 9.5. As information from data and bioinformatics-driven approaches is used to develop and annotate expert-curated AOPs, stakeholder feedback on the quality of the automated content can be used to improve those processes.

9.2.2.2 *Quantitative Modelling*

AOPs, unlike system biology models, do not aim to describe the full dynamics of a system, but rather the essential observable changes in normal biological function, following an initial chemical exposure, for which all compensatory mechanisms are overwhelmed. The choice to represent a sequence of snapshots of the actual dynamics requires the temporal and biological context of the observed changes to be well documented.

Effectopedia uses the biological context in the form of a pathway space as a visual organising principle for AOP descriptions. Each dimension in the pathway space represents a certain aspect of the biological context, *e.g.*, life stage, taxonomic relevance, level of biological organisation, time to effect, *etc.* The space is also user-expandable and can describe factors that are initially introduced as relevant for a specific pathway but subsequently can

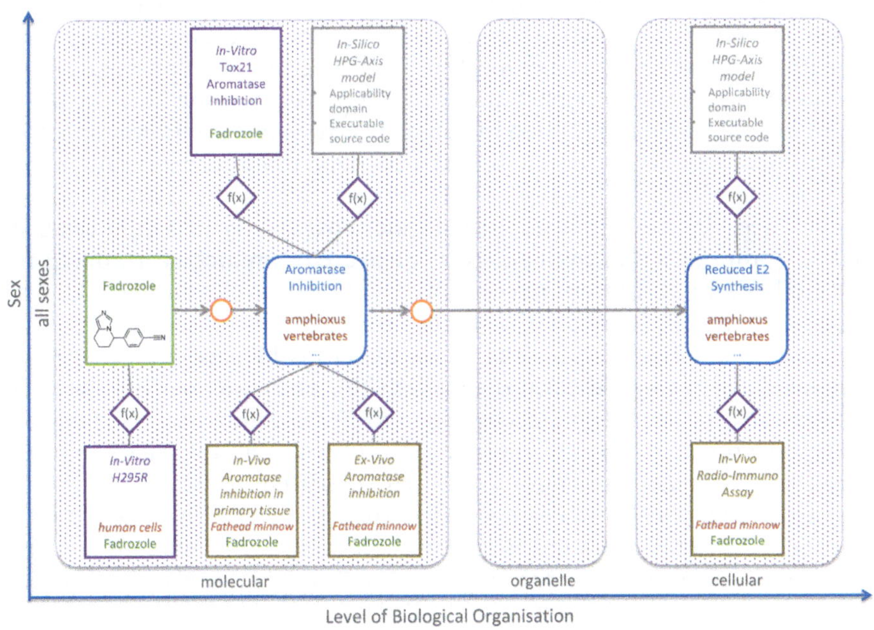

Figure 9.3 Effectopedia example of a pathway fragment. Green rectangular boxes represent the chemical initiators; orange circles represent links or key event relationships (KERs); blue rounded rectangular boxes represent effects; tall rectangular boxes represent supporting evidence; and rhomboids represent transformation functions that encode mathematical equations.

be utilised in all pathways (Figure 9.3). The first element of the pathway description is often one or more reference chemicals that trigger an MIE. Effectopedia includes information on the steps of chemical absorption, distribution, metabolism and excretion (ADME) – while considered to be outside the scope of AOPs since ADME can be chemical specific – which are critically important for describing the relevance, needed biotransformations, and interpretation of various test assays used as supporting evidence. ADME information might also be required to establish how much of the tested chemical is available to interact with the biological target when the pathway is used in a predictive mode. The description of a pathway can also start with a definition of a chemical profile based on a combination of structural alerts and properties.

In Effectopedia's representation of AOPs (Figure 9.3), the element connecting the reference chemical with the MIE is called a link (orange circle, also known as a key event relationship when connecting two KEs). Links encapsulate both the qualitative description of the underlying mechanisms as well as the quantitative dose response relationship. When a link connects two effects, the quantitative relationship can take on a variety of forms: one to one (*i.e.*, $a = b$), proportional, threshold and a general non-linear form characterising the response–response relationship depending on the described biological mechanism. The biological context of the connected upstream and downstream effects and visual location of the link in the pathway space frames the questions that need to be addressed in the description of plausible mechanisms that propagate dysfunction. The position of the effects (blue rounded rectangular boxes) in the pathway space is determined by their biological context and, depending on their role in the pathway, the effects have more specific names: MIE, KE and AO. Supporting evidence (represented by tall rectangular boxes) describes different test methods or models that can be used to measure or predict a certain effect. Test methods can represent *in chemico*, *in vitro*, *ex vivo*, *in vivo* test assays, and *in silico* models (coded with different colours in the pathway diagram). The placement of the test method in the Effectopedia pathway space communicates the biological context in which the assay was performed. Test methods are connected in Effectopedia to effects using a transformation function (represented by a rhomboid) that encodes all the mathematical steps (if any) required to convert an assay media chemical concentration to an *in vivo* chemical concentration, and the measured response to a normalised *in vivo* effect ranging from zero to a maximum value. The *in vivo* chemical concentration used throughout the whole pathway is the same as the one that can be measured at the MIE level. Referencing all effects (this includes KEs) to the same *in vivo* chemical concentration allows the test results from two sequentially connected effects to be used for establishing a quantitative relationship (encoded in the link) between them. The derived relationship between these effects could be correlative or considered causal only when no other pathways are triggered by the chemical at comparable concentrations. Once established in one pathway, this relationship can be reused in all pathways that share the same sequence of events. The use of a transformation function also allows direct comparison of the output of different test assays and *in silico* models. The formulation of the transformation function can be a serious challenge and may require additional models and dedicated experiments. However, a well described biological context of the measured effect and detailed description of the test protocol or assumptions behind an *in silico* model provide clear guidance as to what aspects of the model are different and will need transformation.

There are two distinct sources of uncertainty in the Effectopedia organising framework. First, the ability of the experiments to reproduce and measure an associated *in vivo* effect. This information is stored in the test method description and transformation functions. The second source of uncertainty arises in the confidence of establishing causality between effects and is described in the linked elements of the pathway. The cumulative confidence and reliability of the predictions of a fully quantified pathway is a non-trivial combination of these two sources of uncertainty.

9.2.3 Supporting Decision-making

The ultimate use of structured AOP information is to facilitate decision-making, whether that be in a regulatory context, medical treatment, or other area. Having structured information is valuable because the decision makers then have the required information in a consistent format as they consider a wide variety of biological processes. It is important that the information be summarised concisely with references to more detailed descriptions. The evidence supporting the AOP should be sufficiently described to allow the reader to personally evaluate any assessment of the strength of evidence made by the authors. The OECD AOP Development Program has issued a handbook[103,104] to guide AOP authors in a systematic evaluation of the evidence.[10] By providing information in a consistent manner with regard to both the format and content of the data, the efficiency of decision makers can be increased, who can readily read, understand, and evaluate the evidence provided.

A key component of any knowledge base intended for providing AOP information to inform decision-making is transparency. The provenance of all information in the AOP-KB is captured, including the person who entered the information and the time it was entered. Capturing author information allows the decision maker to check for conflicts of interest when reviewing the information. Capturing time information allows decisions to be made based on current knowledge and then reconsidered later in light of new knowledge without the impression that a "wrong" decision was made. This is further improved when tools are provided that can rapidly show the differences between a previous version and the current version of an AOP.

Finally, having structured descriptions of biological processes allows the development of automated tools to improve decision-making as will be discussed in Section 9.6. More work is needed to determine the level of resolution for the AOP that is required to enable this approach, but the current work suggests that in time this should be feasible if the AOP information is properly documented and stored. By providing open access to all AOP information, the AOP-KB will promote further development of these tools and thereby enhance the use of AOP information in decision-making.

9.3 Integration of High-throughput Screening Assays and Curated Databases for Understanding Chemical–Bioactivity Relationships

Over the last decade, the development of large chemical screening programmes such as ToxCast™ and Tox21 (see Chapter 8) have produced large amounts of biological pathway-based data. The development and progress of HTS assays in generating enormous amounts of data has quickly moved this toxicity testing approach into the big data era. To effectively utilise these datasets, it is important to understand the types of chemical and biological data that are being produced, where these data are stored, and how these data can be integrated to better understand chemical–gene relationships and toxicity.

9.3.1 Data Generated from HTS Assays

9.3.1.1 Chemical Screening Data

Understanding the potential toxicity of chemicals can be supported by read-across approaches, *i.e.*, identifying categories of similar chemicals with similar properties that have documented toxicity effects, to enable prediction of the toxicity and mode of action of the chemicals of interest by reading across from the existing data.[11] Although this approach has been useful for predicting the toxicity of many chemicals, small changes in chemical structure can also result in large differences in toxicity. Therefore, read-across category approaches should not be based on chemical structure alone, but take into account biological similarity. The integration of mechanistic data provided by HTS assays with existing chemical information can thus strengthen the ability to categorise and predict the toxicity of similar chemicals.

To effectively utilise HTS assays for evaluating the impacts of chemicals on biological pathways, the screening of large chemical libraries is needed. HTS assays have long been utilised for drug discovery, providing large amounts of chemical data for potential pharmaceuticals.[12] A growing number of academic centres such as UCLA's Molecular Screening Shared Resources and centres associated with the Molecular Libraries Screening Center Network[13] have produced chemical libraries approaching 200 000 chemical compounds. More recently, large chemical libraries have been developed by government agencies specifically to generate chemical toxicity data. The USEPA's ToxCast™ programme maintains a chemical library of almost 2000 chemicals that includes pesticides, steroid hormones, and pharmaceutical candidates that are not commercially available.[2] A large collaborative group consisting of the USEPA, National Institutes of Health, and Food and Drug Administration launched the Toxicity Testing in the 21st Century (Tox21) programme, which developed a library containing almost 10 000 chemicals, including pesticides, commercial products, and pharmaceuticals.[14] These programmes and centres demonstrate the feasibility of generating data for large inventories of chemicals in an efficient and cost effective manner.

9.3.1.2 *Biological Targets*

The selection of *in vitro* assays for identifying biological targets (*i.e.*, genes or proteins) for toxicity testing is challenging because the molecular mechanisms underlying many types of chemical toxicity are largely unknown. Historically, *in vitro* assays used for toxicity testing have focused on a limited number of gene targets. This created large amounts of chemical–gene target data, but this approach cannot identify all the potential genes and pathways that could be associated with chemical toxicity. HTS technologies for the drug discovery industry provide relevant knowledge and tools to develop assays for a large suite of biological targets to overcome this challenge.[15] The biological targets are representative of various gene families including: protein kinases, G-coupled protein receptors, nuclear receptors, and many others. A large suite of biological targets is needed to capture MIEs and other KEs associated with the majority of toxicity pathways. For example, the ToxCast™ programme currently utilises about 700 different bioassays that include biological targets and responses representative of 21 different gene families; the Tox21 programme utilises about 70 different HTS assays that represent six different gene families. Furthermore, data deposited in publicly available databases from other HTS programmes can augment biological target information to provide greater coverage of relevant toxicity pathways. The identified MIE and KEs from these assays can then be integrated in an AOP to determine the potential for AOs at the organism or population level.[9] The large suite of biological targets associated with HTS assays provides a rich overview of molecular and cellular effects needed to assess chemical toxicity.

9.3.2 **Publicly Available Databases**

To facilitate HTS data analysis and application in scientific and regulatory contexts, the data must be organised and easily accessible. Fortunately, several large-scale, publicly available databases exist for curating and maintaining HTS data, toxicogenomics and other relevant toxicity data (see examples

Table 9.1 Publicly available databases for integrating high throughput, -omics and toxicity data.

Database	Database description	Data description	Database URL
PubChem[16]	A database of biological tests of small molecules generated through screening experiments, medicinal chemistry studies, chemical biology research, and drug discovery programmes.	47 million chemical compounds screened; 700 000 bioassays, over 13 billion data points.	https://pubchem.ncbi.nlm.nih.gov
ChEMBL v20[17]	A manually curated database containing binding, functional, and ADMET (absorption, distribution, metabolism, excretion, toxicity) information for a large number of drug-like bioactive compounds.	Over 1 million chemicals screened; over 1 million assays covering over 10 000 targets; over 13 million bioactivity measurements.	https://www.ebi.ac.uk/chembl
ToxNet[18]	A group of databases covering chemicals and drugs, diseases and the environment, environmental health, occupational safety and health, risk assessment and regulations, and toxicology. Includes information on specific chemicals and mixtures, unknown chemicals and specific toxic effects of chemicals.	*In vitro* and *in vivo* information on over 50 000 environmental chemicals from different sources.	https://toxnet.nlm.nih.gov
USEPA iCSS Dashboard[19]	An interactive database to rapidly explore the high-throughput chemical screening data generated from the Toxicity Forecaster (ToxCast) and Toxicity Testing in the 21st Century (Tox21) collaboration.	ToxCast – over 1800 chemicals screened in over 600 assays; Tox21 – almost 10 000 chemicals screened through 50 assays for Tox 21.	http://actor.epa.gov/dashboard
ToxRefDB[19]	A manually curated database of chronic *in vivo* toxicity data from different organ systems in experimental animals.	Information on 474 environmentally relevant chemicals.	https://catalog.data.gov/dataset/toxcast-toxrefdb

CEBS[20]	The Chemical Effects in Biological Systems is a database that houses several types of study data from academic, industrial, and government laboratories.	6000 chemicals and drugs with over 800 molecular (most microarray) datasets.	https://tools.niehs.nih.gov/cebs3/ui/
DrugMatrix[21]	A database populated with gene expression results from *in vivo* rat or *in vitro* rat hepatocyte exposures to various drugs and chemicals. *In vivo* studies were carried out in multiple target organs.	Over 600 drugs and chemicals with gene expression data (mostly microarray) from over 4000 studies.	https://ntp.niehs.nih.gov/drugmatrix/index.html
cMap[22]	The Connectivity Map is a database containing gene expression results from human cells treated with various chemicals.	Over 1300 chemical compounds with over 7000 gene expression profiles.	https://www.broadinstitute.org/cmap
CTD[23]	The Comparative Toxicogenomics database is a multiple-species, manually curated database containing chemical–gene/protein interactions and chemical–disease relationships from published literature.	Over 10 000 chemicals; over 37 000 genes; over 1 million curated chemical–gene interactions; over 1 million chemical–disease associations.	http://ctdbase.org
Toxic Exposome Database[24]	A database containing chemicals and chemical–gene target information from different databases, government documents, books, and scientific literature.	Over 3000 chemicals including pollutants, pesticides, drugs and food toxins; over 42 000 chemical–gene target associations.	http://www.t3db.ca

in Table 9.1). Many of these databases are routinely updated, providing a growing resource for understanding the different biological mechanisms underlying chemical toxicity (see also Chapter 4).

9.3.3 Current Approaches for Integrating HTS Data and Databases and Computational Modelling to Predict *In Vivo* Toxicity From *In Vitro* Data

A significant challenge of using *in vitro* data for predicting chemical toxicity is extrapolating *in vitro* data results to whole organism responses. This is especially difficult if only a few gene targets are examined rather than a comprehensive suite of targets that cover multiple pathways for a single chemical, such as the approach taken by ToxCast™ and Tox21. A solution to this challenge is the integration of relevant chemical–gene information provided in the databases listed in Table 9.1 to identify a large suite of gene targets and pathway-based perturbations that would likely lead to an AO *in vivo*. The integration of these data provide greater statistical power for data analysis and sufficient information needed to discover patterns, or "signatures", that are correlated with the toxic effects observed in traditional animal toxicity testing.

To integrate HTS data, data mining tools can be used to extract data from the databases listed in Table 9.1. Several commercial software packages such as Spotfire and Vortex are available to mine these databases; however, only a limited number of freely available software packages exist currently, such as HTS Navigator[25] and Chem2Bio2RDF.[26] Thus, there is a great need for further development of freely available software to improve HTS and other relevant toxicity data mining and extraction from these different databases (see Section 9.9 for more information).

Once the data present in these databases has been extracted, the next challenge lies in standardising, structuring, and condensing the data for interpretation. Because the majority of the HTS data have been generated and analysed using different methods (*e.g.*, different experimental set up and biological targets examined), care must be taken before using the integrated data for predicting *in vivo* toxicity. For example, data quality should be assessed to ensure that only the most reliable and adequate data are used in an analysis. Also, depending on the types of data being integrated, it is important that raw data are pre-processed, normalised, or re-scaled together to make sure the data from different sources or experiments are comparable. This is especially important when high-throughput expression data from microarrays or other gene expression platforms are utilised.

After the data are made comparable, computational approaches can be used to identify patterns, or "signatures", predictive of *in vivo* toxicity endpoints or phenotypes.[27] To develop computational models, one starts with a set of chemicals for which *in vitro* data and relevant *in vivo* toxicity endpoints or phenotypes are available. This set of chemicals functions as a training set for which one can develop and validate a predictive model.

Various statistical modelling approaches including linear, non-linear, or logistic-regression-based models can be utilised. Most commonly, supervised machine learning approaches[28] are used for developing models to accurately classify endpoints from *in vitro* data. Examples of machine learning approaches that have been used include weighted voting algorithm, naïve Bayes support vector machine, k-nearest neighbours, and artificial neural networks.[29,30] Each of these approaches have their advantages and disadvantages for developing predictive models and this has been examined extensively.[31] Once a model has been built using the training set, a test set of chemicals can be evaluated to determine if the model results for the test set is better than by chance alone. Before a developed model is acceptable, it is critical to determine if there is support in the literature for a link between the *in vitro* target(s) and the *in vivo* endpoint. This approach has been successfully used to develop predictive models for identifying chemicals related to cancer,[32] developmental,[33] and reproductive toxicity[34] based on the ToxCast™ data.

9.4 Systems Biology: *De Novo* Network Inference of Chemical, Gene, Protein, and Metabolite Relationships Underlying Toxicity

An aim of systems biology is to understand how biological systems maintain homeostasis in the face of a dynamic environment. When a system is exposed to stimuli or stressors, the underlying networks connecting tens of thousands of different molecules transition to different states to adapt to or resist the stress or stimulus. These changes can lead to new stable states that are different from unstressed systems. These stable states can resist perturbations through compensatory feedback mechanisms, until a tipping point is reached and the system moves to a different stable state. Systems biology examines very large datasets that capture changes in gene, protein, metabolite, and chemicals to model and understand these transitions and tipping points in order to predict potential outcomes.

All organisms contain networks of genes, proteins, and biochemical reactions that control responses to stressors and stimuli. Understanding how these networks behave and what tipping points change the state of a system is crucial to understand how the system as a whole works, reacts, and adapts. For example, researchers from Seattle BioMed and the Institute of Systems Biology (Seattle, WA, USA) have deciphered about 200 different molecular switches that *Mycobacterium tuberculosis* uses to sense and respond to the variable landscape of a host by overexpressing different genes.[35,36] They also identified >10 000 "dormant" DNA-binding events that cannot be linked directly with proximal transcriptional control, suggesting that DNA binding could be a common feature that should be considered when developing global gene expression models.[35] Study of these systems can generate large datasets in the order of 10 million to a billion data points in order to properly characterise network dynamics. As a result, the integration, manipulation and analysis of big data is an essential feature in systems biology.

9.4.1 Network Inference

Omics measurements (*e.g.*, transcriptomics, metabolomics, or proteomics among others) can provide a unique understanding of how stressors affect an organism (see Chapter 7). These data have already been used to identify modes of action, or potential adverse effects.[37–41] Traditional bioinformatics approaches rely on prior knowledge of biological pathways and relationships. In the absence of *a priori* information, network biology approaches can be used to infer new connections and interactions by perturbing the system and correlating values to infer the underlying network topology.[42,43] A helpful overview of network biology can be found in *Network Inference in Molecular Biology*.[44] Briefly, network biology allows the inference of significant associations and causal linkages that can be useful in many different applications, disciplines, and biological levels of organisation. To find those significant associations and causal linkages between genes or other molecules, experiments need to be performed that perturb the system under multiple different conditions thereby permitting better inference of connections.

9.4.2 Steady-state Networks

In genomics, network inference has been principally applied to infer causal links among genes, such as gene regulatory networks, but these approaches have also been used to detect network interactions among different molecules, such as genes, proteins, and metabolites.[45,46] A causal assumption must lie behind every causal conclusion.[44,47] For instance, a causal assumption can be that the state of a system at a specific time point will determine its state at a close time point, or that if a perturbation is applied to a steady-state, any changes will be due to that perturbation. Without that external perturbation it is almost impossible to infer useful networks, as biological systems are full of feedback loops that might mask the actual regulatory networks.[44]

One of the challenges of network inference is the large number of genes that can be used, which requires enormous computational power and might be intractable for some algorithms. A possible solution is clustering genes into co-expressed groups, which reduces the data size by replacing individual genes with artificial gene clusters (or supergenes) that can be treated as single nodes, with a minimal amount of information lost. This can reduce the signal-to-noise in the data and therefore improve the accuracy of the inferred network.[44] Biclustering is another useful method to reduce noise and increase the power of the inferred network by clustering across both rows and columns of a matrix, *i.e.*, gene expression values and experimental conditions, using algorithms such as cMonkey, SANDY, or SAMBA.[44,48–51]

Data used to infer networks can come from steady-state or time series experiments, *i.e.*, a stimulus is applied and the next state of the system is measured, or after the stimulus is applied, measurements are taken throughout a time course. This second approach provides extra information as the system evolves toward the next steady-state. Several algorithms have been suggested for steady-state data.[44] For instance, the Median-Corrected Z-Score is a very useful algorithm when mutation data are available, which could be combined into a network inference pipeline when time series data are available.[52] Network Identification by Multiple Regression uses multiple regression to infer networks from steady-state expression data resulting from a known initial perturbation.[53] The Gene Network Inference with Ensemble of trees is an algorithm that uses an ensemble of regression trees to infer networks from multifactorial data.[54] Finally, the Context Likelihood of Relatedness (CLR) algorithm uses mutual information to infer networks from steady-state data.[55]

9.4.3 Dynamic Networks

Network inference from static experiments has provided valuable information on how systems function. However, biological systems are highly dynamic systems that are always responding to environmental or genetic changes.[56] Therefore, to truly understand a system, we also need to understand the system dynamics. While time series data provide more information than steady-state experiments, *i.e.*, can help infer directionality of connections, they also add complexity to the analysis. Furthermore, the increased complexity can result in a higher number of false positive or negative connections, so more care needs to be used when working with time series data.[44] Some of the main algorithms used to work with time series datasets include mutual information approaches, ordinary differential equations, and dynamic Bayesian networks. These algorithms include Time-Delay ARACNE;[57] Time-Lagged Context Likelihood of Relatedness (tlCLR),[52,58] an extension of the CLR algorithm; the Inferelator, based on ordinary differential equations;[52,58] and the Bayesian Network Inference with Java Objects.[59] Time series also enables differential network mapping, which allows for the interrogation of previously unexplored interaction spaces, and is rapidly becoming a standard use of network analysis, just like differential gene expression is standard in genomic analysis.[56] A key challenge in the use and development of these sophisticated computational resources, which are increasingly available to bench biologists, is the need for tailored ongoing education to ensure correct use of these resources.[60]

9.4.4 Networks and Inference of Toxicological Effects

Network inference has been applied to large datasets to connect molecular-level changes to adverse outcomes. For example, Workman *et al.*[61] used steady-state network analysis to combine genome-wide chromatin immunoprecipitation assays, transcriptional effects, phenotyping, and protein interaction databases to identify how yeast responded to the DNA damaging chemical methyl-methanesulfonate. Using this approach, Workman *et al.* were able to identify specific regulatory proteins and signalling mechanisms involved in DNA damage responses. Bandyopadhyay *et al.*[62] expanded this work using 80 000 double mutant yeast genetic interactions to assess differences in static gene interaction networks in the presence and absence of methyl-methanesulfonate. By performing a differential network analysis, Bandyopadhyay *et al.* were able to highlight the participation of genes not previously known to be involved in the yeast DNA damage response. Network analysis integrating blood pressure Genome-Wide Association Studies and blood gene expression profiles identified key molecular mechanisms influencing blood pressure.[63] Network analysis has also been applied in a wide range of toxicological problems including endocrine disrupting chemicals,[45,46] chemical carcinogenicity,[64] and other chemical hazards.[65–67]

While networks can provide key insights into biological systems, and complex interaction networks are being rapidly generated, cells and cellular organisation are fundamentally modular.[68] Recently, the focus has changed from just generating networks to integrating these networks with molecular profiles in order to identify sets of molecules and interactions that participate in a biological function, *i.e.*, "modules" as reviewed in Mitra *et al.*[69] Similarly, network motifs, which are basic building blocks of molecular circuits underpinning cellular functions, have been suggested to provide a basis for understanding thresholds for cellular responses, a key challenge in the AOP framework.[70]

Networks can also be used for cross-species comparison, another key aspect of the AOP framework. To date, comparisons of genetic interactomes (*i.e.*, sets of molecular interactions) are restricted to small sets of genes. But more recently, Ryan *et al.*[71] generated a set of functional modules for the yeast *Schizosaccharomyces pombe* and compared it to an integrated interactome from *Saccharomyces cerevisiae*, revealing a hierarchical model for the evolution of genetic interactions, with the highest conservation within protein complexes, lower conservation among biological processes, and the lowest between distinct biological processes. They also found that despite the extensive rewiring of interactions and evolutionary distance, both networks retained conserved features and displayed similar levels of functional cross-talk between biological processes. Garcia-Reyero *et al.*[72] used a reference network strategy to compare molecular-level effects across species as diverse as rat, bobwhite quail, fathead minnow, and coral. A neuronal gene interaction network was developed using microarray analysis of brain tissue from rats exposed to the munition hexahydro-1,3,5-trinitro-1,3,5-triazine (RDX). Species conservation across this network was then examined by mapping differentially expressed genes from other species onto the rat neuronal network thereby demonstrating conservation of key network attributes.

Ontologies are crucial to "-omics" experiments and data interpretation, but a major challenge to creating ontologies includes the lack of biological knowledge as well as inconsistencies in knowledge curation. Dutkowski *et al.*[73] showed how large networks of gene and protein interactions could be used to infer an ontology, the network-extracted ontology (NeXO), with coverage and power equivalent to the manually curated Gene Ontology (GO), demonstrating yet another use of network biology that could help inform the AOP framework.

9.5 Computationally Predicted and Putative AOP Generation

Development of biological pathways leading to AOs using expert knowledge can be a cumbersome and time-consuming approach for AOP development[74] that is likely to lead to a relatively small number of highly informative pathways. Computational approaches provide new opportunities to utilise big data sources to generate hypothetical biological pathways leading to AOs. A computationally predicted AOP offers great promise in defining the complete range of pathways that may lead to AOs by producing large collections of potential pathways that require validation or expert review. With curation, a cpAOP can be used to generate a putative AOP or pathways that have biological plausibility or are based on statistical evidence but little other support. These computationally driven methods can greatly expedite the numbers of AOPs available by easing the burden of creating a curated AOP. To effectively extract actionable knowledge from an experiment, experimental results are combined with other information. For example, researchers depend on their prior knowledge of relationships between enzymes in a biochemical pathway or disease progression to interpret new experimental data. Algorithms can be useful to drive discovery, but generally they lack expert background knowledge when given a single experiment. Thus, big data and access to knowledge resources in a broader context will enhance toxicologists' ability to identify and extract new knowledge or hypotheses from *in vitro* assays and facilitate predictive toxicology.[76–78]

Most researchers use big data for two distinctly different purposes. The first is hypothesis testing; looking to see if a defined relationship exists using supervised methods and models. The second is knowledge discovery or hypothesis generation, which utilises unsupervised or semi-supervised analyses to identify patterns within the data that meet some predefined measure of interest.[28] Supervised approaches rely on a set of labelled relationships to train an algorithm to make future predictions (*e.g.*, classification approaches like support vector machines). Unsupervised approaches do not use training data but instead look for hidden patterns in unlabelled data (*e.g.*, clustering or self-organising maps). Methods typically used for hypothesis testing or classification in a single or small dataset are also applicable in hypothesis testing of big, integrated data but they are limited in their suitability for knowledge discovery. These limitations arise because many methods cannot accommodate missing, sparse, or incomplete data that are often an artefact of data integration. Additionally, multiple test corrections needed for hypothesis testing methods (such as regression modelling) might further restrict a method's usefulness for initial data exploration. Methods designed for knowledge discovery from data, such as frequent itemset mining,[79] clustering,[80] or graph-based methods[81,82] are robust to these limitations and enable better extraction of new knowledge.

9.5.1 Computational AOP Generation

Large-scale data integration from different sources depends on a few factors, most importantly a shared attribute or key. A cpAOP integrates experimental data with other biological data resources to generate data-driven networks linking an MIE to an AO. Figure 9.4 illustrates data integration for knowledge discovery in the context of cpAOPs, but the general workflow holds across many applications.

Similar data types, *e.g.*, multiple transcriptomic experiments, are combined by concatenation after running applicable scaling algorithms or through merging *via* common attributes such as probe identifiers (Figure 9.4A). The combined "-omics" dataset is enriched by relating experimental treatments to biological pathway perturbations (Figure 9.4B). Corresponding experimental data, *e.g.*, phenotypes (Figure 9.4C), can similarly be processed using literature citations to annotate the observations, or association approaches, *e.g.*, frequent itemset mining[79] can be used to identify associations between the perturbed biological pathways and phenotypes. Independently processed data, *e.g.*, HTS or biomonitoring data, are joined to the network based on a gene/protein target or a literature citation to anchor the data to known phenotypes (Figure 9.4D). Resources that link or translate data from one biological space or identifier to another are essential during data integration, *e.g.*, literature citations for anchoring phenotypes to diseases, and ontologies or protein-to-gene identifiers for relating one identifier to another process or biological entity (Figure 9.4E). Individual data sources are integrated to identify linkages between the datasets and the links and associated properties are carried forward to form a network linking a perturbation to an AO (Figure 9.4F). Associations are inferred from:

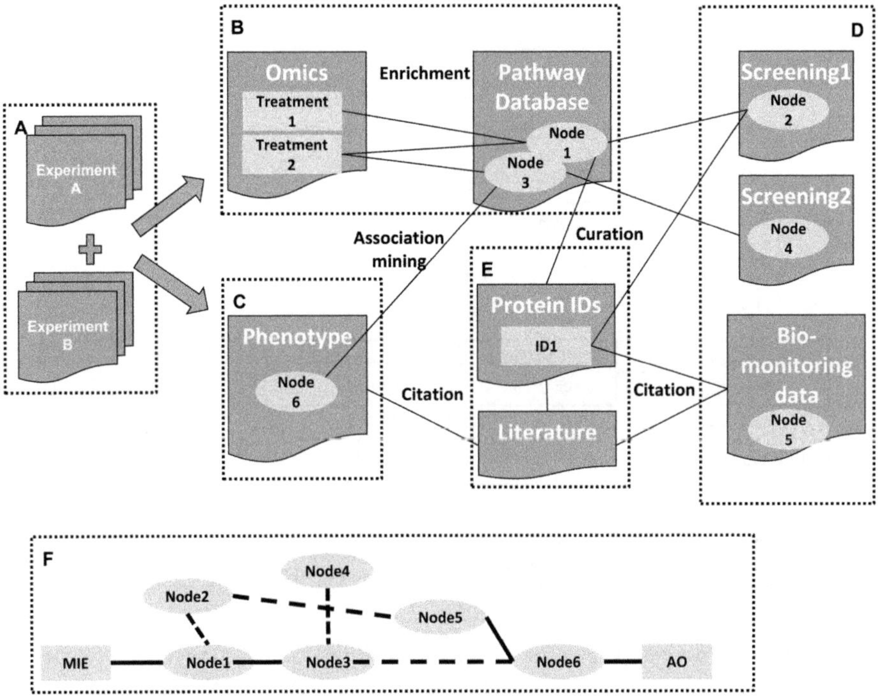

Figure 9.4 Graph-based data integration. Similar datasets may be combined through concatenation (A) prior to data-specific processing for the individual treatments (B, C) to facilitate downstream processing steps. Otherwise independently processed data (D) are brought into the network using identifiers and citations or curations to facilitate the integration (E). The result of the linkage of the different datasets is a network structure (F) that can be used to build hypotheses about the relationships between the different data types. Dashed lines in (F) illustrate the connections made through data integration and inference methods.

(i) direct results of data integration, *e.g.*, a direct link between a gene or assay and its corresponding protein product (Node2 and Node5 to ID1 in Figure 9.4D and E); and
(ii)results from a computational step such as with frequent itemset mining (Node3 to Node6) or an enrichment calculation (Treatment1 to Node1 in Figure 9.4B).

We applied the method outlined in Figure 9.4 to the Toxicogenomics Project-Genomics Assisted Toxicity Evaluation System (TG-GATES) toxicogenomics data[83,84] and generated cpAOP networks for carbon tetrachloride and tamoxifen; two compounds known to result in non-alcoholic fatty liver (*i.e.*, steatosis) with prolonged use.[85–87] The intersection of the

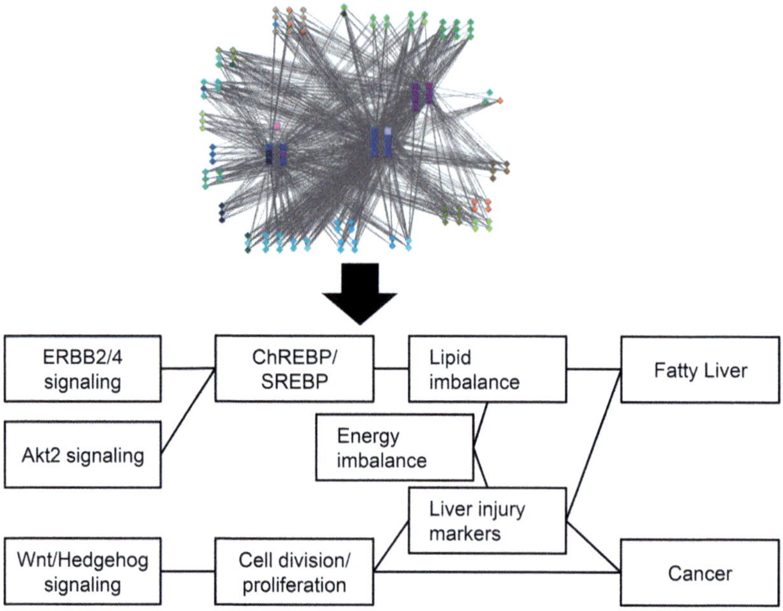

Figure 9.5 The top figure is the computationally predicted AOP (cpAOP) network resulting from the intersection of the cpAOP networks for carbon tetrachloride and tamoxifen using the information on timing and connectivity results in the cpAOPs shown below. Initial stress response is mediated through signalling cascades that activate the master regulator RAC-beta serine/threonine-protein kinase (Akt2). Akt2 interacts with lipid and energy metabolism responding to the lipid imbalance within the system. Specifically, Akt2 activates sterol regulatory element-binding protein (SREBP), which in turn regulates fatty acid, cholesterol and triglyceride metabolism. The result is alteration of liver and serum lipids leading to liver injury. Coincident with the lipid imbalance in the cpAOP network is a cpAOP for cancer. Shown here is the initial stress signalling response through WNT/hedgehog; the increased cell proliferation coincides with phenotypic markers for cell proliferation and liver injury. ERBB2/4 signalling denotes receptor tyrosine-protein kinase (ERBB2) heterodimerisation with other ERBB family members;[88] WNT is the fusion of *wingless* gene from *Drosophila* and its vertebrate homolog *integrated*.[91]

two networks leads to a cpAOP network for non-alcoholic fatty liver (modified summary diagram shown in Figure 9.5). With information on the perturbed biological pathways (based on Reactome v52 pathway annotations[88,89]) and the associated phenotypes, a hypothesised cpAOP was generated. The hypothetical model generated from the cpAOP (Figure 9.5) shows how an initial stress response influences the master regulator RAC-beta serine/threonine-protein kinase (Akt2),[90] leading to metabolic changes. This metabolic response, presumably due to the lipid imbalance, is characterised by changes in fatty acid oxidation and the energy exchange of ketone metabolism–urea cycle–tryptophan metabolism. The lipid dysregulation impacts sterol regulatory element-binding protein (SREBP),[90] ultimately resulting in cholesterol and lipid imbalance at the serum level. The associated liver injury is also apparent based on the cpAOP. The scaffold generated through data integration can now serve as a starting framework for the researcher or can be used to feed into additional workflows for further computational model building.

9.5.2 From Computationally Predicted AOPs to Putative AOPs

Once cpAOPs have been developed, they need to be "upgraded" to putative AOPs prior to use through a verification process where existing literature is consulted for biological plausibility of the pathway, in addition to supporting materials within the pathway databases that were used for cpAOP development. For instance, if the Reactome database (http://www.reactome.org/) were used in the construction of the cpAOP, its supporting documentation would be consulted for the relevant Reactome pathways. Additional literature should also be consulted to assess the completeness of the cpAOP KEs, adding KEs where necessary and trimming out unnecessary predicted KEs. These verification steps will result in a fully developed, putative AOP.

9.6 Use of AOPs to Support Testing and Risk Assessment

The process of identifying potential hazards and assessing risks is time-consuming and expensive. It can take hundreds of person-hours and significant sums of money, resources, and animals to complete studies required for assessments and testing. By using computational tools, including graph and network theory,[81,82] computer science, and mathematics, the efficiency of chemical toxicity evaluations could be significantly improved and provide a better understanding of potential risks in a shorter amount of time.

9.6.1 Identification of Sufficient Key Events in AOP Networks

In reality, AOPs occur as interacting networks of AOPs. By connecting multiple putative AOPs, interacting AOP networks can be generated. This complexity may seem daunting, where thousands of KEs are present that connect to hundreds or thousands of AOs. However, the network approach is particularly useful in understanding how KEs in one AOP could affect another through the identification of intersections where modulation may occur. For instance, the signal propagating through one AOP may be modulated (*i.e.*, increased, decreased, turned off, or turned on) through the action of signals emanating from another AOP.

The intersections of AOPs reflect biological complexity and confound chemical toxicity evaluation. One approach to this problem is to measure every KE in all of the AOPs to obtain a comprehensive snapshot of what is happening across the known biological space. While this would increase knowledge of what happens to signals when AOPs intersect, this approach is impractical due to an inability to measure all the KEs, considering the high cost, and prohibitively large number of assays that would be needed if they were available. Thus, the number of KEs measured must be reduced by strategically designing our experimental testing efforts.

The complexity of an AOP network can be decreased by focusing on only those KEs which are sufficient to infer that an AO will occur. In some cases, the literature can be used to help identify these KEs. For instance, it is well known that activation of tumour protein p53 (p53 – also known as TP53) is sufficient to infer that DNA damage is present. However, our knowledge of biology is incomplete and we do not know all of the sufficient KEs to infer all of the possible AOs that we may care about. We can use the fact that AOPs are causal graphs to our advantage in identifying sufficient KEs. Causal graphs are those where child KEs are direct descendants of their parent KEs, and a child cannot be its own predecessor (*i.e.*, there is no feedback; AOPs with feedback can still be modelled as causal graphs, by using a temporal component). Because of these constraints, causal graphs have certain mathematical properties that can be used to identify sufficient KEs.

We can use Pearl's backdoor algorithm[92] to exploit these mathematical properties and identify sufficient KEs in AOPs. The algorithm requires a starting point and an end point – here we can use an MIE and an AO. The algorithm starts by introducing an edge that connects all parents that share a child KE. Next, the algorithm identifies the shortest path from the AO leading to the MIE. Note that the paths being explored by the algorithm are not directed paths – meaning the directionality of the arrows in the AOP are ignored. The sufficient KE in this path is the one adjacent to the AO. This sufficient KE is removed from the graph, and the path search is performed again, removing the sufficient KEs, until there are no more paths leading from the AO to the MIE. Sufficient KEs identified by the backdoor algorithm are the set of KEs that should be monitored to infer the AO. Note that this method should only be employed when there is insufficient information in the literature to identify sufficient KEs.

In the future, these types of analyses will be facilitated by the AOPXplorer. Computers will use the data encoded according to the AOP Ontology[101] to perform first-order logic queries in order to infer whether a chemical at a given dose may cause any number of AOs. For instance, through a simple query, a user could identify if a chemical causes a deficit in executive function at any exposure level based on the assay data and AOP networks within the system. By using Pearl's backdoor algorithm to identify sufficient KEs and infer executive function of the KEs within an AOP network, a computer running AOPXplorer can easily answer this query. Similarly, AOPXplorer

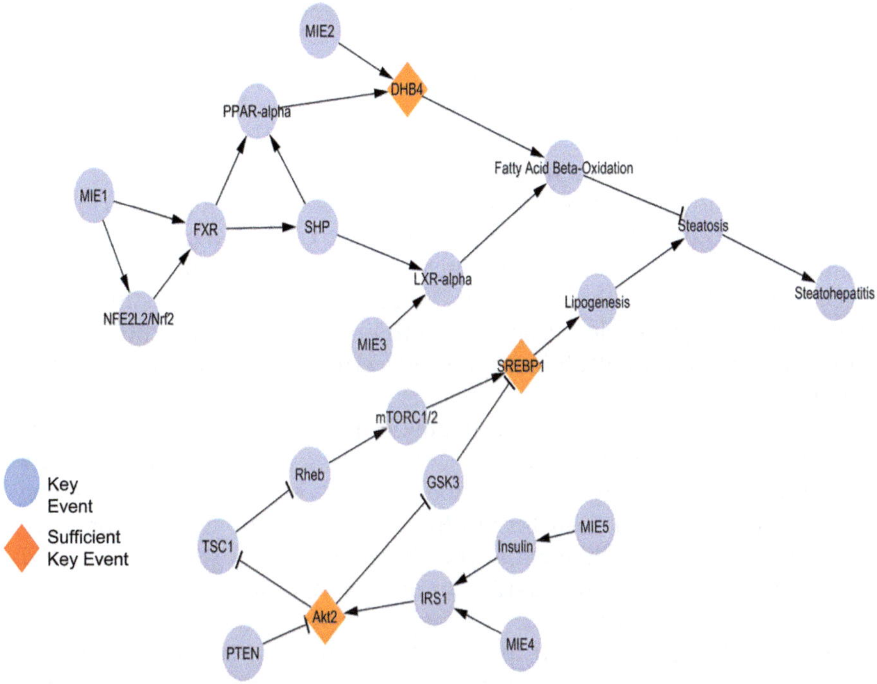

Figure 9.6 Adverse outcome pathway network for steatosis. This is a sample AOP network for steatosis focused on fatty acid beta-oxidation and lipogenesis; all codes represent proteins. It should be noted that a complete AOP network for steatosis is likely to include additional pathways; however, this sample is sufficient for our example. A sufficient key event is one where it has been demonstrated either in the literature (Akt2, SREBP1, and DHB4) or through causal network analysis (DHB4 and SREBP1) for activation/deactivation of the key event to be sufficient to infer that the adverse outcome (*e.g.*, steatosis and steatohepatitis) will occur. By definition, all key events in the network are necessary for the adverse outcome to occur.

could be queried for combinations of sufficient assays across one or more AOP networks to create a test battery for one or more AOs.

9.6.2 Case Study: Steatosis

A putative AOP network for steatosis can be assembled from several individual AOPs connected together based on the current understanding of steatosis mediated through protein regulatory control of fatty acid beta-oxidation and lipogenesis (Figure 9.6). Using the cpAOP as a starting point, additional information was included from pathway databases and the literature to facilitate assembly of the AOP network. In this case, literature information identifies peroxisomal multifunctional enzyme type 2 (DHB4), RAC-beta serine/threonine-protein kinase (Akt2), and SREBP1 as sufficient

KEs that can determine whether or not a chemical exposure is likely to result in steatosis.

The evidence for each of these proteins is based on either a pharmacological, biochemical, or genetic knockout study. DHB4 is a protein involved in fatty acid beta-oxidation. If this gene is knocked out in rodents or non-functional in humans, this leads to D-bifunctional deficiency, which leads to steatosis, among other AOs (http://omim.org/clinicalSynopsis/261515). In mice, Akt2 activation is sufficient to induce steatosis based on a study using constitutively active Akt2. A lack of functional Akt2 is sufficient to block lipogenesis in the Lep$^{ob/ob}$ obesity model.[93,94] SREBP1 activation is sufficient to result in hypertriglyceridemia secondary to lipogenesis, while knocking out SREBP1 is sufficient to result in a hypotriglyceridemia secondary to blockage of lipogenesis.[94]

Using the backdoor algorithm available in the computing language Pearl and network theory, it can be seen that monitoring DHB4 and SREBP1 activity is all that is necessary to infer whether or not a chemical exposure may result in steatosis. If an assay for SREBP1 is activated, or an assay for DHB4 is inactivated, then it is likely that the chemical may cause steatosis. Modulation of both SREBP1 and DHB4 is not necessary to infer steatosis. The network analysis also demonstrates the assumptions that are required for monitoring of Akt2 activation to be sufficient to infer steatosis. Although the literature reports that constitutive activation of Akt2 is sufficient to lead to steatosis, secondary to increased lipogenesis, this requires a functional SREBP1. Thus, for information about Akt2 to be sufficient to infer steatosis, SREBP1 activity should also be monitored.

The steatosis case study demonstrates the importance of understanding the network, and how network theory can be used to identify a minimal set of sufficient KEs to monitor for toxicity assessment. By starting with a hypothesis derived from cpAOPs, incorporating literature and pathway database information to upgrade the cpAOP into a pAOP, and then applying causal network theory, the minimal set of KEs required to infer an AO was identified. This approach will allow investigators to save time and money in future toxicity assessments, and to streamline the testing required for risk assessments.

9.7 Translating Experimental Data into Model Structure: Integration of High-throughput Datasets or Expert Elicitation?

To rapidly assess the potential hazards and risks of chemicals found in the environment, researchers have sought high-throughput approaches to characterise chemical effects through standardised screening assays as described in Section 9.3. For example, the ToxCast™ database provides concentration–response curves for thousands of unique substances, and aims to support characterisation of *in vivo* chemical effects by providing automated *in vitro* endpoint data. However, molecular data gathered from other biological scales, such as microarray and sequencing data, provide yet another dimension of toxicological insight. Integrating such disparate data into a mechanistic understanding of the molecular regulation of xenobiotic compounds and substances within an organism remains a great challenge. One solution involves using mechanistic mathematical modelling to evaluate and assess individual-level AOs predicted from these molecular level data by causally linking KEs of an AOP. We approach this subject by examining two exemplary models developed in the context of the hypothalamus-pituitary-gonad (HPG) axis of teleost fish, a highly conserved regulatory network among vertebrates, wherein two different yet complementary model development strategies were utilised in an effort to explain experimental data.

9.7.1 Mechanistic Modelling Methods

An iterative procedure can be employed to develop mathematical models that predict biological data. First, the structure of the model is conceptualised from the underlying regulatory biology, with a goal of capturing the salient processes (*e.g.*, rate-limiting fluxes) that affect measured endpoints (*e.g.*, plasma hormone concentrations). Next, this conceptual model is translated into a mathematical model using standard methods determined by both biology and experimental details, *e.g.*, reaction-limited chemical kinetic equations may be used if organ tissues are homogenised to measure concentrations, which amounts to a molecular averaging over the cellular population of the tissue. Model parameter values can be assigned through a combination of literature sourcing and computational methods, such as model "training"/parameter estimation from the available datasets. Finally, model fidelity can be evaluated through comparison of its predictions with data, and disparities can be addressed by revisiting the conceptual model, thus completing the model-building cycle.

There are at least two approaches to developing a conceptual model, which, hypothetically, captures the necessary regulatory biology of the system. One is to develop a model structure based on expert elicitation. In this case, expert knowledge, which represents the sum of experiences gained during a professional career, is used to filter information available in the literature to provide specific, implementable guidance on the model structure. An example of this is a model for aromatase inhibition described in Section 9.7.2.1. Alternatively, when larger, more complex "-omics" datasets are available, formal bioinformatics analyses may be used to develop detailed hypotheses by interrogating the regulatory biology. This type of analysis can result in a visual "hairball", which as a whole is uninformative for model development, but can be used to develop hypotheses about a regulatory network. In this case, some level of expert elicitation is still needed in order to distil the complex network of regulatory interactions into a level of detail that balances the need for parsimony with literature documentation for development into a computational model (see Section 9.7.2.2).

9.7.2 Case Study: Aromatase Inhibition Model Development

9.7.2.1 Model Development by Expert Elicitation

Perturbation of the HPG axis by environmental chemicals can alter sex steroid levels, leading to reproductive dysfunction. In female fathead minnows (*Pimephales promelas*, abbreviated to FHM), inhibition of aromatase (CYP19A), an enzyme that converts testosterone into 17β-estradiol (E2), impairs oogenesis and reproduction. The key anatomical and regulatory components of the FHM HPG axis were identified by expert elicitation, and coded into a differential equation-based computational model.[95] In this model, an FHM is represented by six tissue compartments: gill, brain (as a tissue including the hypothalamus and pituitary glands), ovary, liver, venous blood, and the rest of body. The "rest of body" compartment accounts for blood flow and tissue volume not directly associated with the HPG axis or reproductive function. The key input from the expert elicitation was identification of a regulatory loop that senses decreased levels of E2 and upregulates aromatase activity in an attempt to compensate for its inhibition. This regulatory loop identified through expert elicitation involves

(i) *sensing* changes in the circulating E2 level in the brain compartment;
(ii) release from the brain of a signalling molecule (analogous to luteinising hormone (LH) and follicle stimulating hormone (FSH));
(iii) interaction of these signalling molecules with receptors at the ovary–blood interface;
(iv) an increased production of ovarian aromatase mRNA as a function of the amount of signal–receptor complex; and, finally,
(v) *mRNA* translation into new aromatase protein.

In parallel with this model development, experimental data were obtained from female FHMs exposed to a non-steroidal aromatase inhibitor, fadrozole.[96] The data consist of plasma E2 and ovarian aromatase mRNA concentrations from two dose–response and time-course experiments, each of which included exposure and depuration phases. The combination of dose–response and time-course data is ideal for model calibration/training, and good descriptions of the data were obtained. Since the increase in aromatase activity is due to a transcriptionally mediated process, the activity does not return to control levels immediately upon cessation of exposure to the aromatase inhibitor. Thus, in agreement with the data, the model predicts plasma E2 concentrations higher than control levels, *i.e.*, a rebound effect, for several days post-exposure. Computational models of the type described here are useful in two ways – as part of the research process, where the model can aid in data analysis and experimental design, and as part of the regulatory process, where model-generated predictions can support regulatory decision-making. This latter application requires, however, confidence on the part of regulators that the uncertainties associated with the predictions are no greater than the uncertainties associated with decision-making by more traditional approaches.

9.7.2.2 *Model Development by Integration of High-throughput Microarray Analysis and Network Reconstruction*

In contrast to the "top down" approach of the expert elicitation method, which aims to develop computational models predictive of key biological endpoints by interrogating subject matter experts, a "bottom up" approach may be employed when suitable experimental data are available. In this approach, "-omics" technologies may be used for direct identification of key regulatory pathways altered in response to toxicant exposure. This method has been used before to develop a dynamic computational model of the HPG axis[97] and, similar to the expert elicitation model explained earlier, used to predict plasma E2 concentrations in addition to transcriptional induction curves for CYP19A, steroidogenic acute regulatory protein (StAR), and follicle stimulating hormone receptor (FSHR).

In this approach, tissue samples obtained from the same fadrozole study[96] used in Section 9.7.2.1 were analysed using one-colour microarray analysis. The Context Likelihood of Relatedness method[55] was used to reconstruct a gene-regulatory network mapping the expression patterns involved in the responses to aromatase inhibition. Expert elicitation was then used to reduce this large network of regulatory interactions into a smaller network more amenable to mathematical modelling and analysis. Results suggested that testosterone, which accumulates in ovary tissues if conversion to E2 is suppressed by aromatase inhibition, interacts with the androgen receptor, to associate with expression levels of key steroidogenic genes (*i.e.*, transport proteins, enzymes). A mechanistic feedback loop between testosterone and aromatase was therefore identified, and a mathematical model of the additional regulatory loops was incorporated into a previously developed model of the HPG axis.[98] Predictions from this new androgen-receptor-based model exhibited high fidelity with the experimental data for plasma E2 concentrations in addition to data obtained from quantitative polymerase chain reaction (qPCR) for the CYP19A, StAR, and FSHR genes. This model was predictive of both an "overshoot" and rebound observed in the plasma E2 data, similar to performance of the expert elicitation based model described in Section 9.7.2.1. This overshoot could be tracked to a mechanism wherein aromatase–fadrozole complex, unable to bind with testosterone, accumulated under the fadrozole exposure. When the exposure ceased, fadrozole dissociated from aromatase and the larger pool of previously inhibited aromatase combined with the accumulated testosterone to generate a surge in E2 concentrations as observed in the plasma data. In this model, the dynamics of aromatase inhibition result from the regulatory feedback loops that are programmed into it, and as such should apply to a broad range of exposure concentrations. While the available data hints at dose-dependent levels of overshoot and rebound in the plasma E2 concentrations, the experimental sampling rate used by Villeneuve *et al.*[96] was insufficient to resolve the issue. In the future, experiments could be designed to address this issue by increasing the number of time points and further resolving the overshoot dynamics of the plasma E2 concentrations.

9.7.2.3 Comparison of the Two Model Development Approaches

"Omics" technologies can generate large amounts of data, so much data in fact that no longer is the acquisition of data rate-limiting, but rather, its analysis. A significant challenge in this analysis is identifying a reduced dataset that directly supports development of a computational model of the system under study. A "hairball" of affected genes and their connections, while often visually impressive, is typically not useful for model development – there is a need to extract a reduced dataset from the hairball to provide a limited set of genes and connections for which development of a computational model is practical. This second step requires expert consultation. Thus, the model reported by Mayo *et al.*[97] was developed from a conceptual framework that combined bioinformatics-based network reconstruction analysis with expert-driven interpretation of the resulting regulatory interactions. In contrast, the model reported by Breen *et al.*[95] employed an approach driven primarily by expert elicitation to arrive at a quite different regulatory structure. Nevertheless, both models were developed with the goal of better understanding the dynamic regulation of proteins and hormones after exposure to fadrozole, which perturbs reproduction of the FHM and potentially other teleost fish.

The two different models of the FHM HPG axis and its perturbation through inhibition of aromatase likely reflect the complexity of the structure of the HPG axis and the multiple regulatory circuits embedded within it. Both are plausible until additional data about the molecular regulatory dynamics are obtained to validate one model over the other. We suspect that the two models are in fact complementary, with the different approaches to model development having identified different parts of the overall regulatory circuitry to integrate transcriptional-regulatory, protein-interaction, and metabolic network layers. Kafri *et al.*[99] noted that functional redundancy increases the robustness of organisms to mutations and other stresses. We would thus anticipate that with additional exploration of the regulatory dynamics in the FHM HPG axis, the two models could be merged into a single, more sophisticated description of the regulatory biology governing teleost reproduction.

9.8 Conclusions

Big data enables a number of new approaches to understand and predict how chemicals cause toxicological effects. Within this chapter, a variety of approaches and tools are described that enable aggregation, querying, and analysis of big data sources to infer biological networks and their regulatory dynamics. Aggregating knowledge of chemicals, toxicological assays, HTS data, and other sources permits efficient data mining, modelling, and decision-making.

The AOP-KB is an example of how to compile and organise big data in toxicology so that people unfamiliar with sophisticated data analytics have access to the current state of knowledge. Sophisticated analysis of HTS assays and curated databases can drive development of toxicological relationships and facilitate discovery of new pathways that may lead to adverse effects. Big datasets for chemical effects that include many different parameters including phenotype, gene and protein expression, metabolites, and DNA binding sites can be used to explore systems relationships and discover new relationships from toxicological data. Finally, different modelling approaches can utilise large datasets to create dynamic models of the toxicological mechanisms leading to chemical effects.

The development of the structured data knowledge base, networks, computationally predicted AOPs, and dynamic models serve a valuable purpose as resources for assessing chemical effects and both generating and testing hypotheses. These resources extend our basic knowledge and improve the efficiency of ongoing research; however, the main justification for the activity must come from its end use, *e.g.*, regulatory toxicology and risk assessment. The structuring of knowledge in an AOP framework directly enables translation of data to support regulatory toxicology and risk assessments.

The use of data integration tools (*e.g.*, databases and online resources) and quantitative/computational modelling and analytic methods to access the rich information in big data has become integral to successfully addressing toxicology questions. Knowledge bases and models continue to evolve as additional data become available. Likewise, the case studies may also evolve, and readers should keep in mind that these are examples meant to demonstrate different approaches to big data integration and inference in predictive toxicology. As better tools become available to facilitate data integration from a wide variety of sources, our knowledge and understanding of toxicity and AOPs will improve. Analyses and interpretation of big datasets has only just begun.

9.9 Future Directions

Advances in biotechnology and HTS have helped transform toxicity testing today. With screening programmes like the ToxCast™ and Tox21 programmes, enormous amounts of chemical–gene data from HTS assays will continue to be produced. These large HTS datasets have led to a big data scenario, leading to a need for the development of large databases to share and utilise these data. Integration of HTS data from multiple databases can provide a robust dataset for understanding chemical toxicity. Computational models and bioinformatics approaches have provided a means to utilise these HTS data for predicting *in vivo* toxicity. However, there is still an urgent need to develop other novel approaches to access, integrate, and use high quality and well-curated data for chemical toxicity. As more of these data are generated from various sources, strategies will need to be developed to ensure the most relevant and comparable HTS data are utilised.

Software development can greatly enhance data interpretation. For example, the development of software consisting of an automated pipeline to extract raw data from available HTS, toxicogenomics, and *in vivo* effects (*e.g.*, ToxRefdb) databases and pre-process values according to quality assurance guidelines to standardise the data would reduce the number of person-hours required for this task. Following data quality assessment, the software could cluster similar biological activity profiles for chemicals, and allow use of different machine-learning approaches to predict the bioactivity responses of a reported *in vivo* effect. This type of free software would facilitate the ability to utilise big data from *in vitro* screening and toxicogenomics studies for predicting toxicologically relevant outcomes. Software is available to do this, with some being better than others, though they are not free. GeneData in Switzerland is working on trying to develop this type of software, but it will likely require purchase.

While some of the potential applications of network inference are still under development, the future holds great promise. As suggested by a very innovative idea,[75] the actual concept of network generation and interpretation as well as currently used gene ontologies, which are static, might evolve toward more predictive or "executable" ontologies that could be supported by recent progress in artificial intelligence. The future of network inference and the AOP framework can be intrinsically linked, where each field benefits from developments in the other complementary fields.

Lastly, dynamic models within an AOP context can provide quantitative predictions of the time-course and dose–response for chemical effects related to perturbations of normal biological function (*e.g.*, HPG axis).[105] Model-generated predictions with no/low uncertainties would have obvious relevance to support regulatory decision-making. The main obstacle to such an application rests not only with a formal evaluation of confidence in their predictions, but also with how this confidence compares to that associated with decisions made in the absence of the models. This need for "comparisons of confidences" is an area that requires more investigation. The routine use of structurally sophisticated, biologically based models for support of regulatory decision-making may not occur without significant progress on this issue.

References

1. J. G. Hengstler, H. Foth, R. Kahl, P. J. Kramer, W. Lilienblum, T. Schulz and H. Schweinfurth, The REACH concept and its impact on toxicological sciences, *Toxicology*, 2006, **220**, 232–239.
2. R. Kavlock, K. Chandler, K. Houck, S. Hunter, R. Judson, N. Kleinstreuer, T. Knudsen, M. Martin, S. Padilla, D. Reif, A. Richard, D. Rotroff, N. Sipes and D. Dix, Update on EPA's ToxCast Program: Providing high throughput decision support tools for chemical risk management, *Chem. Res. Toxicol.*, 2012, **25**, 1287–1302.
3. T. Hartung and C. Rovida, Chemical regulators have overreached, *Nature*, 2009, **460**, 1080–1081.
4. M. J. W. A. Schiffelers, B. J. Blaauboer, C. F. M. Hendriksen and W. E. Bakker, Regulatory acceptance and use of 3R models: a multilevel perspective, *Altex*, 2012, **29**, 287–300.
5. National Research Council, *Toxicity Testing in the 21st Century: A Vision and a Strategy*, The National Academies Press, Washington, D.C, 2007.
6. M. E. Andersen and D. Krewski, Toxicity testing in the 21st century: bringing the vision to life, *Toxicol. Sci.*, 2009, **107**, 324–330.
7. D. L. Villeneuve and N. Garcia-Reyero, Vision & strategy: Predictive ecotoxicology in the 21st century, *Environ. Toxicol. Chem.*, 2011, **30**, 1–8.
8. I. Cote, P. T. Anastas, L. S. Birnbaum, R. M. Clark, D. J. Dix, S. W. Edwards and P. W. Preuss, Advancing the next generation of health risk assessment, *Environ. Health Perspect.*, 2012, **120**, 1499–1502.
9. G. T. Ankley, R. S. Bennett, R. J. Erickson, D. J. Hoff, M. W. Hornung, R. D. Johnson, D. R. Mount, J. W. Nichols, C. L. Russom, P. K. Schmieder, J. A. Serrrano, J. E. Tietge and D. L. Villeneuve, Adverse outcome pathways: A conceptual framework to support ecotoxicology research and risk assessment, *Environ. Toxicol. Chem.*, 2010, **29**, 730–741.
10. R. A. Becker, G. T. Ankley, S. W. Edwards, S. W. Kennedy, I. Linkov, B. Meek, M. Sachana, H. Segner, B. Van Der Burg, D. L. Villeneuve, H. Watanabe and T. S. Barton-Maclaren, Increasing scientific confidence in Adverse Outcome Pathways: Application of tailored Bradford-Hill considerations for evaluating weight of evidence, *Regul. Toxicol. Pharmacol.*, 2015, **72**, 514–537.

11. S. J. Enoch, Chemical Category Formation and Read-Across for the Prediction of Toxicity, in *Recent Advances in QSAR Studies: Methods and Applications*, ed. T. Puzyn,J. Leszczynski and M. T. Cronin, Springer Science+Business Media B.V., 2010, pp. 209–219.
12. A. Carnero, High throughput screening in drug discovery, *Clin. Transl. Oncol.*, 2006, **8**, 482–490.
13. A. McCarthy, The NIH Molecular Libraries Program: Identifying Chemical Probes for New Medicines, *Chem. Biol.*, 2010, **17**, 549–550.
14. R. R. Tice, C. P. Austin, R. J. Kavlock and J. R. Bucher, Transforming public health protection: A U.S. Tox21 progress report, *Environ. Health Perspect.*, 2013, **121**, 756–765.
15. K. A. Houck and R. J. Kavlock, Understanding mechanisms of toxicity: Insights from drug discovery research, *Toxicol. Appl. Pharmacol.*, 2008, **227**, 163–178.
16. Y. Wang, T. Suzek, J. Zhang, J. Wang, S. He, T. Cheng, B. A. Shoemaker, A. Gindulyte and S. H. Bryant, PubChem BioAssay: 2014 update, *Nucleic Acids Res.*, 2014, **42**, D1075–D1082.
17. A. P. Bento, A. Gaulton, A. Hersey, L. J. Bellis, J. Chambers, M. Davies, F. A. Krüger, Y. Light, L. Mak, S. McGlinchey, M. Nowotka, G. Papadatos, R. Santos and J. P. Overington, The ChEMBL bioactivity database: an update, *Nucleic Acids Res.*, 2014, **42**, D1083–D1090.
18. G. C. Fonger, D. Stroup, P. L. Thomas and P. Wexler, Toxnet: A computerized collection of toxicological and environmental health information, *Toxicol. Ind. Health*, 2000, **16**, 4–6.
19. R. S. Judson, M. T. Martin, P. Egeghy, S. Gangwal, D. M. Reif, P. Kothiya, M. Wolf, T. Cathey, T. Transue and D. Smith, *et al.*, Aggregating data for computational toxicology applications: The U.S. Environmental Protection Agency (EPA) Aggregated Computational Toxicology Resource (ACToR) System, *Int. J. Mol. Sci.*, 2012, **13**, 1805–1831.
20. M. Waters, S. Stasiewicz, B. A. Merrick, K. Tomer, P. Bushel, R. Paules, N. Stegman, G. Nehls, K. J. Yost and C. H. Johnson, CEBS—Chemical Effects in Biological Systems: a public data repository integrating study design and toxicity data with microarray and proteomics data, *Nucleic Acids Res.*, 2008, **36**, D892–D900.
21. B. Ganter, R. D. Snyder, D. N. Halbert and M. D. Lee, Toxicogenomics in drug discovery and development: mechanistic analysis of compound/class-dependent effects using the DrugMatrix database, *Pharmacogenomics*, 2006, **7**, 1025–1044.
22. J. Lamb, E. D. Crawford, D. Peck, J. W. Modell, I. C. Blat, M. J. Wrobel, J. Lerner, J.-P. Brunet, A. Subramanian and K. N. Ross, *et al.*, The Connectivity Map: using gene-expression signatures to connect small molecules, genes, and disease, *Science*, 2006, **313**, 1929–1935.
23. A. P. Davis, C. J. Grondin, K. Lennon-Hopkins, C. Saraceni-Richards, D. Sciaky, B. L. King, T. C. Wiegers and C. J. Mattingly, The Comparative

Toxicogenomics Database's 10th year anniversary: update 2015, *Nucleic Acids Res.*, 2015, **43**, D914–D920.
24. D. Wishart, D. Arndt, A. Pon, T. Sajed, A. C. Guo, Y. Djoumbou, C. Knox, M. Wilson, Y. Liang and J. Grant, T3DB: the toxic exposome database, *Nucleic Acids Res.*, 2015, **43**, D928–D934.
25. D. Fourches, M. F. Sassano, B. L. Roth and A. Tropsha, HTS navigator: freely accessible cheminformatics software for analyzing high-throughput screening data, *Bioinformatics*, 2014, **30**, 588–589.
26. B. Chen, X. Dong, D. Jiao, H. Wang, Q. Zhu, Y. Ding and D. J. Wild, Chem2Bio2RDF: a semantic framework for linking and data mining chemogenomic and systems chemical biology data, *BMC Bioinf.*, 2010, **11**, 255.
27. R. J. Kavlock, D. J. Dix, K. A. Houck, R. S. Judson, M. T. Martin and A. M. Richard, ToxCast: Developing predictive signatures for chemical toxicity, *AATEX J.*, 2008, **14**, 623–627.
28. C. S. Greene, J. Tan, M. Ung, J. H. Moore and C. Cheng, Big data bioinformatics, *J. Cell. Physiol.*, 2014, **229**, 1896–1900.
29. A. Lavecchia, Machine-learning approaches in drug discovery: methods and applications, *Drug Discovery Today*, 2015, **20**, 318–331.
30. J. B. O. Mitchell, Machine learning methods in chemoinformatics, *Wiley Interdiscip. Rev.: Comput. Mol. Sci.*, 2014, **4**, 468–481.
31. R. Judson, F. Elloumi, R. W. Setzer, Z. Li and I. Shah, A comparison of machine learning algorithms for chemical toxicity classification using a simulated multi-scale data model, *BMC Bioinf.*, 2008, **9**, 241.
32. N. C. Kleinstreuer, D. J. Dix, K. A. Houck, R. J. Kavlock, T. B. Knudsen, M. T. Martin, K. B. Paul, D. M. Reif, K. M. Crofto and K. Hamilton, *et al.*, In vitro perturbations of targets in cancer hallmark processes predict rodent chemical carcinogenesis, *Toxicol. Sci.*, 2013, **131**, 40–55.
33. N. S. Sipes, M. T. Martin, D. M. Reif, N. C. Kleinstreuer, R. S. Judson, A. V. Singh, K. J. Chandler, D. J. Dix, R. J. Kavlock and T. B. Knudsen, Predictive models of prenatal developmental toxicity from ToxCast High-Throughput Screening Data, *Toxicol. Sci.*, 2011, **124**, 109–127.
34. M. T. Martin, T. B. Knudsen, D. M. Reif, K. A. Houck, R. S. Judson, R. J. Kavlock and D. J. Dix, Predictive model of rat reproductive toxicity from ToxCast high throughput screening, *Biol. Reprod.*, 2011, **85**, 327–339.
35. K. J. Minch, T. R. Rustad, E. J. R. Peterson, J. Winkler, D. J. Reiss, S. Ma, M. Hickey, W. Brabant, B. Morrison, S. Turkarslan, C. Mawhinney, J. E. Galagan, N. D. Price, N. S. Baliga and D. R. Sherman, The DNA-binding network of Mycobacterium tuberculosis, *Nat. Commun.*, 2015, **6**, 5829.
36. T. R. Rustad, K. J. Minch, S. Ma, J. K. Winkler, S. Hobbs, M. Hickey, W. Brabant, S. Turkarslan, N. D. Price, N. S. Baliga and D. R. Sherman, Mapping and manipulating the Mycobacterium tuberculosis transcriptome using a transcription factor overexpression-derived regulatory network, *Genome Biol.*, 2014, **15**, 502.

37. N. Garcia-Reyero, B. L. Escalon, E. Prats, J. K. Stanley, B. Thienpont, N. L. Melby, E. Barón, E. Eljarrat, D. Barceló, J. Mestres, P. J. Babin, E. J. Perkins and D. Raldúa, Effects of BDE-209 contaminated sediments on zebrafish development and potential implications to human health, *Environ. Int.*, 2014, **63**, 216–223.
38. N. Garcia-Reyero, A. J. Kennedy, B. L. Escalon, T. Habib, J. G. Laird, A. Rawat, S. Wiseman, M. Hecker, N. Denslow, J. A. Steevens and E. J. Perkins, Differential effects and potential adverse outcomes of ionic silver and silver nanoparticles in vivo and in vitro, *Environ. Sci. Technol.*, 2014, **48**, 4546–4555.
39. K. Hallén, J. Björkegren and J. Tegnér, Detection of compound mode of action by computational integration of whole-genome measurements and genetic perturbations, *BMC Bioinf.*, 2006, **7**, 51.
40. C. M. Warner, K. A. Gust, J. K. Stanley, T. Habib, M. S. Wilbanks, N. Garcia-Reyero and E. J. Perkins, A Systems Toxicology approach to elucidate the mechanisms involved in RDX species-specific sensitivity, *Environ. Sci. Technol.*, 2012, **46**, 7790–7798.
41. R. L. Wang, D. Bencic, D. L. Villeneuve, G. T. Ankley, J. Lazorchak and S. Edwards, A transcriptomics-based biological framework for studying mechanisms of endocrine disruption in small fish species, *Aquat. Toxicol.*, 2010, **98**, 230–244.
42. K. Basso, A. A. Margolin, G. Stolovitzky and U. Klein, *et al.*, Reverse engineering of regulatory networks in human B cells, *Nat. Genet.*, 2005, **37**, 382–390.
43. G. Della Gatta, T. Palomero, A. Perez-Garcia, A. Ambesi-Impiombato, M. Bansal, Z. W. Carpenter, K. De Keersmaecker, X. Sole, L. Xu, E. Paietta, J. Racevskis, P. H. Wiernik, J. M. Rowe, J. P. Meijerink, A. Califano and A. A. Ferrando, Reverse engineering of TLX oncogenic transcriptional networks identifies RUNX1 as tumor suppressor in T-ALL, *Nat. Med.*, 2012, **18**, 436–440.
44. *Network Inference in Molecular Biology: A Hands-on Framework*, ed. J. M. Lingeman and D. Shasha, Springer, New York, 2012.
45. N. Garcia-Reyero, D. R. Ekman, T. Habib, D. L. Villeneuve, T. W. Collette, D. C. Bencic, G. T. Ankley and E. J. Perkins, Integrated approach to explore the mechanisms of aromatase inhibition and recovery in fathead minnows (Pimephales promelas), *Gen. Comp. Endocrinol.*, 2014, **203**, 193–202.
46. E. J. Perkins, J. K. Chipman, S. Edwards, T. Habib, F. Falciani, R. Taylor, G. Van Aggelen, C. Vulpe, P. Antczak and A. Loguinov, Reverse engineering adverse outcome pathways, *Environ. Toxicol. Chem.*, 2011, **30**, 22–38.
47. D. Koller and N. Friedman, *Probabilistic Graphical Models*, MIT Press, 2009.
48. N. M. Luscombe, M. M. Babu, H. Yu, M. Snyder, S. A. Teichmann and M. Gerstein, Genomic analysis of regulatory network dynamics reveals large topological changes, *Nature*, 2004, **431**, 308–312.

49. S. C. Madeira and A. L. Oliveira, Biclustering algorithms for biological data analysis: a survey, *IEEE/ACM Trans. Comput. Biol. Bioinf.*, 2004, **1**, 24–45.
50. D. J. Reiss, N. S. Baliga and R. Bonneau, Integrated biclustering of heterogeneous genome-wide datasets for the inference of global regulatory networks, *BMC Bioinf.*, 2006, 7, 280.
51. A. Tanay, R. Sharan, M. Kupiec and R. Shamir, Revealing modularity and organization in the yeast molecular network by integrated analysis of highly heterogeneous genomewide data, *Proc. Natl. Acad. Sci. U. S. A.*, 2004, **101**, 2981–2986.
52. A. Greenfield, A. Madar, H. Ostrer and R. Bonneau, DREAM4: Combining genetic and dynamic information to identify biological networks and dynamical models, *PLoS One*, 2010, **5**, e13397.
53. T. S. Gardner, D. di Bernardo, D. Lorenz and J. J. Collins, Inferring genetic networks and identifying compound mode of action *via* expression profiling, *Science*, 2003, **301**, 102–105.
54. V. A. Huynh-Thu, A. Irrthum, L. Wehenkel and P. Geurts, Inferring regulatory networks from expression data using tree-based methods, *PLoS One*, 2010, **5**, e12776.
55. J. J. Faith, B. Hayete, J. T. Thaden, I. Mogno, J. Wierzbowski, G. Cottarel, S. Kasif, J. J. Collins and T. S. Gardner, Large-scale mapping and validation of Escherichia coli transcriptional regulation from a compendium of expression profiles, *PLoS Biol.*, 2007, **5**, 0054–0066.
56. T. Ideker and N. J. Krogan, Differential network biology, *Mol. Syst. Biol.*, 2012, **8**, 1–9.
57. P. Zoppoli, S. Morganella and M. Ceccarelli, Time-Delay-ARACNE: Reverse engineering of gene networks from time-course data by an information theoretic approach, *BMC Bioinf.*, 2010, **11**, 154.
58. A. Madar, A. Greenfield, E. Vanden-Eijnden and R. Bonneau, DREAM3: Network inference using dynamic context likelihood of relatedness and the inferelator, *PLoS One*, 2010, **5**, e9803.
59. J. Yu, V. A. Smith, P. P. Wang, A. J. Hartemink and E. D. Jarvis, Advances to Bayesian network inference for generating causal networks from observational biological data, *Bioinformatics*, 2004, **20**, 3594–3603.
60. H. Bolouri, Modeling genomic regulatory networks with big data, *Trends Genet.*, 2014, **30**, 182–191.
61. C. T. Workman, H. C. Mak, S. McCuine, J.-B. Tagne, M. Agarwal, O. Ozier, T. J. Begley, L. D. Samson and T. Ideker, A systems approach to mapping DNA damage response pathways, *Science*, 2006, **312**, 1054–1059.
62. S. Bandyopadhyay, M. Mehta, D. Kuo, M. K. Sung, R. Chuang, E. J. Jaehnig, B. Bodenmiller, K. Licon, W. Copeland, M. Shales, D. Fiedler, J. Dutkowski, A. Guénolé, H. Van Attikum, K. M. Shokat, R. D. Kolodner, W. K. Huh, R. Aebersold, M. C. Keogh, N. J. Krogan and T. Ideker,

Rewiring of genetic networks in response to DNA damage, *Science*, 2010, **330**, 1385–1389.
63. T. Huan, Q. Meng, M. A. Saleh, A. E. Norlander, R. Joehanes, J. Zhu, B. H. Chen, B. Zhang, A. D. Johnson, S. Ying, P. Courchesne, N. Raghavachari, R. Wang, P. Liu, C. J. O'Donnell, R. Vasan, P. J. Munson, M. S. Madhur, D. G. Harrison, X. Yang and D. Levy, Integrative network analysis reveals molecular mechanisms of blood pressure regulation, *Mol. Syst. Biol.*, 2015, **11**, 799.
64. R. Thomas, R. S. Thomas, S. S. Auerbach and C. J. Portier, Biological networks for predicting chemical hepatocarcinogenicity using gene expression data from treated mice and relevance across human and rat species, *PLoS One*, 2013, **8**, e63308.
65. S. Chowbina, Y. Deng, J. Ai, X. Wu, X. Guan, M. S. Wilbanks, B. L. Escalon, S. A. Meyer, E. J. Perkins and J. Y. Chen, A new approach to construct pathway connected networks and its application in dose responsive gene expression profiles of rat liver regulated by 2,4DNT, *BMC Genomics*, 2010, **11**, S4.
66. Y. Deng, D. R. Johnson, X. Guan, C. Y. Ang, J. Ai and E. J. Perkins, In vitro gene regulatory networks predict in vivo function of liver, *BMC Syst. Biol.*, 2010, **4**, 153.
67. Y. Deng, S. A. Meyer, X. Guan, B. L. Escalon, J. Ai, M. S. Wilbanks, R. Welti, N. Garcia-Reyero and E. J. Perkins, Analysis of common and specific mechanisms of liver function affected by nitrotoluene compounds, *PLoS One*, 2011, **6**, e14662.
68. U. Alon, Biological networks: The tinkerer as an engineer, *Science*, 2003, **301**, 1866–1867.
69. K. Mitra, A.-R. Carvunis, S. K. Ramesh and T. Ideker, Integrative approaches for finding modular structure in biological networks, *Nat. Rev. Genet.*, 2013, **14**, 719–732.
70. Q. Zhang, S. Bhattacharya, R. B. Conolly, I. Clewell, J. Harvey, N. E. Kaminski and M. E. Andersen, Molecular signaling network motifs provide a mechanistic basis for cellular threshold responses, *Environ. Health Perspect.*, 2014, **122**, 1261–1270.
71. C. J. Ryan, A. Roguev, K. Patrick, J. Xu, H. Jahari, Z. Tong, P. Beltrao, M. Shales, H. Qu, S. R. Collins, J. I. Kliegman, L. Jiang, D. Kuo, E. Tosti, H.-S. Kim, W. Edelmann, M.-C. Keogh, D. Greene, C. Tang, P. Cunningham, K. M. Shokat, G. Cagney, J. P. Svensson, C. Guthrie, P. J. Espenshade, T. Ideker and N. J. Krogan, Hierarchical modularity and the evolution of genetic interactomes across species, *Mol. Cell*, 2012, **46**, 691–704.
72. N. Garcia-Reyero, T. Habib, M. Pirooznia, K. A. Gust, P. Gong, C. Warner, M. Wilbanks and E. Perkins, Conserved toxic responses across divergent phylogenetic lineages: a meta-analysis of the neurotoxic effects

of RDX among multiple species using toxicogenomics, *Ecotoxicology*, 2011, **20**, 580–594.
73. J. Dutkowski, M. Kramer, M. A. Surma, R. Balakrishnan, J. M. Cherry, N. J. Krogan and T. Ideker, A gene ontology inferred from molecular networks, *Nat. Biotechnol.*, 2013, **31**, 38–45.
74. D. Villeneuve, D. Crump, N. Garcia-Reyero, M. Hecker, T. H. Hutchinson, C. A. Lalone, B. Landersmann, T. Lettieri, S. Munn, M. Nepelska, M. A. Ottinger, L. Vergauwen and M. Whelan, Adverse Outcome Pathway (AOP) development I: Strategies and principles, *Toxicol. Sci.*, 2014, **142**, 312–320.
75. A.-R. Carvunis and T. Ideker, Siri of the cell: what biology could learn from the iPhone, *Cell*, 2014, **157**, 534–538.
76. J. P. Bai and D. R. Abernethy, Systems pharmacology to predict drug toxicity: integration across levels of biological organization, *Annu. Rev. Pharmacol. Toxicol.*, 2013, **53**, 451–473.
77. M. D. Ritchie, E. R. Holzinger, R. Li, S. A. Pendergrass and D. Kim, Methods of integrating data to uncover genotype–phenotype interactions, *Nat. Rev. Genet.*, 2015, **16**, 85–97.
78. H. Zhu, J. Zhang, M. T. Kim, A. Boison, A. Sedykh and K. Moran, Big data in chemical toxicity research: the use of high-throughput screening assays to identify potential toxicants, *Chem. Res. Toxicol.*, 2014, **27**, 1643–1651.
79. R. Agrawal, T. Imielinski and A. Swami, Mining association rules between sets of items in large databases, Proceedings of the 1993 ACM SIGMOD International Conference on Management of Data, Washington DC, 25–28 May 1993, *ACM SIGMOD Record*, 1993, **22**, 207–216.
80. A. K. Jain, Data clustering: 50 years beyond K-means, *Pattern Recognit. Lett.*, 2010, **31**, 651–666.
81. T. Aittokallio and B. Schwikowski, Graph-based methods for analysing networks in cell biology, *Briefings Bioinf.*, 2006, **7**, 243–255.
82. D. Conte, P. Foggia, C. Sansone and M. Vento, Thirty years of graph matching in pattern recognition, *Int. J. Pattern Recognit. Artif. Intell.*, 2004, **18**, 265–298.
83. Y. Igarashi, N. Nakatsu, T. Yamashita, A. Ono, Y. Ohno, T. Urushidani and H. Yamada, Open TG-GATEs: a large-scale toxicogenomics database, *Nucleic Acids Res.*, 2015, **43**, D921–D927.
84. T. Uehara, A. Ono, T. Maruyama, I. Kato, H. Yamada, Y. Ohno and T. Urushidani, The Japanese toxicogenomics project: Application of toxicogenomics, *Mol. Nutr. Food Res.*, 2010, **54**, 218–227.
85. A. Aiyar, P. Fatterpaker and A. Sreenivasan, Lipid metabolism in liver injury caused by carbon tetrachloride in the rat, *Biochem. J.*, 1964, **90**, 558–563.

86. A. de Conti, V. Tryndyak, M. I. Churchwell, S. Melnyk, J. R. Latendresse, L. Muskhelishvili, F. A. Beland and I. P. Pogribny, Genotoxic, epigenetic, and transcriptomic effects of tamoxifen in mouse liver, *Toxicology*, 2014, **325**, 12–20.
87. T. Saphner, S. Triest-Robertson, H. Li and P. Holzman, The association of nonalcoholic steatohepatitis and tamoxifen in patients with breast cancer, *Cancer*, 2009, **115**, 3189–3195.
88. D. Croft, A. F. Mundo, R. Haw, M. Milacic, J. Weiser, G. Wu, M. Caudy, P. Garapati, M. Gillespie and M. R. Kamdar, *et al.*, The Reactome pathway knowledgebase, *Nucleic Acids Res.*, 2014, **42**, D472–D477.
89. M. Milacic, R. Haw, K. Rothfels, G. Wu, D. Croft, H. Hermjakob, P. D'Eustachio and L. Stein, Annotating cancer variants and anti-cancer therapeutics in reactome, *Cancers*, 2012, **4**, 1180–1211.
90. A. Bateman, M. J. Martin, C. O'Donovan, M. Magrane, R. Apweiler, E. Alpi, R. Antunes, J. Arganiska, B. Bely, M. Bingley, C. Bonilla, R. Britto, B. Bursteinas, G. Chavali, E. Cibrian-Uhalte, A. Da Silva, M. De Giorgi, T. Dogan, F. Fazzini, P. Gane, L. G. Castro, P. Garmiri, E. Hatton-Ellis, R. Hieta, R. Huntley, D. Legge, W. Liu, J. Luo, A. MacDougall, P. Mutowo, A. Nightingale, S. Orchard, K. Pichler, D. Poggioli, S. Pundir, L. Pureza, G. Qi, S. Rosanoff, R. Saidi, T. Sawford, A. Shypitsyna, E. Turner, V. Volynkin, T. Wardell, X. Watkins, H. Zellner, A. Cowley, L. Figueira, W. Li, H. McWilliam, R. Lopez, I. Xenarios, L. Bougueleret, A. Bridge, S. Poux, N. Redaschi, L. Aimo, G. Argoud-Puy, A. Auchincloss, K. Axelsen, P. Bansal, D. Baratin, M. C. Blatter, B. Boeckmann, J. Bolleman, E. Boutet, L. Breuza, C. Casal-Casas, E. De Castro, E. Coudert, B. Cuche, M. Doche, D. Dornevil, S. Duvaud, A. Estreicher, L. Famiglietti, M. Feuermann, E. Gasteiger, S. Gehant, V. Gerritsen, A. Gos, N. Gruaz-Gumowski, U. Hinz, C. Hulo, F. Jungo, G. Keller, V. Lara, P. Lemercier, D. Lieberherr, T. Lombardot, X. Martin, P. Masson, A. Morgat, T. Neto, N. Nouspikel, S. Paesano, I. Pedruzzi, S. Pilbout, M. Pozzato, M. Pruess, C. Rivoire, B. Roechert, M. Schneider, C. Sigrist, K. Sonesson, S. Staehli, A. Stutz, S. Sundaram, M. Tognolli, L. Verbregue, A. L. Veuthey, C. H. Wu, C. N. Arighi, L. Arminski, C. Chen, Y. Chen, J. S. Garavelli, H. Huang, K. Laiho, P. McGarvey, D. A. Natale, B. E. Suzek, C. R. Vinayaka, Q. Wang, Y. Wang, L. S. Yeh, M. S. Yerramalla and J. Zhang, UniProt: a hub for protein information, *Nucleic Acids Res.*, 2015, **43**, D204–D212.
91. Y. Komiya and R. Habas, Wnt signal transduction pathways, *Organogenesis*, 2008, **4**, 68–75.
92. J. Pearl, *Causality*, Cambridge University Press, 2000.
93. K. F. Leavens, R. M. Easton, G. I. Shulman, S. F. Previs and M. J. Birnbaum, Akt2 Is Required for Hepatic Lipid Accumulation in Models of Insulin Resistance, *Cell Metab.*, 2009, **10**, 405–418.

94. H. Ono, H. Shimano, R. Katagiri, N. Yahagi, H. Sakoda, Y. Onishi, M. Anai, T. Ogihara, M. Fujishiro, A. Y. I. Viana, Y. Fukushima, M. Abe, N. Shojima, M. Kikuchi, N. Yamada, Y. Oka and T. Asano, Hepatic Akt activation induces marked hypoglycemia, hepatomegaly, and hypertriglyceridemia with sterol regulatory element binding protein involvement, *Diabetes*, 2003, **52**, 2905–2913.
95. M. Breen, D. L. Villeneuve, G. T. Ankley, D. C. Bencic, M. S. Breen, K. H. Watanabe, A. L. Lloyd and R. B. Conolly, Developing predictive approaches to characterize adaptive responses of the reproductive endocrine axis to aromatase inhibition: II. Computational modeling, *Toxicol. Sci.*, 2013, **133**, 234–247.
96. D. L. Villeneuve, N. D. Mueller, D. Martinoviś, E. A. Makynen, M. D. Kahl, K. M. Jensen, E. J. Durhan, J. E. Cavallin, D. Bencic and G. T. Ankley, Direct effects, compensation, and recovery in female fathead minnows exposed to a model aromatase inhibitor, *Environ. Health Perspect.*, 2009, **117**, 624–631.
97. M. Mayo, T. Habib, N. Garcia-Reyero, D. L. Villeneuve, G. T. Ankley and E. J. Perkins, Androgen receptor mediated compensation of estradiol in response to aromatase inhibition: a mathematical model, in *SETAC North America 33rd Annual Meeting*, Long Beach, CA, 2012, p. 362.
98. J. Shoemaker, K. Gayen, N. Garcia-Reyero, E. Perkins, D. Villeneuve, L. Liu and F. Doyle, Fathead minnow steroidogenesis: in silico analyses reveals tradeoffs between nominal target efficacy and robustness to cross-talk, *BMC Syst. Biol.*, 2010, **4**, 89.
99. R. Kafri, M. Levy and Y. Pilpel, The regulatory utilization of genetic redundancy through responsive backup circuits, *Proc. Natl. Acad. Sci. U. S. A.*, 2006, **103**, 11653–11658.
100. European Parliament and Council, Regulation (EC) No 1907/2006 of the European Parliament and of the Council of 18 December 2006 concerning the Registration, Evaluation, Authorisation and Restriction of Chemicals (REACH), establishing a European Chemicals Agency, amending Directive 1999/4, L396/1 Official Journal of the European Union 849 (2006), EC: Official Journal of the European Union.
101. C. Ives, I. Campia, R.-L. Wang, C. Wittwehr and S. Edwards, Creating a structured AOP knowledgebase *via* ontology-based annotations, *Appl. In Vitro Toxicol.*, 2018, **3**, 298–311.
102. National Academies of Sciences, Engineering, and Medicine, *Using 21st Century Science to Improved Risk-Related Evaluations*, National Academies Press, Washington DC, 2017.
103. OECD, Revised Guidance Document on Developing and Assessing Adverse Outcome Pathways, Series on Testing & Assessment No. 184, 2017, ENV/JM/MONO(2013)6.

104. Users' Handbook supplement to the Guidance Document for developing and assessing Adverse Outcome Pathways (AOPs), Series on Testing & Assessment No. 233, Series on Adverse Outcome Pathways No. 1, 2018, ENV/JM/MONO(2016)1.
105. R. B. Conolly, G. T. Ankley, W. Cheng, M. L. Mayo, D. H. Miller, E. J. Perkins, D. L. Villeneuve and K. H. Watanabe, Quantitative Adverse Outcome Pathways and Their Application to Predictive Toxicology, *Environ. Sci. Technol.*, 2017, **51**(8), 4661–4672.

CHAPTER 10

Chemometrical Analysis of Proteomics Data

MARJAN VRAČKO*

Kemijski Inštitut/National Institute of Chemistry, Hajdrihova 19, 1000 Ljubljana, Slovenia
*E-mail: marjan.vracko@ki.si

10.1 Introduction

In the last decades we have been witnessing an explosion of data and a discussion of the four principal attributes of big data: the 4 Vs – volume, velocity, variety and veracity. However, the data *per se* is useless; we do need to pass the next two steps on the path to knowledge: data – information – knowledge. When we talk about big data in drug research and toxicology, we often think of -omics data, such as genomics, proteomics, secretomics, and metabolomics (see also Chapter 7).[1–4]

In drug discovery research, the effect of a (new) drug on the organism is investigated. In our mechanistic understanding, the response of an organism is a cascade of events starting from an initial event at a chemical level leading to the final effect. Current emphases in drug research are placed on ADMET (absorption, distribution, metabolism, excretion, toxicity) properties, physiology-based pharmacokinetic (PBPK) studies and drug–target proteins interactions.[4] In toxicology, the adverse outcome pathway (AOP) concept was also introduced as a concept[5,6] describing the sequence of

Issues in Toxicology No. 41
Big Data in Predictive Toxicology
Edited by Daniel Neagu and Andrea-Nicole Richarz
© The Royal Society of Chemistry 2020
Published by the Royal Society of Chemistry, www.rsc.org

events that leads at the end to dis-functioning of the biological system (see also Chapter 9). It is schematically shown in Figure 10.1. Prerequisites to postulating an AOP are on the one side the understanding of the normal functioning of biological systems, and on the other side, an understanding of the response of the system to stress situations. The question is how to integrate the available and newly produced -omics data into this scheme and to develop algorithms that are able to extract useful information from the data.

In this chapter, we focus on proteomics data. 'Proteomics' describes the protein status of the cell, *i.e.*, the presence and abundance of all proteins. When an organism is exposed to a xenobiotic this may influence the metabolism and thus change the status of proteins in cells. Cells of an organism subjected to a disease may show a different protein status in comparison to a healthy organism. On the other hand, even in healthy cells, the protein status depends on the life cycle of the cells, on their origin, or on the environmental conditions. A part of proteomics – secretomics – is focused on proteins outside the cells, which are mostly responsible for signalling pathways. In any case, we must be aware that a proteomic measurement provides 'a snap-shot' in time. Proteomics data are usually obtained by two-dimensional (2D) electrophoresis and/or with mass spectrometric techniques.

A typical proteomic study consists of several steps: experimental design, cell and probe preparation, data acquisition, and chemometrical analysis. The protein status depends on the biological stage of the cell, *i.e.*, in the stage of the reproductive cycle of the cell. Moreover, the measurements are very complex processes. Thus, the intra-laboratory and inter-laboratory

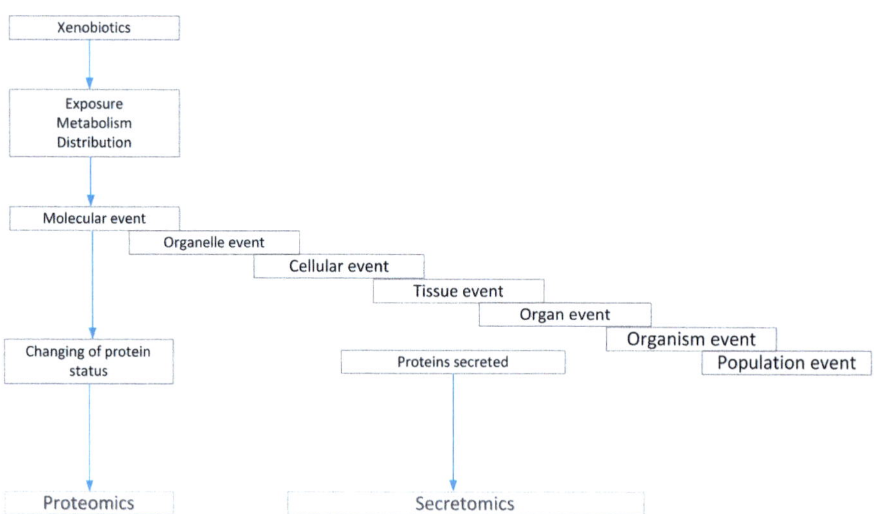

Figure 10.1 Schematic description of an adverse outcome pathway (AOP) and the role of protein research.

reproducibility of the results is a problem. The question arises whether a unique moment was recorded or a reproducible result obtained. Therefore, identification of systematic errors and uncertainties in all stages of research is essential.[7]

The basic technique for proteomics measurements is two-dimensional (2D) electrophoresis. The cell proteins are resolved on the basis of their content of acidic and basic amino acids (isoelectric points) and their molecular weights. A result of such a measurement is a 2D plate with spots, which correspond to condensed proteins. The intensity of a spot measures the abundance of a particular protein. Using high-resolution techniques, thousands of proteins can be resolved in a single proteomics map. The method is particularly effective when applied in a comparative way. This means that proteomics maps of cells in normal conditions are compared with diseased or treated cells. The differences indicate the changes of the status of particular proteins. Thus, these proteomics maps are an important tool for identifying biomarkers, potential drug targets, or indicators for toxicity of xenobiotics. Pattern recognition software is applied to the 2D patterns to present the results in numerical tabular form. Other methods to investigate the proteome are based on mass spectroscopy for detection of the proteins after their separation.[8,9] These techniques can be divided into two categories. In the bottom-up approach, the proteins are digested into peptides, which then undergo spectroscopic analysis.[8] In the top-down approach, the entire protein is ionised and analysed in the spectrometer.[9] The spectra are analysed with bio-informatics tools to identify the proteins.

As results, the statistician or chemometrics expert receives a list of proteins with their abundances, recorded at specific biological conditions. The task is to extract the useful information that may lead to new knowledge, *i.e.*, to find the relationship to the biological endpoint.[7] In Section 10.2, we present some established chemometrical methods that have been applied for analysis of proteomics data. Emphasis is placed on the definition of similarity between proteomics maps, clustering and classification and, in particular, on searching for the relationship between information obtained from the proteomics analysis and the biological response (for example, toxicity). Examples of these approaches have been reported under the term bio-descriptors.[10,11]

2D maps can be encoded following the idea of a topological representation of molecular structures. Topological descriptors deduced from 2D maps have been described in several studies on the characterisation of proteomics maps.[12–15] A selected pattern of spots produced in 2D electrophoresis maps can be graphically presented as two-dimensional graphs. Randić *et al.* proposed partial ordering for clustering the proteins.[16] Partial ordering is a mathematical technique that enables ordering of elements into a series with respect to several parameters. In 2D electrophoresis, the proteins are separated relative to their mass and isoelectric point and, thus, the proteins can be ordered according to both these parameters. Such series produce patterns of spots, which can be embedded into geometrical polyhedrons (for example, Voroni's polygons) and can be, in further steps, described with numerical invariants. Some pilot studies have been performed on this.[13–16] Briefly, the entire mathematical formalism, which was developed and successfully applied in the description of molecules, was also applied for a numerical description of proteomics data and, further, proteins.[17–19]

When proteins are isolated, the next step is to determine the sequences of amino acids. The question is how far the sequences determine the three-dimensional structure of proteins and their role in the cells' life. As an example showing the usefulness of the numerical encoding of protein sequences, we refer to the models of PredαTM and PredβTM, which were developed to predict the α- and β-transmembrane regions of proteins (http://transpred.ki.si/odseki/d03-odsek-za-kemijsko-informatiko/transmembrane-region-predictors).[18] The protein is encoded as a matrix expressing the frequency of amino acid pairs. A Support Vector Machines algorithm has been trained on a dataset of known transmembrane sequences. An application example is the protein bilitranslocase, which is responsible for transport of bilirubin.[19]

This chapter will describe chemometrical methods and tools that can be used for the analysis of proteomics data, with a particular focus on the methods of clustering and classification as well as genetic algorithms to search for 'important' proteins, *i.e.*, which can be correlated to a biological activity or adverse outcome. Furthermore, the concept of bio-descriptors and biomarkers will be presented. Overall, application examples of the methods and tools are shown, demonstrating how knowledge can be extracted from proteomics data to support the understanding of toxicity mechanisms and help build predictive models.

10.2 Chemometrical Methods for Data Analysis

10.2.1 Measure of Similarity

A way to express the similarity relationship between proteomics data is given with the similarity index.[20] The similarity index directly compares the intensities between two spots that belong to the same protein, and thus the change in the abundance of it. It can be extended to any combination of proteins or to entire proteomics. Defining similarities and thus comparing the abundance of proteins under different physiological or pathological conditions enables the identification of proteins that might be specifically correlated to biological events in the organism caused, for example, by toxicity.

The similarity between two proteomics maps a and b can be determined with the similarity index defined as:

$$S^{a,b} = \frac{1}{N}\sum_{i=1}^{N}\frac{\left(z_i^a z_i^b\right)}{\max\left(z_i^a, z_i^b\right)^2}$$

N is the number of spots, z is the intensity of the i-th spot, in nominator is the product of both spots and in denominator the square of the maximal value. From the definition, the following properties of the similarity index are straightforward:

1. For two identical maps the index is 1.
2. The index is defined only for spots that occur in both maps on the same positions.
3. It is not sensitive about the absolute value of spots, *i.e.*, changes of small spots can contribute in the same way to the index as changes of the largest spots.
4. It is not sensitive about whether the intensity spots are increasing or decreasing for untreated and treated samples, respectively.

As an example, the similarity index has been applied in a recently published report to study the proteomic data of Caco-2/HT29-MTX cells treated with two kinds of nanosubstances (multi-walled carbon nanotubes and TiO_2 nanobelts). Out of approximately 3000 proteins the authors identified eight proteins strongly affected by multi-walled carbon nanotubes, 25 proteins affected by TiO_2 nanobelts and 14 proteins affected by either nanosubstance.[21]

10.2.2 Clustering and Classification

Analysing a cluster structure of a dataset reveals whether or not a dataset exhibits an intrinsic (natural) grouping structure. It should be emphasised that it is not known whether or not such a structure exists before the clustering analysis is performed. Several methods have been proposed for such cluster analysis; some of them are: the nearest neighbour method, the furthest neighbour method, the centroid method, the group average method, and the median method.[22–25] On the other hand, the classification approach sorts objects into defined categories (classes). In other words, it is sorting elements according to classification (or allocation) rules. The rules must be known before the classification procedure starts. Thus, it is evident that the classification is conceptually different from the clustering. To visualise the difference we can imagine a virtual experiment of classification and clustering on a hypothetic dataset. The classification analysis could show that some classes are empty, but a cluster can never be empty. Classification is often performed in combination with clustering, where the cluster analysis is used to define the classification function, or in other words to define the rules for classification.[24]

10.2.2.1 Hierarchical Clustering (Dendrograms)

A dendrogram shows the relationship between the sequence of consecutive fusions of dataset elements and the clustering coefficient. The clustering coefficient is a numerical parameter that expresses the similarity relationship between objects or between groups of objects. It can be calculated in different ways. Usually, it is calculated as the distance between objects or distance between centres of groups.[25] Examples of dendrograms are given in Section 10.3.1.

10.2.2.2 Principal Component Analysis and Partial Least Squares Method

The main idea of the Principal Component Analysis (PCA) is to reduce the number of variables in data vectors and at the same time to preserve the informational content of the system. Basically, the PCA is a linear transformation of original variables into new (latent) variables with the linear transformation matrix. The basis for the transformation matrix represents the matrix of variances, and as the first step its eigenvalues and eigenvectors must be calculated. An eigenvalue explains how much information the corresponding eigenvector carries. They are sorted in the transformation matrix according to their eigenvalues, *i.e.*, the vector with the largest eigenvalue is located on the top and so on. The same is true for the latent variables; they are also sorted according to their informational content. Therefore, it is sufficient to consider only a few latent variables that carry the essential part of the information. If an intrinsic grouping structure in the dataset exists, it is expressed as a cluster structure (groups of objects well separated from others) in PCA score plots.[25,26] Thus, the basic idea of PCA is the concentration (compression) of information into a few new variables and therefore PCA represents a powerful tool for the reduction of the number of variables. In proteomics studies, the individual protein spots represent independent variables, thus we have to consider systems with a large number of independent variables.

The Partial Least Squares (PLS) method represents a generalisation of PCA, which requires also the responses (dependent variables or output variables). Basically, the PLS method finds a linear regression model between transformed variables. When the output variables are categories (for example: toxic or non-toxic), the method is referred to as Partial Least Squares-Discriminant Analysis (PLS-DA). Logistic regression is a type of linear regression function where the dependent variable y is coded as a probability that the object belongs to a particular class. Instead of a single regression function for k classes, one obtains k-1 regression equations known also as logit functions.

10.2.2.3 Self-Organising Maps for Clustering and Classification

Self-Organising Maps (SOM) or Kohonen neural networks can be used to analyse the clustering structure of multi-dimensional datasets, as described in the literature.[27,28] Briefly, the architecture of SOM represents a two-dimensional array of neurons, which are vectors of weights. The learning algorithm is, essentially, the topology preserving projection of data points from the multi-dimensional space to a two-dimensional grid of points. Contrary to PCA, the learning algorithm is non-linear. After the training, the objects are arranged in a two-dimensional array (network) in such a way that similar objects are located close to each other. With visual inspection, one recognises clusters and other similarity relationships among data points, which are organised in a map. In our case, the multi-dimensional objects are proteomics measurements, and abundances of individual proteins are elements of representation vectors.

10.2.3 Genetic Algorithm

Genetic algorithm (GA) is a machine learning technique that is used to identify optimal models. The philosophy of GA mimics Darwin's evolution theory, where only the best subjects survive. The concept and the name of GA was introduced in 1961, the first mathematical formulation was given in 1975.[29] Optimal models are found within a multi-dimensional variable space. The GA technique is based on three pillars.

Setting of generation. Inside the multi-dimensional space, numerous configurations (chromosomes) are created, which are low numbered combinations of variables. In the first run of the algorithm these are set randomly, but in the ongoing cycles they are modified following the natural principle of sexual reproduction; they exchange the variables and include some new variables as mutations.

Modelling technique. For each configuration a model is built, which expresses the relationship between configuration and property. Different techniques can be used, such as linear regression, neural networks, PLS, decision trees, *etc.*[30]

Evaluation function. A criterion must be defined after the models are evaluated. At each step the models are evaluated and the fittest models are promoted to the next generation.

GA was successfully applied in many problems in molecular and protein modelling. Examples are: descriptor selection in (quantitative) structure–activity relationship modelling, prediction of protein structures, analysis of spectral data, *etc.*[31] In the example described in Section 10.3.2, the GA was applied to select proteins that may play a crucial role in different cell toxicities.

10.2.4 Chemical Descriptors and Bio-descriptors

QSAR (Quantitative Structure–Activity Relationship) modelling plays an important role in rational drug design and in predictive toxicology. This modelling approach is based on the assumption that similar molecules show similar properties and that a molecular property depends on the chemical structure. The development of a QSAR model consists of several steps: the selection of training data, descriptors, modelling techniques and testing and validation of models. The final model can be used to predict a property for new (or hypothetical) compounds. One central question is how to represent the chemical structures (see also Chapter 3). Usually, the molecules are represented with a set of descriptors. An ideal representation shall be: unique, uniform, reversible, and invariant on rotation and translation of molecules. Unique means that different structures give different representations, uniform means that the dimension of representation is the same for all structures, reversible means that the structure can be unambiguously reconstructed from the representation. Furthermore, invariant means that the representation is not sensitive if a molecule is rotated or translated. Unfortunately, no representation fulfils all requirements simultaneously.[14] In our context of chemical structures, the descriptors can be ordered in a hierarchical manner.[32]

Fragments as descriptors. A molecular structure can be encoded in terms of its fragments (substructures). In this representation an individual element of a representation vector shows the presence or absence of a particular fragment.[32]

One-dimensional (constitutional) descriptors. To this class belong the discrete numerical descriptors that describe the basic molecular structure; for example, the number of atoms, number of rings, *etc.*

Two-dimensional descriptors. They are deduced from the topological image (2D image) of molecules, which shows how the atoms are connected and the nature of bonds ('structural formula of a molecule'). Mathematically, the topology is described with topological indices.

Electrotopological (E-state) descriptors summarise topological information of a molecule with atomic properties.[32]

Three-dimensional descriptors. They are calculated from the 3D molecular structure, which is defined with coordinates of all atoms in the molecule. It is a crucial step from the 2D to the 3D description of molecules. The 3D structure is an ambiguously defined entity, which depends on the molecular environment, *i.e.*, it is different in crystal structures, in solutions, or in a vacuum. To this group belong the *geometrical descriptors* and *quantum chemical descriptors*.

Molecular structures in the environment of receptors. Molecules are often described in their environment as, for example, in the environment of receptors, *i.e.*, describe the molecules interacting with receptors.

Physico–chemical descriptors. To this group belong different physico–chemical properties like log P, which measures the hydrophobicity (or lipophilicity) of compounds, solubility, molecular weight, *etc.*

Generally, a QSAR equation can be expressed as:

$$\text{activity} = f(\text{molecular structure})$$

Furthermore, bio-descriptors can be considered that describe the biological status of cells (proteomics) after treatment with a chemical. This conceptually means that in addition to the chemical information, the biological information is also taken into account in QSAR equations. Thus, the equation can be written as:

$$\text{activity} = f(\text{molecular structure, status of cells})$$

The concept of bio-descriptors has been often discussed in the literature,[33,34] addressing the crucial question of how to encode the large amount of biological data in a way suitable for QSAR modelling.

10.3 Application Examples of Chemometrical Analyses of Proteomics Data

In the following, application examples of chemometrics to analyse proteomics data will be described. They show how chemometrics can contribute to extract information and build knowledge from proteomics big data, such as by identifying proteins that could be causally related to events in the biological pathways leading from exposure to chemicals to adverse outcomes in the organism.

10.3.1 Clustering and Classification of Proteomics Data

Cluster analyses investigate whether the objects of a dataset are assembled in groups that are separated from each other. Specifically, the analysis of the cluster structure of proteome data may allow the identification of proteins that are associated to particular biological effects.

Clustering of 2D electrophoresis maps of pancreatic ductal carcinoma cells. Marengo *et al.* studied a set of eighteen 2D electrophoresis map images of pancreatic ductal carcinoma cells. The dataset was divided into four classes: two cell-lines of pancreatic ductal carcinoma cells, which were untreated (control samples) or drug-treated with Trichostatin A.[35] The 2D proteomics images were projected to the space of Legendre polynomials applying the standard procedure known from the functional theory. Since the images are well structured, *i.e.*, the intensity has a lot of uncorrelated peaks and valleys, up to 100 moments were necessary to get a good representation. The hierarchical clustering method, which was performed with the Legendre moments, confirmed the classification of samples into four classes. In addition, the linear discriminant analysis confirmed this classification scheme, which was achieved under consideration of only six selected Legendre moments. The methods were able to identify four clusters, which clearly separated between the two cell-lines and the controls from the treated samples.

Analysis of proteomics maps from peroxisome-proliferator-treated hepatocytes. In a study of peroxisome proliferators reported by Anderson *et al.*,[36] mouse liver cells were treated with five peroxisome proliferators: LY171883, DEHP,

Clofibric acid, WY14643, and Nafenopin. One group of animals was treated with LY163443, which is not a proliferator (negative control) and the control group was not treated at all. The chemicals were added to the mice's diet. The data consisted of three groups, the proteomics data was taken after five days and 35 days of treatment with the chemicals, plus a dose-related study. In the latter, the proliferator LY171883 was administrated in seven doses: 0%, 0.003%, 0.01%, 0.03%, 0.1%, 0.3%, and 0.6% in diet. After the treatment of animals, the hepatocytes were analysed with 2D electrophoresis, where 107 proteins were assigned. Peroxisome proliferation is a complex biochemical process, which involves a number of proteins, but is regulated by a single gene that may affect the proteins in a coordinated way. In the paper, the authors analysed the 107 proteins with PCA (principal component analysis). The first principal axis PC1 carried 51% of the total variance and was related to the strength of the peroxisome effect. PC2 and PC3 carried 6.7% and 5% of the total variance, respectively. The second axis separated the five- and 35 day exposure groups while the third axis separated peroxisome proliferators from the control group.

The data of this study were further analysed by calculating the similarity indices between proteome maps taking into account all 107 assigned proteins.[20] It should be emphasised that in this case a lot of information is condensed into a single index, which can describe the general behaviour of the protein status under different biological conditions. The similarity indices are graphically presented in Figure 10.2. The dose–response shows a surprising behaviour, a local minimum at low doses. It can be speculated that this is a case of a biphasic process, or hormesis, *i.e.*, that the biological response is stronger at a lower dose.[37]

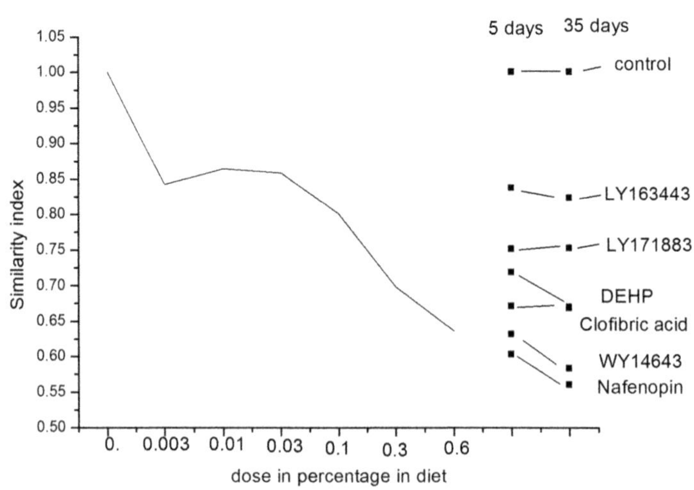

Figure 10.2 Similarity index dependent on administrative dose of LY171883. The right side shows similarity indices for five- and 35 day exposures. Dose is given as the percentage in diet. Reproduced from ref. 20 with permission from Elsevier, Copyright 2003.

Figure 10.3 represents the dendrograms of the three datasets. The *x* axis shows the treated samples while the *y* axis shows the Euclidian distance between 107 protein abundances (spots). Figure 10.3A shows the five-day study. Two clusters with proliferators LY171883, DEHP, WY14643, and Nafenopin are evident. On the other hand, the control and negative control (LY163443) samples are well separated from the others. A similar situation can be observed for the 35 day study (Figure 10.3B), the only difference being that DEHP and Clofibric acid form a cluster. Figure 10.3C shows the situation for the dose-related study. It confirms the curve in Figure 10.2, where the low doses from 0.003% to 0.03% in diet show a similar response.

From the Self-Organising Map, which is shown in Figure 10.4, the following conclusions can be made.

- The dose-related study is separated from the five- and 35 day studies.
- Control and negative control samples form a cluster.
- WY 14643 and Nafenopin are on the same neuron in the five- and 35 day studies.
- In the dose-related study, the low dose samples (0.003%, 0.01% and 0.03% in diet) are separated from the 0.1%, 0.3% and 0.6% samples.

Overall, these examples show the power of chemometrical analyses to process a large amount of data points, order the results and extract information. Analysing proteomics data, the different clustering methods can be used to find similar samples and to support making conclusions about the biological status of cells.

10.3.2 Genetic Algorithm Applied to Proteomics Data

Vracko *et al.* reported a proteome study of rat hepatocytes exposed to 14 halocarbons.[38] The list of compounds and the concentration range is given in Table 10.1. With a 2D electrophoresis of the rat hepatocytes, 92 2D maps were recorded in total, whereof 27 from untreated cells (controls). In each proteomics map, 1401 spots were assigned. For each spot, the average value and the standard deviation over 27 recordings were calculated in the data pre-processing. With the aim of selecting only reliable spots, only spots with a standard deviation less than 0.39 were selected, which applied to 188 spots.

In addition, six toxicological endpoints were measured in the hepatocytes with *in vitro* techniques: cell viability (MTT assay), membrane integrity measured as a decrease in lactate dehydrogenase (LDH), total cellular thiols, lipid peroxidation, reactive oxygen species production, activity of the enzyme catalase. The biological responses were expressed as logarithms of molar concentrations that cause a defined biological effect: EC_{50MTT} = effective concentration that causes 50% decrease of cell viability (MTT assay), EC_{50LDH} = concentration that causes 50% decrease in lactate dehydrogenase (LDH), EC_{20} = 20% decrease in the thiol level, LEC_{LP} = lowest concentration

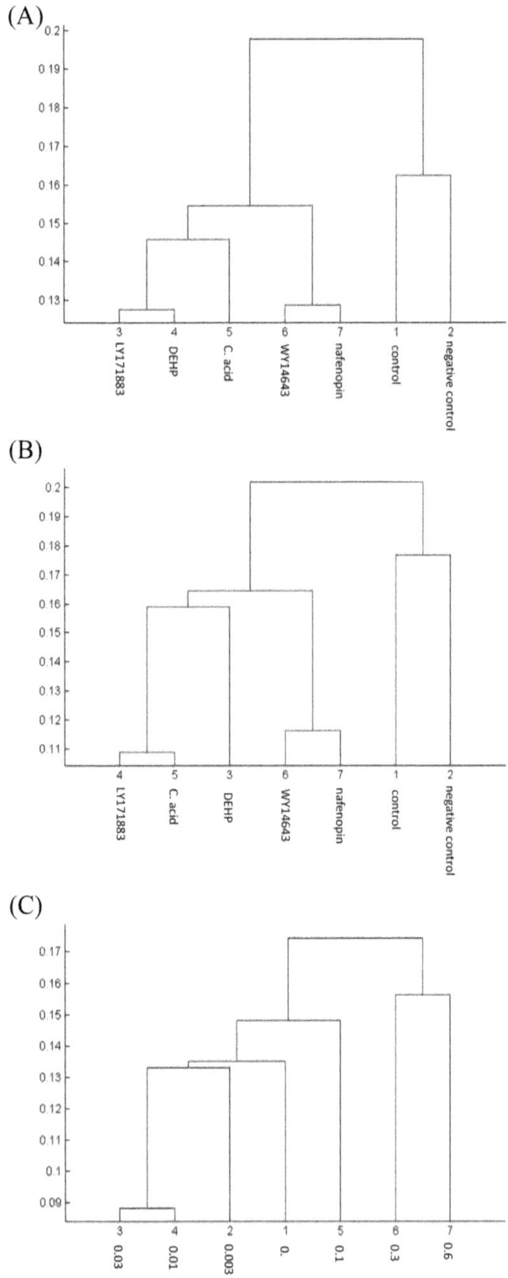

Figure 10.3 Dendrograms of the datasets. (A) Five days' exposure, (B) 35 days' exposure, (C) dose-related study. The *y* axis shows the Euclidean distance between intensities of proteins.

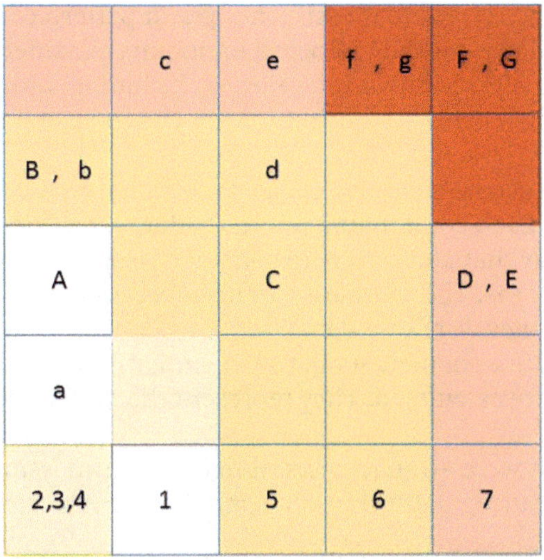

Figure 10.4 Self-Organising Map (Kohonen network) of 21 proteomic maps. a–g five days' study, A–G 35 days' study. Reproduced from ref. 20 with permission from Elsevier, Copyright 2003. (a,A): control, (b,B): LY163443 (negative control), (c,C): LY171883, (d,D): DEHP, (e,E): Clofibric acid, (f,F): WY14643, (g,G): Nafenopin, 1–7 dose-related study.[20] The intensity of colour indicates the intensity of changes in protein abundances (dark colours – higher change.).

Table 10.1 14 halocarbons used in the reported example proteome study (see ref. 38).

	Compound	Dose range (mM)
1	Control	
2	C_2Cl_4	0.100–2.000
3	CH_2Br_2	2.500–25.000
4	CH_2BrCl	10.000–100.00
5	CH_2Cl_2	10.000–100.000
6	$CHCl_3$	1.000–10.000
7	$CHBr_3$	0.143–4.580
8	1,2-$C_2H_4Br_2$	0.070–8.630
9	1,1,2-$C_2H_3Br_3$	0.310–2.460
10	1,2-C_2H_4BrCl	0.420–3.364
11	1,2-$C_2H_4Cl_2$	3.160–38.000
12	C_2HCl_3	0.277–2.666
13	1,1,1-$C_2H_3Cl_3$	0.063–0.501
14	1,1,2-$C_2H_3Cl_3$	1.080–6.480
15	$C_2H_2Cl_4$	1.188–4.750

that causes lipid peroxidation, LEC_{ROS} = lowest concentration causing reactive oxygen species production, LEC_{CAT} = lowest concentration causing inhibition of catalase. The electrophoresis was performed for concentrations causing defined biological effects.

The aim of the study was to identify the spot or pattern of spots (proteins), which were significant for the biological endpoints considered. For this purpose, a searching algorithm was developed, including similarity indices as defined earlier and a genetic algorithm. Briefly, it consisted of five steps:

1. Five thousand combinations of spots (patterns) were created as the first generation. Each combination was a random selection of 10 to 20 spots in such a way that all spots were considered equivocally.
2. For all patterns, the similarity indices between control sample and treated samples were calculated.
3. The correlation coefficients between similarity indices and biological activities were calculated. They represent the evaluation criteria for the patterns.
4. The patterns were arranged according to the correlation coefficients.
5. The neighbouring patterns exchanged their spots (crossover).

Steps 2–5 were repeated many times. After 50 generations saturation was achieved.

At the end all patterns with correlation coefficients higher than 0.75 were analysed. From the pool of patterns, for each toxic response, 10 spots were selected, which were present in different patterns and may give indications about the mechanism of activity. Figure 10.5 shows the positions of spots as they appear in the 2D gels for the six biological activities. Two spots were associated with the activities EC_{50MTT}, LEC_{ROS} and EC_{20SH}, and one spot was associated with five activities (with the exemption of LE_{CLP}). The authors performed a comprehensive testing of selection with exclusions of spots. When significant spots were excluded, the correlations dropped, or, in many cases, no correlations were found. Three spots are common for all six activities (labelled as 15, 70 and 108 in Figure 10.5).

The example shows the power of chemometrics to gain information from proteomics data: the genetic algorithm has searched thousands of possible patterns (combinations) of spots in order to identify important proteins for further consideration.

10.3.3 Proteomics as Bio-descriptors

In the following, we continue the example of the study presented in Section 10.3.2. From 1401 spots in the 2D proteomics map, 188 spots were selected, which seemed to be statistically reliable. The genetic algorithm was run on this subset and identified the ten 'best spots'. After the 'best' subsets of spots were identified, the abundances of individual spots were taken as bio-descriptors to build a 'proteomic map/spots–toxicity relationship' model. Counter propagation neural networks were applied for modelling. The models were tested using a leave-one-out (LOO) procedure. The results are presented in Table 10.2. The best correlation coefficient in the LOO test was found for toxic property EC_{50MTT} with $r = 0.78$, and after removing one outlier, $r = 0.88$ (see Figure 10.6A and B). The model shows moderate correlation for the other biological

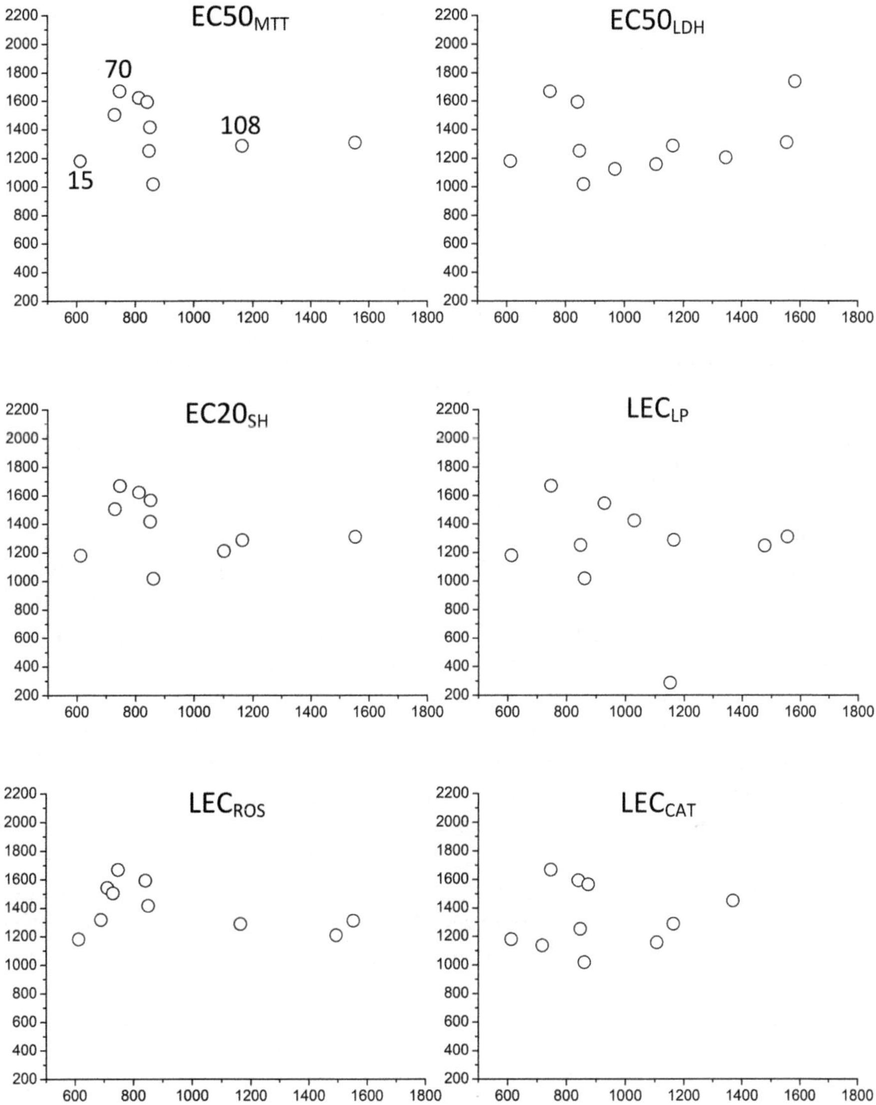

Figure 10.5 The ten most important spots for six different toxicological endpoints as they appeared in the 2D gels. Reproduced from ref. 38 with permission from American Chemical Society, Copyright 2006.

parameters measured. It is to speculate that different mechanisms are involved and thus different spots, *i.e.*, proteins, are important.

Another study performed with the same dataset of proteomics maps is reported by Hawkins *et al.*[26] The abundances of proteins were included as bio-descriptors in a QSAR model. The set of compounds were the same 14 halocarbons. Eight toxicity measures were considered: the lowest concentration arresting conidial development (ARR) or inducing one lethal hit per cell (D_{37}), as well as the same assays determining cytotoxicity and

Table 10.2 Leave-one-out (LOO) correlation coefficients for the 'proteomic map/spots–toxicity relationship' model.

Toxicity	LOO corr. coeff.	LOO corr. coeff.
EC_{50MTT}	**0.78**	0.88[a]
EC_{50LDH}	0.48	0.82[b]
EC_{20SH}	**0.74**	0.76[a]
LEC_{LP}	0.38	0.60[c]
LEC_{ROS}	0.56	0.62[a]
LEC_{CAT}	0.16	0.66[d]

[a]Compound no. 13 ($C_2Cl_3H_3$) was excluded as an outlier.
[b]Compounds no. 13 ($C_2Cl_3H_3$) and 11 ($C_2Cl_2H_4$) were excluded as outliers.
[c]Compounds no. 13 ($C_2Cl_3H_3$), 11 ($C_2Cl_2H_4$) and 7 ($CHBr_3$) were excluded as outliers.
[d]Compounds no. 13 ($C_2Cl_3H_3$) and 2 (C_2Cl_4) were excluded as outliers.

oxidative stress as earlier: cell viability (EC_{50MTT}), membrane integrity (EC_{50LDH}), total cellular thiols (EC_{20SH}), lipid peroxidation (LEC_{LP}), reactive oxygen species (LEC_{ROS}), and catalase activity (LEC_{CAT}). The halocarbons were represented by chemo-descriptors as described earlier. The protein abundances were considered as bio-descriptors. In the gels, 1401 individual proteins (spots) were determined and summarised into 17 markers after individual analysis. The markers were condensed into seven important (PCA) canonical variables. In the next step, QSAR models were developed taking the chemical descriptors, the seven canonical variables as bio-descriptors and the eight toxicity measures as predicted variables. The results of the predictions were summarised by cross validation correlation coefficients, which were comparable for chemo- and bio-descriptors. The use of both chemo- and bio-descriptors significantly improved the models for the endpoints EC_{50MTT} and EC_{50LDH}. A moderate improvement was observed for the endpoint EC_{20SH}; for the other five endpoints, no significant improvements were observed. It was concluded though that both types of descriptors give different information and using both information sources may provide more insights. In addition, the authors reported the clustering structure of compounds considering the first two most important canonical variables. They concluded that compounds differing only by exchanging the Cl and Br form clusters and thus show similar proteomic effects.

Alternatively, proteomics data can be encoded *via* graphical representation.[12–17] Randić *et al.* introduced topological descriptors to study the changes in proteomics maps investigating data reported by Anderson *et al.*[36,39] The descriptors were developed on the basis of ordering the proteins into a series according to their abundances. The corresponding spots were connected to a zigzag line, which was described by numerical invariants (deduced from distance matrices). It was expected that the order of proteins changes when the cells are treated with peroxisome proliferators, and thus, the zigzag line changes. It was shown that changes of these invariants are correlated with dose in the diet.[39]

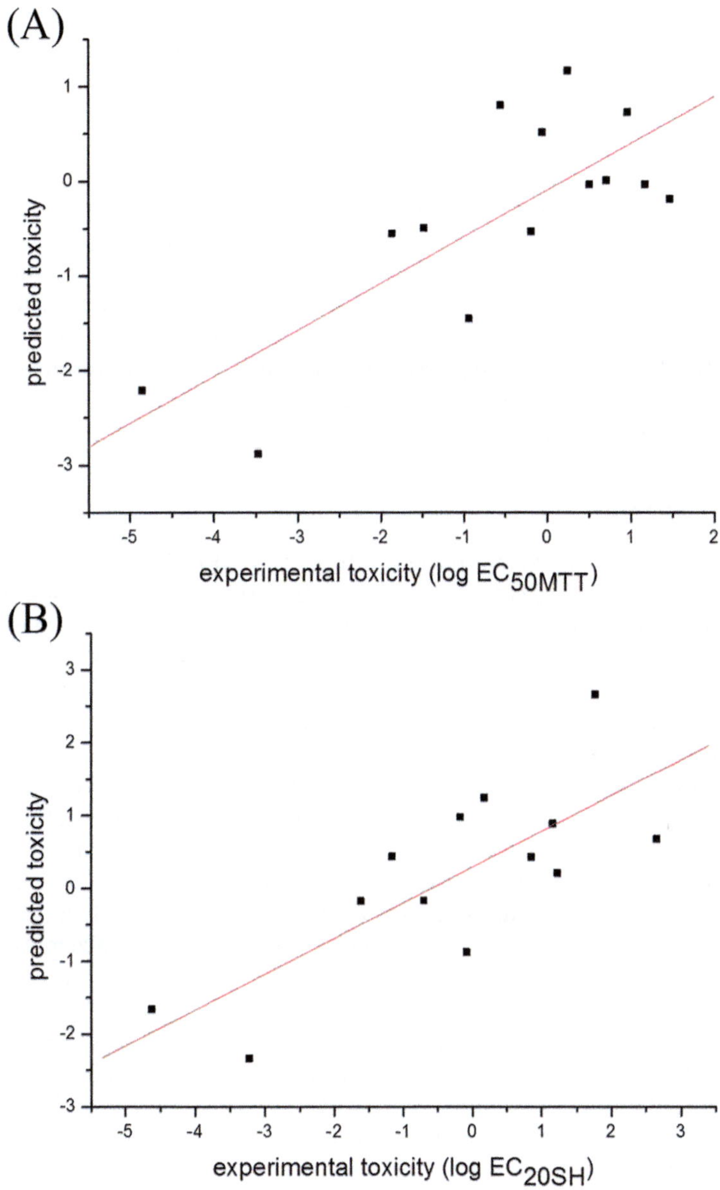

Figure 10.6 Linear regression curves for the leave-one-out test for (A) EC_{50MTT} and (B) EC_{20SH}. (In the second case, compound no. 13 ($C_2Cl_3H_3$) was excluded as an outlier.)[32]

In related work, the topological indices deduced from proteomic maps were used to develop a 'Quantitative Proteome–Property Relationship' model for early detection of drug-induced cardiac toxicity.[40] Linear discriminant analysis (LDA) was applied as a modelling technique. For the selected models, the authors reported accuracy, sensitivity and specificity values up to 87%.

In conclusion, we would like to stress the value of the concept of bio-descriptors, namely, the integration of chemical and biological information into one single model. Chemical information is embedded in structural descriptors while the proteomics descriptors carry the biological part of information. During modelling, an entire set of data is reduced to the model and extracted knowledge. In our case, this means that entire proteomics, comprising data on all proteins, are included in the model, which by means of statistical or chemometrical analysis, selects the proteins that seem important for the biological activity with the most significant chemo-descriptors. Once the chemometrical analysis is completed, the role of bio-chemists and biologists is to evaluate and interpret the results.

10.3.4 Biomarkers

In medical diagnostics, biomarkers are used as indicators of a particular physiological/pathological status in humans. In (eco)toxicological science they can provide information on relationships between a chemical exposure and a particular adverse effect on organisms or populations.[41] The chemometrical analysis of proteomics in some cases leads to the discovery of biomarkers, *i.e.*, proteins that control the physiological status of organisms. In the following, a few examples will be given.

Yeoh *et al.* compared samples from colorectal cancer tissue and normal tissue.[42] The samples were taken in pairs from 26 patients who suffered from colorectal cancer, one from cancerous and one from healthy tissues. In 2D electrophoresis maps, they identified 37 aqueous proteins and 24 membrane-associated proteins, using Tris and thiourea lysis buffer (TLB) extracting methods, respectively. The proteins were analysed with electrospray ionisation ion trap mass spectrometry and identified in the MASCOT protein database (http://www.matrixscience.com). The intensities of the 37 proteins were considered for PCA and LDA. For Tris buffer samples, the results of PCA and LDA classification were that 12 principal axes carried 76.43% of the total variance. The PCA model was built with six PCs carrying 53.97% of the variance. From 52 samples (26 from cancer tissues, 26 from normal tissues), 26 cancer and 22 normal samples were correctly classified. In TLB samples, seven PCs carried 72.46% of the total variance. The LDA model built with six PC axes (76.61%) classified 22 normal tissues and 19 cancerous tissues correctly. PCA and LDA results showed that six from 37 proteins were reliable to discriminate the colorectal cancer tissue from normal tissue, or six from 24 proteins, when the Tris or TLB extraction was applied, respectively. They can be considered as potential biomarkers.

Shkoda analysed the changes of proteome in patients with Crohn's disease and ulcerative colitis.[43] Proteins were identified with 2D electrophoresis and peptide mass fingerprints *via* matrix-assisted laser desorption ionisation time of flight (MALDI-TOF) mass spectrometry and Western blot analysis. A two-fold change was taken as a criterion for considering the changes in expression levels. 21 proteins were identified. To identify the statistically significant proteins, the individual peaks were analysed using the Mann–Whitney *U* test. Nine proteins were identified that could serve as biomarkers, including the Rho-GDP dissociation inhibitor. Furthermore, 40 proteins with significant changes in the expression were identified as potential biomarkers to separate Crohn's disease patients' tissues from non-inflamed ones. The article focuses on proteins and their role in the cells' life and concludes that the proteins associated with signal transduction, stress response and energy metabolism show the most significant changes in the expression.

Forshed *et al.* studied proteomics data of patients with acute myeloid leukemia.[44] The aim of the study was to find biomarkers that could indicate a complete remission. The authors present a complete workflow, including a description of the applied chemometrical methods. A chemometrical experimental design was conceived to optimise the number of measurements necessary to make conclusions. Furthermore, the peaks obtained with surface-enhanced laser desorption ionisation time of flight (SELDI-TOF) mass spectrometry were pre-processed with chemometrical methods considering the variability of data and repeatability, and criteria for outliers were discussed. The 'cleaned' data were then analysed by means of PCA and PLS to identify the important proteins and potential biomarkers. 21 potential candidates were selected, which seemed to be significant and could contribute to a decision on whether remission is achieved or not.

Klenø *et al.* studied the mechanism(s) of hydrazine toxicity in rat liver.[45] The Sprague Dawley rats were divided into six groups: two control groups, two groups with an administered dose of 30 mg hydrazine per kg body weight, and two groups with a dose of 90 mg kg^{-1} body weight. The animals were sacrificed after 48 hours or after 168 hours, respectively, in each pair of groups. The hepatocytes were analysed with 2D electrophoresis and the separated proteins were identified with MALDI-TOF mass spectrometry. 431 proteins were identified and analysed by means of PCA and PLS. PC1 and PC4 carried 17% and 9% of the total variance, respectively. PC1 separated the control group from groups treated with 30 mg kg^{-1} and 90 mg kg^{-1}, PC4 separated the 48- and 168 hour groups. However, it was not possible to correlate individual proteins with toxic effects. It was concluded that the intoxication not only affects single proteins, but many proteins simultaneously. In a refined partial least squares regression (PLSR) study, the dose-dependence was considered (0, 30 mg kg^{-1} and 90 mg kg^{-1}). A model with 117 selected proteins was built, which clearly separated samples according to the administered dose. In this case, the first two principal axes PC1 and PC2 carried 94% and 4% of the total variance, respectively. Following Jack-knife (leave-out) analysis and Student's *t*-test, twenty significant proteins were selected and analysed. These proteins are related to lipid metabolism, Ca^{2+} homeostasis, stress response and thyroid hormone pathways. The most perturbed protein by hydrazine treatment was alpha-2μ-microglobulin.

In conclusion, the section has shown further examples to highlight how chemometrics analyses can contribute to make sense and extract information from proteomics big data.

10.4 Conclusions

In different areas of biology, we are witnessing the explosion of available data, in particular -omics data. They can make essential contributions to our knowledge and understanding of the living world and support research in medicine and toxicology.

In order to extract the information of the large amount of data points in proteomics and help transform them into knowledge, chemometrical tools are very useful. This chapter focused on the numerical representation of proteomics data and described examples of chemometrical methods, in particular approaches for clustering and classification (categorisation). A caveat for the chemometrical analysis is that proteomics (and in general -omics) data, as biological data, have an inherent variability and are very diverse. This should be taken into account for the selection of chemometrical techniques and evaluation of the results.

The examples summarised in this chapter show the utility of the chemometric approaches to successfully distill information out of a large amount of data points, such as to identify protein changes associated with specific toxicological endpoints and conclude on potential biomarkers. Although dealing with proteomics, the examples shown here were using relatively small datasets. The techniques, however, are applicable to taking on the challenges of the big data era and supporting predictive toxicology.

References

1. R. A. Fasani, C. B. Livi, D. R. Choudhury, A. Kleensang, M. Bouhifd, S. N. Pendse, P. D. McMullen, M. E. Andersen, T. Hartung and M. Rosenberg, The Human Toxome Collaboratorium: A shared environment for multi-omic computational collaboration within a consortium, *Front. Pharmacol.*, 2016, **6**, 322.
2. G. S. Karagiannis, M. P. Paulou and E. P. Diamandis, Cancer secretomics reveal pathophysiological pathways in cancer molecular oncology, *Mol. Oncol.*, 2010, **4**, 496–510.
3. D. Chen, Y. Xu, W. Zheng, D. Huang, W. Wong, W. C. Tai, Y. Cho and A. T. Y. Lau, Proteomic analysis of secreted proteins by human bronchial

epithelial cells in response to cadmium toxicity, *Proteomics*, 2015, **15**, 3075–3086.

4. S. C. Basak, M. Vračko and A. K. Bhattacharjee, Big data and new drug discovery: tackling "big data" for virtual screening of large compound databases, *Curr. Comput.-Aided Drug Des.*, 2015, **11**, 197–201.
5. G. T. Ankley, R. S. Bennett, R. J. Erickson, D. J. Hoff, M. W. Hornung, R. D. Johnson, D. R. Mount, J. W. Nichols, C. L. Russom, P. K. Schmieder, J. A. Serrano, J. E. Tietge and D. L. Villeneuve, Adverse outcome pathways: a conceptual framework to support ecotoxicology research and risk assessment, *Environ. Toxicol. Chem.*, 2010, **29**, 730–741.
6. T. W. Schultz, Adverse outcome pathways: A way of linking chemical structures to *in vivo* toxicological hazard, in *In Silico Toxicology: Principles and Applications*, ed. M. T. D. Cronin and J. C. Madden, Royal Society of Chemistry, Cambridge, 2010, pp. 346–371.
7. C. H. Spiegelman, R. Pfeiffer and M. Gail, Using chemometrics and statistics to improve proteomics biomarker discovery, *J. Proteome Res.*, 2006, **5**, 461–462.
8. A. A. M. Heemskerk, A. M. Deelder and O. A. Mayboroda, CE–ESI-MS for bottom-up proteomics: Advances in separation, interfacing and applications, *Mass Spectrom. Rev.*, 2016, **35**, 259–271.
9. F. Lermyte, M. K. Lacki, D. Valkenborg, G. Baggerman, A. Gambin and F. Sobott, Understanding reaction pathways in top-down ETD by dissecting isotope distributions: A mammoth task, *Int. J. Mass Spectrom.*, 2015, **390**, 146–154.
10. S. C. Basak and B. Gute, Mathematical biodescriptors of proteomic maps: background and applications, *Curr. Opin. Drug Discovery Dev.*, 2008, **11**, 320–326.
11. B. D. Gute, S. C. Basak, K. Balasubramanian, K. Geiss and D. M. Hawkins, Chemodescriptors *versus* biodescriptors for toxicity predictions on halocarbons, *Environ. Toxicol. Pharmacol.*, 2004, **16**, 121–129.
12. M. Randić, F. Witzmann, M. Vračko and S. C. Basak, On characterization of proteomics maps and chemically induced changes in proteomes using matrix invariants: Application to peroxisome proliferators, *Med. Chem. Res.*, 2001, **10**, 456–479.
13. M. Randić, A graph theoretical characterization of proteomics maps, *Int. J. Quantum Chem.*, 2002, **90**, 848–858.
14. M. Randić and S. C. Basak, A comparative study of proteomics maps using graph theoretical biodescriptors, *J. Chem. Inf. Sci.*, 2002, **42**, 983–992.
15. M. Randić, F. A. Witzmann, V. Kodali and S. C. Basak, On the dependence of a characterization of proteomic maps on the number of protein spots considered, *J. Chem. Inf. Model.*, 2006, **46**, 116–122.
16. M. Randić, M. Novič, M. Vračko and D. Plavšić, Study of proteome maps using partial ordering, *J. Theor. Biol.*, 2010, **266**, 21–28.

17. M. Randić, M. Novič, A. Roy Choudhury and D. Plavšić, On graphical representation of trans-membrane proteins, *SAR QSAR Environ. Res.*, 2012, **23**, 327–343.
18. A. Roy Choudhury and M. Novič, PredβTM: A novel β-transmembrane region prediction algorithm, *PLoS One*, 2015, **10**, e0145564.
19. A. Perdih, A. Roy Choudhury, Š. Župerl, E. Sikorska, I. Zhukov, T. Šolmajer and M. Novič, Structural analysis of a peptide fragment of transmembrane transporter protein bilitranslocase, *PLoS One*, 2012, **7**, e38967.
20. M. Vračko and S. C. Basak, Similarity study of proteomic maps, *Chemom. Intell. Lab. Syst.*, 2004, **70**, 33–38.
21. S. C. Basak, M. Vracko and F. A. Witzmann, Mathematical nanoproteomics: quantitative characterisation of effects of multi-walled carbon nanotubes (MWCNT) and TiO_2 nanobelts (TiO_2-NB) on protein expression patterns in human intestinal cells, *Curr. Comput.-Aided Drug Des.*, 2016, **12**, 259–264.
22. J. H. Ward Jr, Hierarchical grouping to optimize an objective function, *J. Am. Stat. Assoc.*, 1963, **58**, 236–244.
23. V. Drgan, J. Panek, M. Vračko and M. Novič, Encoding and Clustering of Proteins in Mycobacterium Tuberculosis Proteome, *Int. J. Chem. Model.*, 2014, **6**, 377–389.
24. A. P. Worth, A. Bassan, A. Gallegos, T. Netzeva, G. Patlewicz, M. Pavan, I. Tsakovska and M. Vračko, *The characterisation of (quantitative) structure-activity relationships: preliminary guidance*, JRC Report EUR 21866 EN, 2005. p. 85.
25. K. Varmuza, Multivariate data analysis in chemistry, in *Handbook of Chemoinformatics*, ed. J. Gasteiger, Wiley-VCH, Weinheim, 2003, pp. 1098–1133.
26. D. M. Hawkins, S. C. Basak, J. Kraker, K. T. Geiss and F. A. Witzmann, Combining chemodescriptors and biodescriptors in quantitative structure-activity relationship modeling, *J. Chem. Inf. Model.*, 2006, **46**, 9–16.
27. J. Zupan and J. Gasteiger, *Neural Networks in Chemistry and Drug Design*, Wiley-VCH, Weinheim, 1999.
28. M. Vračko, Kohonen artificial neural network and counter propagation neural network in molecular structure-toxicity studies, *Curr. Comput.-Aided Drug Des.*, 2005, **1**, 73–78.
29. B. T. Luke, Genetic algorithms and beyond, in *Nature-inspired Methods in Chemometrics: Genetic Algorithms and Artificial Neural Networks*, ed. R. Leardi, Elsevier, Amsterdam, 2003, pp. 3–54.
30. V. Drgana, Š. Župerla, M. Vračko, F. Comob and M. Novič, Robust modelling of acute toxicity towards fathead minnow (Pimephales promelas) using counter-propagation artificial neural networks and genetic algorithm, *SAR QSAR Environ. Sci.*, 2016, **27**, 501–519.

31. A. Maiocchi, Genetic algorithms in molecular modeling: a review, in *Nature-inspired Methods in Chemometrics: Genetic Algorithms and Artificial Neural Networks*, ed. R. Leardi, Elsevier, Amsterdam, 2003, pp.109–139.
32. M. Vračko, Mathematical (structural) descriptors in QSAR: Applications in drug design and environmental toxicology, in *Advances in Mathematical Chemistry and Applications*, ed. S. C. Basak, Bentham Science Publishers, 2014, vol. 1, pp. 222–250.
33. S. C. Basak, Mathematical descriptors for the prediction of property, bioactivity, and toxicity of chemicals from their structure: A chemical-cum-biochemical approach, *Curr. Comput.-Aided Drug Des.*, 2013, **9**, 449–462.
34. S. C. Basak, Role of mathematical chemodescriptors and proteomics-based biodescriptors in drug discovery, *Drug Dev. Res.*, 2011, **72**, 225–233.
35. E. Marengo, M. Bobba, M. C. Liparota, E. Robotti and P. G. Righette, Use of Legendre moments for the fast comparison of two-dimensional polyacrylamide gel electrophoresis maps images, *J. Chromatogr. A*, 2005, **1096**, 86–91.
36. N. L. Anderson, R. Esquer-Blasco, F. Richardson, P. Foxworthy and P. Eacho, The effects of peroxisome proliferators on protein abundances in mouse liver, *Toxicol. Appl. Pharmacol.*, 1996, **137**, 75–89.
37. E. J. Calabrese and L. A. Baldwin, Toxicology rethinks its central belief, *Nature*, 2003, **421**, 691–692.
38. M. Vracko, S. C. Basak, K. Geiss and F. Witzmann, Proteomic maps-toxicity relationship of halocarbons studied with similarity index and genetic algorithm, *J. Chem. Inf. Model.*, 2006, **46**, 130–136.
39. M. Randić, M. Novič and M. Vračko, On characterization of dose variations of 2-D proteomics maps by Matrix Invariants, *J. Proteome Res.*, 2002, **1**, 217–226.
40. M. Cruz-Monteagudo, C. R. Munteanu, F. Borges, M. N. D. S. Cordeiro, E. Uriarte and H. Gonzalez-Diaz, Quantitative Proteome-Property Relationships (QPPRs). Part 1: Finding biomarkers of organic drugs with mean Markov connectivity indices of spiral networks of blood mass spectra, *Bioorg. Med. Chem.*, 2008, **16**, 9684–9693.
41. A. Jemec, D. Drobne, T. Tišler and K. Sepčić, Biochemical biomarkers in environmental studies–lessons learnt from enzymes catalase, glutathione S-transferase and cholinesterase in two crustacean species, *Environ. Sci. Pollut. Res.*, 2010, **17**, 571–581.
42. L.-C. Yeoh, S. Dharmaraj, B.-H. Gooi, M. Singh and L.-H. Gam, Chemometrics of differentially expressed proteins from colorectal cancer patients, *World J. Gastroenterol.*, 2011, **17**, 2096–2103.
43. A. Shkoda, T. Werner, H. Daniel, M. Gunckel, G. Rogler and D. Haller, Differential protein expression profile in the intestinal epithelium from patients with inflammatory bowel disease, *J. Proteome Res.*, 2007, **6**, 1114–1125.

44. J. Forshed, M. Pernemalm, C. S. Tan, M. Lindberg, L. Kanter, Y. Pawitan, R. Lewensohn, L. Stenke and J. Lehtiö, Proteomic data analysis workflow for discovery of candidate biomarker peaks predictive of clinical outcome for patients with acute myeloid leukemia, *J. Proteome Res.*, 2008, 7, 2332–2341.
45. T. G. Klenø, L. R. Leonardsen, H. Ø. Kjeldal, S. M. Laursen, O. N. Jensen and D. Baunsgaard, Mechanism of hydrazine toxicity in rat liver investigated by proteomics and partial least squares regression, *Proteomics*, 2004, **4**, 868–908.

CHAPTER 11

Big Data and Biokinetics

MIYOUNG YOON*[a], GINA SONG[b], HARVEY CLEWELL[c] AND BAS BLAAUBOER[d]

[a]ToxStrategies, Research Triangle Park, 1249 Kildaire Farm Road 134, Cary, NC 27511, USA; [b]ScitoVation, 6 Davis Drive, Research Triangle Park, NC 27709, USA; [c]Ramboll Environ, Research Triangle Park, NC 27709, USA; [d]Utrecht University, Institute of Risk Assessment Sciences, Yalelaan 2, 3584, CM Utrecht, The Netherlands
*E-mail: yoon.m.work@gmail.com

11.1 Introduction

There is an increasing need to be able to rapidly deal with large numbers of compounds and large volumes of exposure, mechanistic and toxicity data to assess the toxicological risk of being exposed to chemicals. An important aspect of this process is information on the kinetic behaviour of the chemicals in organisms. There is therefore a need to be able to rapidly evaluate this multitude of compounds in terms of their biokinetics. In light of the shifting paradigm in toxicology,[1] this is even more evident. This shift replaces the emphasis from being based on apical endpoints in animal studies to an approach incorporating knowledge of modes of action in human cellular (*in vitro*) systems together with biokinetics.[2]

Results obtained from *in vitro* studies are often not directly applicable to the *in vivo* situation. One of the most obvious differences between the situations *in vitro* and *in vivo* is the absence *in vitro* of processes regarding absorption,

Issues in Toxicology No. 41
Big Data in Predictive Toxicology
Edited by Daniel Neagu and Andrea-Nicole Richarz
© The Royal Society of Chemistry 2020
Published by the Royal Society of Chemistry, www.rsc.org

distribution, metabolism and excretion (ADME; *i.e.*, biokinetics) that govern the exposure of the target tissue in the intact organism. The concentrations to which *in vitro* systems are exposed may not correspond to the actual situation at the target tissue after *in vivo* exposure. In addition, the occurrence of metabolic activation and/or saturation of specific metabolic pathways or absorption and elimination mechanisms may also become relevant for the toxicity of a compound *in vivo*. This may lead to misinterpretation of *in vitro* data if such information is not taken into account. Therefore, predictive studies on the biological activity of compounds require the integration of data on the mechanisms of action with data on biokinetic behaviour. Over the last decade, the feasibility of using mathematical models for prediction of *in vivo* biokinetics has grown substantially.[3]

The application of physiologically-based kinetic models in these approaches in principle allows the estimation of concentration–time profiles of compounds at any place in the organism, after any exposure scenario. Likewise, it also allows the interpretation of *in vitro*-derived concentration–effect relationships for its relevance for an *in vivo* exposure. This is done by the process of reverse dosimetry: calculating a dose that can be kinetically related to, for example, the concentration giving 10% of the maximum effect *in vitro*, thus allowing a quantitative *in vitro*-to-*in vivo* extrapolation (IVIVE) and the interpretation of *in vitro* data in risk evaluations.[4]

The consequence of this new possibility is an even greater need for an ability to incorporate biokinetic modelling in the day-to-day work of toxicological risk assessors. This incorporation, described, for example, in integrated approaches of toxicity assessment (IATA) will be the challenge for toxicology in the future. In the future, data from different domains will need to be rapidly integrated, *i.e.*,

- data on exposure scenarios;
- biokinetic and toxicodynamic data from multiple *in vitro* systems;
- data to support development and parameterisation of modelling approaches for *in vitro* biokinetics and dynamics;
- physiological data for the construction of *in vivo* physiologically-based biokinetic (PBBK) models: blood flows, organ volumes, *etc.*
- chemical-specific data relevant for incorporation in the PBBK models (partitioning, metabolism, transporter affinity, *etc.*)

Key problem areas in predicting kinetics of chemicals include: (1) biotransformation (*in vitro* systems for determining biotransformation, interpretation of *in vitro* data, scaling up to the *in vivo* situation) and (2) the passage across special barrier systems (*e.g.*, in the gastrointestinal (GI) tract, the blood–brain barrier, and the kidney). It will be obvious that the quality of the evaluation of a compound's toxic profile making use of the data given earlier will be highly dependent on the quality of the data as well as on the appropriate incorporation in the different models.

To fully incorporate biokinetics into modern toxicity testing and risk assessment processes based particularly on a large amount of information from high-throughput *in vitro* assay results in the big data era, easy to use and yet robust simulation tools are necessary. To support both screening and assessment phases of risk assessment, these tools need to be flexible in terms of data requirements. Previous efforts in biokinetic modelling have been conducted on a chemical by chemical basis, which is entirely too slow for dealing with large numbers of compounds. Instead, biokinetic modelling in the future must focus on defining a comprehensive physiological/biochemical modelling framework, encompassing human life stages from conception to senescence, in which the *in vivo* biokinetics of chemicals can be rapidly predicted. Chemical-specific data generated from *in silico* prediction methods, available database resources, and experimental methods can be placed in the framework to simulate human exposure to a given chemical in an exposure condition of human safety concerns. This chapter will describe ways in which biokinetic data and modelling can be used to support rapid interpretation of high-volume toxicity data on large numbers of chemicals, from simple approaches that can be used in high-throughput screening to the more complicated approaches that are necessary to provide more accurate risk assessments.

11.2 Simple Biokinetic Modelling Approaches for Supporting the Use for Large Data Amounts

11.2.1 Biokinetic Approach for Supporting High-throughput Testing

High-throughput *in vitro* toxicity assays promise to rapidly screen and identify the potential biological activity (*i.e.*, hazard) of chemicals. However, in order to evaluate the safety of potential *in vivo* exposures based on the concentrations at which effects are observed in the *in vitro* assays, bioavailability and clearance of the chemicals must be considered.[2] Considering the extensive data requirement to build compound-specific biokinetic or physiologically-based biokinetic models, the possibility of applying a simple steady-state *in vitro*-to-*in vivo* extrapolation (IVIVE) approach to interpret the results of high-throughput assays conducted under the US Environmental Protection Agency (EPA) ToxCast programme[5,6] was evaluated. In these studies, hepatic metabolic clearance and plasma protein binding were measured *in vitro* for the ToxCast Phase I chemicals. The experimental data were used to parameterise a simple IVIVE equation to estimate the human oral equivalent doses that would produce steady-state *in vivo* blood concentrations equivalent to *in vitro* AC50 (concentration at 50% of maximum activity) or LEC (lowest effective concentration) values in the *in vitro* high-throughput assays.

Briefly, a simple clearance equation[7] was used to estimate expected steady-state blood concentrations. The equation assumes zero-order uptake of a daily dose from the gut (assuming 100% oral bioavailability). Both renal and hepatic clearance are described. The steady-state concentration in the blood (C_{ss}) is calculated as eqn (11.1) (see discussion in next section):

$$C_{ss} = k_o/[\text{GFR} \times f_{ub} + \text{QL} \times f_{ub} \times \text{CLint}/(\text{QL} + f_{ub} \times \text{CLint})] \quad (11.1)$$

In this equation, the term GFR × f_{ub} represents the renal excretion of unbound (free) parent compound in the blood by glomerular filtration, where GFR is the glomerular filtration rate, which is about 6.7 L h^{-1} in human adults.[8] f_{ub} is the fraction of the compound in the blood that is unbound. The second term in the denominator is hepatic clearance (CLh), where QL is liver blood flow (typically on the order of 90 L h^{-1} in adults) and CLint is the intrinsic metabolic clearance for first-order conditions of metabolism in the liver at low concentrations, *i.e.*, below k_m. Hepatocellular clearance in this study was experimentally measured at 1 μM and the slope of the disappearance of the chemical over time was measured. Metabolic intrinsic clearance measured in hepatocyte suspension was normalised to cell number, with the units μL/min/10^6 cells. *In vivo* intrinsic clearance was estimated by multiplying the *in vitro* clearance by the number of cells per gram of liver (roughly 137 × 10^6), referred to as hepatocellularity, and the weight of the liver (about 1820 g in an adult). An arbitrary daily dose of 1 mg kg^{-1} day^{-1} was used for C_{ss} calculations in those ToxCast dosimetry studies. In their analysis, the SimCyp simulation platform[9] was used to perform Monte Carlo analysis to simulate variability across a population of 100 healthy individuals of both sexes from 20–50 years of age. A coefficient of variation of 30% was used for intrinsic and renal clearance. Reverse dosimetry was then performed to generate oral equivalent doses according to eqn (11.2):

$$\text{Oral Equivalent Dose (mg kg}^{-1}\text{ day}^{-1}) = \text{AC50 or LEC}/C_{ss} \quad (11.2)$$

For a small number of these chemicals, it was possible to find *in vivo* biokinetic data to estimate a steady state concentration at an exposure of 1 mg kg^{-1} day^{-1} for comparison with the *in vitro* predictions. The results of the comparison are shown in Table 11.1.

Two alternative hepatic clearance assumptions[7] were compared: (1) restrictive hepatic clearance (assuming only unbound chemical is available for clearance), using f_{ub} determined experimentally and (2) non-restrictive hepatic clearance (assuming all of the chemical is available for clearance), where the f_{ub} was set to one. Picloram appears to have non-restrictive clearance, while triclosan is an example of a chemical that appears to have restrictive clearance. The behaviour of lindane appears to be intermediate between the two extremes. In general, the assumption of restrictive clearance produces a higher estimate of C_{ss}. This may be a conservative assumption if the parent compound is largely responsible for biological activity. These two

Table 11.1 Comparison of *in vitro*-to-*in vivo* extrapolation modelling results with *in vivo* -based results (adapted from ref. 6 with permission from Oxford University Press, Copyright 2011).

Chemical	*In vivo* derived $C_{ss}{}^{a}$ (μM)	IVIVE $C_{ss}{}^{a,b}$ (μM)	IVIVE, Caco-2[c] $C_{ss}{}^{a,b}$ (μM)	IVIVE, Fu=0.99 $C_{ss}{}^{a,b}$ (μM)	IVIVE, Fu = 0.99, Caco-2[c] $C_{ss}{}^{a,b}$ (μM)
2,4-Dichlorophenoxyacetic acid	9.05–90.05	39.25	40.34	39.25	40.34
Bisphenol-A	<0.13	0.35	0.40	0.06	0.07
Cacodylic acid	1.80	3.06	—	3.06	—
Carbaryl	0.03	0.07	0.07	0.03	0.03
Fenitrothion	0.03	17.91	—	0.10	—
Lindane	0.46	13.21	—	0.07	—
Oxytetracycline dihydrate	0.36	2.00	0.44	2.00	0.44
Parathion	0.17	24.63	—	0.14	—
Perfluorooctane sulfonic acid	19990	160.78	179.96	160.78	179.96
Perfluorooctanoic acid	20120	55.34	58.19	0.40	0.40
Picloram	0.27	57.63	32.01	0.37	0.19
Thiabendazole	0.45	13.76	15.20	13.76	15.20
Triclosan	2–10	1.56	1.59	0.01	0.01

[a] C_{ss}, concentration at a steady state for 1 mg kg^{-1} day^{-1} dose.
[b] Predicted using the 1 μM metabolic clearance rate.
[c] IVIVE performed incorporating Caco-2 data into the simulation, if available.

hepatic clearance assumptions represent extremes of the range of the possible relationship between chemical disposition/transport and hepatocellular metabolism that can result in C_{ss} estimates that differ by several orders of magnitude. Hepatic clearance is complexly determined by a number of factors, including liver blood flow, the association and dissociation rates for binding of the chemical to plasma proteins such as albumin, the kinetics of hepatocellular uptake of the chemical, and the kinetics of hepatocellular metabolism. Indeed, no approaches have yet been demonstrated to predict the fraction of compound available for metabolism in liver, even in the case of drugs. In general, the comparison result indicates that non-restrictive hepatic clearance appears to be more consistent with the *in vivo* data for environmental chemicals.

While this simple approach demonstrated the need to consider biokinetics in interpreting high-throughput screening data, interpretation of the simple IVIVE approach requires caution considering the limitations of the simple approach.[10] From a human health standpoint, the assumption of 100% oral bioavailability is conservative because lower absorption results in a higher oral dose required to achieve a specific C_{ss}. However, incorporation of Caco-2 assay data on bioavailability into the steady-state IVIVE equation can increase the predictivity of the C_{ss} determination, as in the case of oxytetracycline dihydrate and picloram in Table 11.1. On the other hand, the assumption that renal clearance is solely a function of f_{ub} and the GFR is not necessarily conservative, since active renal resorption would result in a higher C_{ss} at a given dose, as in the case of the two perfluorinated chemicals in Table 11.1.

The other two important limitations that need to be noted when interpreting the simple IVIVE results are the use of nominal concentration *in vitro* and lack of consideration of metabolite-mediated activity. The analysis is performed on the assumption that blood concentrations equivalent to the nominal *in vitro* AC50 or LEC values would produce equivalent responses *in vivo*. However, the concentration of free chemical in an *in vitro* assay that elicits a certain biological response may differ from the nominal effective concentrations due to factors such as protein–lipid composition of the media and binding of the chemical to surfaces of *in vitro* assay systems.[2,11] In addition, the biokinetics and bioactivity in those ToxCast analyses were only evaluated for the parent compound. No attempt was made to evaluate biological activities and dosimetry of metabolites.

11.2.2 Biokinetic Approach for Predicting Acute Toxicity

The diagram presented in Figure 11.1 illustrates a conceptual structure for the use of kinetic information in the estimation of acute oral toxicity. Under this scheme, available *in vitro* data on the absorption, tissue partitioning, metabolism, and excretion of a test material would be used to parameterise

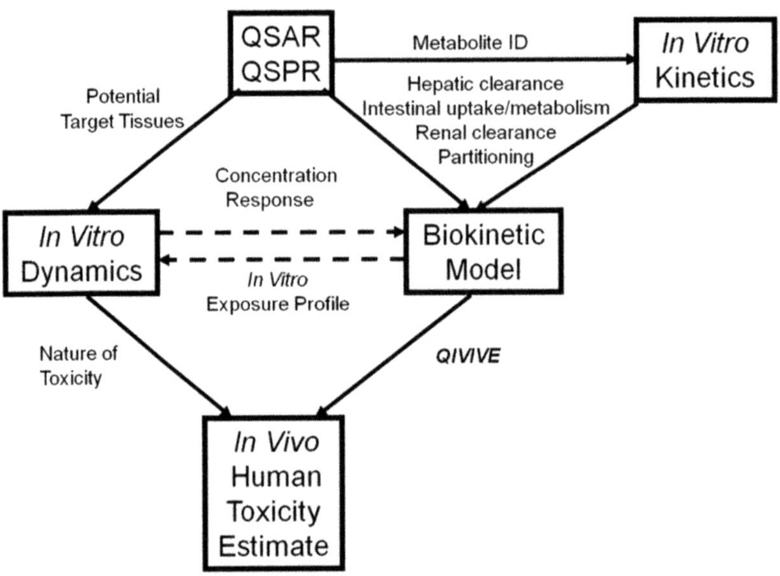

Figure 11.1 A scheme for incorporation of QSAR (QSPR) information, *in vitro* metabolism data, and biokinetic modelling in the estimation of human toxicity from *in vitro* assays. (Reprinted with permission ref. 18.)

a chemical-specific biokinetic model.[3] In many cases, currently available Quantitative Structure Property Relationship (QSPR)/Quantitative Structure Activity Relationship (QSAR) techniques could be used to estimate chemical properties and kinetics when the specific data for that chemical is lacking. For example, simple empirical correlations have been developed for estimating the tissue partitioning of a chemical from its water solubility, vapor pressure, and octanol/water partitioning.[12–14] More physiologically-based approaches to predict tissue partitioning based on tissue composition data, including the case of highly lipophilic chemicals in addition to drug-like chemicals, have become available in the past few years.[15,16] Emerging QSAR techniques (*e.g.*, knowledge-based systems) may eventually prove useful in predicting potential target tissues for toxicity so that the appropriate assays of *in vitro* dynamics (response) could be selected. These target tissue assays would, in turn, provide information on the nature and location of the toxicity produced by the chemical.[17]

The complexity of the biokinetic model would depend on the physico–chemical and biochemical characteristics of the chemical. In the specific case of acute toxicity, a simple one-compartment description of the administered chemical may suffice for many chemicals. The volume of distribution for such a model could be estimated from the volume-weighted average of the estimated partitioning into various tissues, and estimates of fractional absorption and rate of clearance could be based on data for structurally similar compounds. Each of these assumptions or predictions, however, introduces its own associated uncertainty into the result of the lethality risk estimate. Even with such a simple model, it may be possible to estimate the systemic concentrations that could be expected to result from an *in vivo* exposure to a given dose.[18] Thus, the model could be used to relate the concentrations at which toxicity is observed in an *in vitro* toxicity assay to the equivalent dose that would be expected to be associated with toxicity for *in vivo* exposure. These models can also provide information on the temporal profile for tissue exposure *in vivo*, which can then be used in the design of the most appropriate *in vitro* experimental protocol.[19]

There are chemical classes for which a one-compartment description would not be expected to be adequate. However, the physiological mammalian structure (tissue volumes, blood flows, ventilation rate, glomerular filtration rate, *etc.*) is well characterised, and there is no difficulty in describing tissues separately. As mentioned, techniques exist for estimating tissue-specific partitioning. Other data required would depend on the class of chemical. For volatile chemicals, ventilatory clearance can be estimated from the blood–air partition. For water-soluble chemicals, urinary clearance can be estimated from the glomerular filtration rate or the renal blood flow (for secreted compounds). For some classes of chemicals, it would also be necessary to determine the fractional binding of the chemical to plasma proteins or the partitioning of the chemical into red blood cells.

The greatest challenge in parameterising the biokinetic model remains the estimation of metabolic clearance. The possibility to use *in-vitro*-determined metabolic parameters (Vmax and Km) in order to accurately predict total body metabolic clearance is increasing.[20] Currently, it would be necessary to perform *in vitro* assays of the dose–response (capacity and affinity) for metabolic clearance.[21–23] These assays are generally more expensive than the dynamic (toxicity) assays, since they necessarily involve the development of an analytical method for quantifying the concentration of the parent compound and its metabolite(s) in each tissue of interest over time. Quantification of the concentration of compound in the dynamic assays should also be preferred or at least estimated as the *in vitro* kinetics is also critical in accurately determining the *in vitro* dose–response relationship.[12,24] Eventually, as data accumulate for a large number of structurally diverse materials, it might be possible to predict metabolism and disposition using knowledge-based systems.

An important underpinning of this process is that the kind of information necessary for a particular test material depends on its structure and physico-chemical properties. It seems reasonable to expect that chemicals could be categorised into classes based on their properties, and that this categorisation would simplify the process of determining the data needed for a particular compound. As noted earlier, the key physico-chemical properties of a test material involve its volatility (reflected in its blood–air partition, Hb/g), its water solubility (S_w), and its lipophilicity (reflected in its octanol–water partition, $k_{o/w}$). Compounds with similar properties can be grouped, and data from similar compounds can be used to fill gaps in the knowledge of a particular compound.

In the first step, estimates of key kinetic parameters can be obtained either from data available on the chemical or from the use of QSPR techniques (which are based on the physico-chemical properties of the compound). QSPR techniques can be used as a first approximation of key kinetic parameters such as absorption, partition, *etc.* If one can use kinetic data that are actually measured, then these data will prevail. Chemical properties to group chemicals based on similarities include:

- octanol/water partition coefficient
- water solubility
- saturation vapor pressure or blood–air partition
- pK_a
- molecular weight/volume (for estimating gastrointestinal absorption)
- hydrogen bond donors/acceptors (for estimating gastrointestinal absorption).

This prior knowledge on kinetic parameters or the estimation on the basis of QSPR data can then be used to evaluate the *in vitro* LC50 values for a chemical. The assumption is that this LC50 value is equal to the concentration in the intact organism at which 50% of the cells die *in vivo*. Depending on the chemical's physico-chemical properties, the kinetic model to be used for this estimation may be simple or more complex. For many (*e.g.*, water-soluble compounds), a simple one-compartment model can be used to estimate the oral dose that would result in an average systemic exposure equivalent to the *in vitro* LC50 value over the time period of interest. The key factors needed for the model would be estimates of the oral bioavailability, tissue partitioning (to obtain the volume of distribution), and total clearance. Depending on the properties of the compound, the clearance could be dominated by metabolism, urinary excretion, or pulmonary ventilation. In most cases, metabolic clearance will have to be determined empirically.

11.2.3 Biokinetics Supporting Chemical Grouping

One issue that could create some order in the large amount of data is the grouping of chemicals. There are different ways to group chemicals. Established procedures are used to group chemicals in the "chemical space", making use of chemical descriptors (see Chapter 3) based on structural, molecular orbit, physico–chemical properties, *etc.* However, approaches for grouping chemicals based on similarities in molecular structure and chemical properties alone are not sufficient for read-across predictions unless supported by biological properties,[25] with grouping referring to "biological space". This concept is less well-defined since it can be approached in different ways. One possibility is to group compounds on the basis of their biokinetic behaviour. This could include parameters linked to all ADME characteristics of the compound, *e.g.*, the rate of dermal absorption, the distribution over the different organs and tissues, the metabolism (biotransformation) pattern and the excretion *via* exhalation or kidney clearance. There is an obvious relationship between the physico–chemical characteristics and these ADME parameters. However, there is also a clear biological aspect to them, *e.g.*, for absorption in the skin or gut, the biological characteristics certainly will play a role. This is even more obvious for the biotransformation of compounds – given the link with the activities of the enzyme systems involved – as well as for the distribution over barriers (blood–brain, blood–placenta, blood–testis) – here the involvement of transporter systems is of importance.[26]

These considerations would then give the opportunity to place chemicals within a "chemical-biological space" based on a set or sub-set of the parameters involved in the processes mentioned earlier, and depending on the issue(s) at stake. For instance, one can group chemicals according to the rate of dermal absorption. This would then allow making distinctions between compounds for which the need to take systemic toxicity into account is eminent or not. Other groupings might have other purposes: lipophilicity in combination with biotransformation rate would give clues to the ability of compounds to bio-accumulate.

Another grouping to be distinguished is on the basis of a toxicological mode of action. Here again, there is a link with the physico–chemical properties, given the paradigmal axioma that the basis of each toxicological effect of a compound is related to a chemico–biological interaction (also called a molecular initiating event). This is obvious, for example, for reactive compounds such as electrophiles, that easily react with DNA or proteins; heavy metal ions, reacting to SH groups in proteins; organophosphates, reacting with acetylcholinesterase (AChEs) *etc.* Since knowledge of such reactivity can easily be linked to a mode of action or adverse outcome pathway, grouping of chemicals on this basis will be very helpful in selecting the appropriate system to quantify this, thus helping in building the hypothesis for the development of fit-for-purpose *in vitro* assays for understanding the toxic mechanisms and hence the need to perform well-targeted *in vitro* experiments. In other words: the aim of an exercise determines the way in which a grouping – in a chemical or a biological space – can have obvious advantages to perform.

11.2.4 Minimalist Biokinetics Approach

A key problem for this near-term application is that for many high-production-volume chemicals, adequate analytical methods may not have been developed yet. Therefore, metabolism assays may be too expensive and time-consuming for high-throughput LD50 estimation. However, a simple, conservative estimate for the oral dose resulting in systemic exposure equivalent to an *in vitro* LC50 value could be obtained by assuming 100% bioavailability, ignoring metabolic clearance, and simply estimating tissue partitioning to obtain the volume of distribution (Vd). For example, a commonly used default for the volume of distribution for water-soluble chemicals as a function of body weight (BW) is Vd = 0.65 × BW.

In this simple approximation, the relationship between the *in vivo* and *in vitro* assays could be described by eqn (11.3):

$$\mathrm{LD50} = \mathrm{LC50} \times \mathrm{Vd}/\mathrm{BW} \tag{11.3}$$

Other adjustments could be made to this approach for chemicals where ventilatory or urinary clearance would be important, as described in the previous section. In addition, if data on bioavailability are available, such information could be factored in to obtain a more accurate LD50 estimate. An additional benefit of this approach is that similar calculations could be used to convert the *in vitro* LC50 value to an *in vivo* LC50 value for acute inhalation.[10] These assumptions, however, introduce inherent uncertainties into the resulting calculation of the oral LD50 value and depending upon the material of concern, may result in substantial inaccuracies.

Another useful exercise would be to identify the compounds that represent the outliers in the Registry of Cytotoxicity (RC) correlations of *in vitro* basal cytotoxicity assays with LD50 values. By determining the physico–chemical properties of these compounds, and knowing their target tissues, it might be possible to identify factors that could improve the correlation (*e.g.*, consideration of blood–brain barrier penetration) between predicted oral LD50 values in rodents and empirical values. In this way, it might be possible to define a "predictive range" for various chemical properties over which the *in vitro* assay might be expected to provide reasonable LD50 estimates. Also, exclusion rules for identifying compounds for which the results of the *in vitro* assay should not be relied upon might be defined.

A particular problem area in terms of the predictive value of the currently available *in vitro* toxicity assays is where toxicity is secondary to metabolic activation. In particular, it is possible that rapid oxidative or reductive metabolism could result in acute liver toxicity from oral exposure. Examples of such toxicity are the production of phosgene by the oxidative metabolism of chloroform and the acute liver necrosis seen after carbon tetrachloride exposure. Such toxicity would not be observed in *in vitro* assays using basal cells with little or no metabolic competence.

One possible approach for dealing with this problem is illustrated in Figure 11.3. The first step would be to estimate hepatocyte metabolism at a relatively low concentration (*e.g.*, 10 μM). If the rate of metabolism (V_{max}/K_m) observed is low, then the basal cell LC50 value could be relied upon. If, however, the rate is high, then it would be necessary to identify the responsible enzyme system. This identification could be performed, for example, by using a microsomal (S9) fraction with selective addition of cofactors or inhibitors. If these studies indicate that the primary enzyme system is oxidative or reductive, then the possibility of toxicity associated with metabolic activation exists. In this case, it would be necessary to perform a hepatocyte cytotoxicity assay. If the LC50 value for the hepatocytes was much lower than for the basal cells, it would be necessary to characterise the concentration-response for metabolism in order to predict the *in vivo* doses that might be associated with toxicity. On the other hand, if the primary metabolism represents detoxication (conjugation, sulfation, *etc.*), then the (acute) toxicity of the metabolites will generally be much lower and, therefore, the basal cell assay results for the parent compound could be used with some confidence to calculate the LD50 value. Of course, the metabolic clearance *in vivo* needs to be considered during this *in vitro* to *in vivo* translation of LD50.

An alternative approach would be to begin with a basal cell cytotoxicity assay (to screen out highly toxic compounds) and then perform a toxicity assay with a hepatocyte primary culture. If similar LC50 values were obtained in both assays, the concern for toxicity secondary to metabolic activation could be effectively ruled out. In such cases, a much less extensive characterisation of metabolism would be needed to support an estimate of clearance. On the other hand, if the toxicity in the hepatocyte assay was strikingly greater than that for the basal cells, the more complete characterisation of metabolism discussed earlier would be justified.

Caution is needed when interpreting *in vitro* toxicity assay results for repeated dose effect predictions. *In vitro* biokinetic factors may create different microenvironments only applicable for *in vitro* assays hampering proper *in vitro* to *in vivo* extrapolation. For example, *in vivo* target tissue concentrations due to cellular accumulation of highly lipophilic compounds would not have been able to be predicted if for biokinetics of cellular uptake and equilibration of this compound only *in vitro* media concentrations were measured and used for *in vitro* to *in vivo* extrapolation.[27]

11.3 Big Data Facilitating the Parameterisation of Biokinetic Modelling to Support the Use of Big Data for Risk Assessment

Both the adverse and beneficial responses to compounds are related to the free concentrations of active compounds at the target rather than the amounts of compound at the site of absorption. Therefore, the internal exposure at the target tissue is the appropriate dose metric for use in safety assessment. Biokinetic modelling, especially physiologically-based such as physiologically-based pharmacokinetic (PBPK) modelling, offers a scientifically sound framework to integrate mechanistic data for physiology and biochemical processes and serves as a tool to predict internal exposure at the target tissue for a wide range of exposure conditions in animal or human. The mechanistic basis of PBPK models enhances their predictive power, allowing for various applications of this tool in a risk assessment context. These applications include inter-species extrapolation, route-to-route extrapolation, high to low dose extrapolation, and the recent application area of quantitative *in vitro* to *in vivo* extrapolation, supporting the new safety assessment paradigm based on *in vitro* and computational methods.[3,28] Quantitative IVIVE is a good example of the utility of PBPK modelling from the perspective of the changing paradigm in toxicology and the increasing needs for the use of big data for risk assessment. PBPK models provide an effective way to integrate independent knowledge from diverse tools including big data to support mode of action, or adverse outcome pathway-based risk assessment frameworks. Beyond their applications for quantitative risk assessment, PBPK models can be used to interpret human biomonitoring data[29] and epidemiological data,[30–32] being considered as part of a weight of evidence evaluation by regulatory agencies in support of exposure and hazard assessments.

The physiological, mechanistic basis of the models is both their strength (the mechanistic basis provides exceptional utility) and their weakness (PBPK models can be expensive and time-consuming to construct). Obtaining chemical specific parameters, metabolism parameters in particular, has been the biggest challenge in expanding the use of PBPK models to a wide range of chemicals as well as in gaining acceptance by regulatory agencies. The recent advances in *in vitro* and *in silico* technologies for predicting chemical absorption, distribution, metabolism, and excretion and their variability in humans, along with the availability of "generic" PBPK modelling software like SimCyp, has significantly contributed to a rise in the application of PBPK modelling, particularly in drug development.[33] The validity of the *in vitro*- and *in silico*-based parameterisation strategies has been shown with a number of environmental chemicals (reviewed in[3]) and can be applied to build a generic PBPK modelling tool for chemicals.[34,35] Key parameters for PBPK models, *e.g.*, metabolism parameters, are provided from *in vitro* assays based on human-derived or human-relevant systems, increasing confidence in using the model predictions for human health risk assessment.

11.3.1 Example of an Approach for Rapid PBPK Modelling: PLETHEM

PLETHEM stands for Population Lifecourse Exposure-To-Health-Effects Modelling.[34,36] It is being developed by ScitoVation under a Memorandum of Understanding with EPA's National Exposure Research Laboratory and National Center for Computational Toxicology (NCCT).[34,36] PLETHEM is being developed to provide a freely available, open-source, user-friendly platform for rapid modelling across the source-to-outcome continuum (Figure 11.2), using only *in silico* and *in vitro* data.

PLETHEM is being developed with a modular design approach that makes it possible to support a variety of exposure and PBPK models in order to provide a wide range of capabilities:

- Provide quantitative dosimetry to bridge between exposures and potential health outcomes
- Population level simulations
- Lifetime exposure modelling capability
- Test alternative exposure estimates against human biomonitoring data (Reverse Dosimetry)
- Support rapid quantitative risk assessments from *in vitro* studies (high-throughput (HT)-IVIVE)
- Replace studies in neonatal animals with *in vitro* human metabolism and PBPK (Early-Life Sensitivity Modelling)
- Provide a user-friendly run-time platform for PBPK models developed in various languages

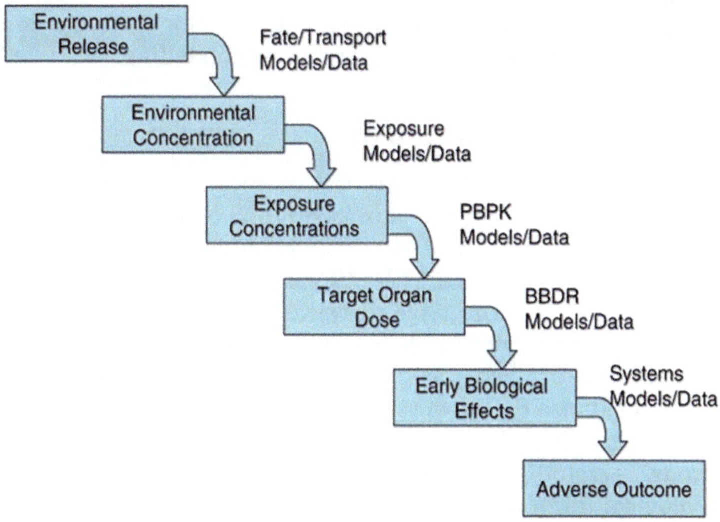

Figure 11.2 The source-to-outcome continuum. (Reprinted from ref. 37 with permission of Elsevier, Copyright 2004.)

PLETHEM includes both generic PBPK models and chemical-specific (*i.e.*, bespoke) PBPK modelling capabilities. Currently, two generic PBPK models and one HT-IVIVE model are included:

- CEFIC-LRI's IndusChemFate PBPK model extended from the published version,[38] which supports rapid estimation of tissue dose for new chemicals
  - Single or multiple exposures (oral, dermal and inhalation)
  - 11 Physiological compartments, including metabolites
  - Database of chemicals and physiological parameters
  - Partition coefficients estimated using QSAR models
  - Accepts a variety of *in vitro* metabolism data
- US EPA NCCT's High-Throughput ToxicoKinetic (httk) Model[35]
  - Rapid, conservative PBPK modelling for chemicals with limited data
- HT-IVIVE Model
  - Supports rapid *in vitro* to *in vivo* extrapolation of high-throughput toxicity testing results
  - Implements approach developed by The Hamner Institutes for ToxCast dosimetry[5,6]

PLETHEM is being designed to accept inputs from multiple EPA and Organisation for Economic Cooperation and Development (OECD) exposure estimation programmes, including EPA's Consumer Exposure Model (CEM), ECETOC's Targeted Risk Assessment (TRA) tool, Semi Empirical Evaluation of Models (SEEM2), and EPA's High-Throughput Stochastic Human Exposure and Dose Simulation (SHEDS)-HT.[36] This exposure information can then be used to drive PBPK model-based estimates of resulting internal tissue exposures to support risk assessments. In the future, PLETHEM will also be able to link these internal exposures to cellular pathway models to provide quantitative adverse outcome pathway dose–response modelling.

11.3.2 Building *In silico* Humans

Given the fact that risk assessments become more and more probabilistically-based rather than relying on the standard individual, simulation of inter-individual variability has become increasingly important for the assessment of pharmacokinetics in humans. This poses a significant challenge in implementing the *in vitro*- and *in silico*-based risk assessment paradigm. Use of big data for parameterisation of PBPK models can provide a way to overcome such challenges. The mechanistic framework and physiological basis of PBPK models facilitate the process to predict inter-individual variability in pharmacokinetics when the necessary information is provided from independent sources including computational predictions and *in vitro* data.[39]

A physiological framework based on the available data for human physiology and biochemical processes determining *in vivo* biokinetic behaviors of chemicals is included in PLETHEM. The physiological framework forms a generic, not chemical specific, predictive pharmacokinetic framework to characterise the chemical internal exposure (*e.g.*, blood concentrations) in various human populations of different ages and genders increasing efficiency of PBPK modelling. The "life-stage" PBPK model simulates the time-dependence of physiological and biochemical parameters due to growth and aging leading to a creation of various *in silico* humans in different life stages. Either equations or linear interpolation between discrete data points based on available age-related quantitative information is used in the PLETHEM database to describe changes of physiological and biochemical parameters with time for average humans. Clewell and colleagues demonstrated this approach[40] and showed that the predictions of average blood concentrations for several environmental chemicals across life stages using this approach were in general within a factor of two, although larger transient variations were predicted during early childhood periods. Applying new *in vitro* and *in silico* approaches to PBPK model parameterisation is a promising direction to move forward to increase confidence in risk assessment for early life stages.[28]

Together with the Monte Carlo (MC) approach, the life stage model framework has been extended to simulate human populations in various life stages. The perfluoroalkyl substances (PFAS)-PBPK modelling work by Wu and colleagues[32] is one of the examples of population life stage PBPK modelling works. They expanded the previouly published human growth models[40–42] to match age-specific distributions of body height, weight, and body mass index (BMI) in females reported in the National Health and Nutrition Examination Survey (NHANES) 2003–2004 data. In these models, growths in height and weight were described as a function of age and a parameter controlling the age of the growth spurt during adolescence. By conserving the physiological correlations in the model, inter-individual differences in pharmacokinetics can be better recapitulated. This appraoch enables linking internal dose to physiological parameters that are considered as potential markers for health effects such as age at menarche (Figure 11.3). A few examples of the life-stage curves in the Wu *et al.* study[32] are included in the Appendix.

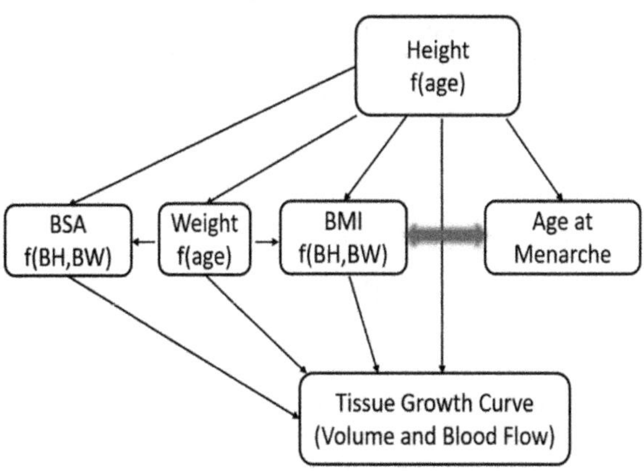

Figure 11.3 Schematic diagram of the growth dynamic model. (Reprinted from ref. 32 with permission from Elsevier, Copyright 2015.)

11.3.3 Obtaining Chemical Specific Parameters

The recent rise in the development of generic PBPK modelling platforms for chemicals including PLETHEM and httk is attributable to the ability to rapidly parameterise the PBPK models for ADME processes. Among those, computational prediction of tissue partition partitioning is one of the most prominent examples of the use of big data to support the incorporation of biokinetics in the new testing paradigm based on alternative approaches.

11.3.3.1 Partition Coefficients

Several mechanistic algorithms have been proposed to predict tissue/blood or blood/air partition coefficients (PCs) based on physico-chemical properties of compounds (*e.g.*, water solubility, vapor pressure, and octanol/water partitioning).[12,15,38,43] These QSPR algorithms can facilitate the computation of PCs for developing organ- and cellular-level PBPK model for drugs and chemicals.

In the PLETHEM generic model (the extended IndusChemFate model), PC prediction is based on the QSPR algorithm for blood/air PC with independent variables, a dimensionless Henry coefficient and the octanol/air partition coefficient.[38] For the tissue/blood partitioning, the QSPR algorithm proposed by DeJongh *et al.*[12] was applied, describing tissue partitioning on the basis of the water and lipid content of the two matrices (tissue and blood) and the octanol/water partition coefficient at the physiological pH of 7.4 considering the speciation of the extent of ionisation of a compound.[38]

In the future, other *in silico* approaches will be incorporated as well for PC prediction within PLETHEM. Peyret *et al.* have recently reported a unified algorithm for prediction of the tissue/blood PC for both environmental chemicals and drugs by integrating all mechanistic algorithms previously published.[15] The model is characterised by the capability to predict the distribution of organic compounds at the macro and micro levels in the four matrices including tissue cells, tissue interstitial fluid, erythrocyte and plasma. To improve the predictivity for the domain of highly lipophilic chemicals, a simplified tissue composition-based model using the tissue–plasma ratio of neutral lipids equivalent[16] will be included.

11.3.3.2 Metabolism Parameters

The challenge in parameterising the PBPK models largely lies in the estimation of metabolic clearance,[44] especially for non-drug-like compounds. Most of the computational prediction tools for metabolism have been developed for drug-like compounds, questioning the applicability of those for a broader chemical space including environmental chemicals. For pharmaceutical compounds, application of alternative approaches for PBPK model parameterisation primarily involves evaluation of whether the candidate compound possesses drug-like properties and largely focuses on estimating oral bioavailability and metabolic stability of parent compound screening to evaluate the drug's half-life.[45,46] However, there are challenges in applying the same principle to environmental chemicals due to a wider range of chemical properties compared to drugs. The role of metabolism in determining chemical toxicity should be considered in the context of the formation and clearance of toxic metabolite(s)[32] and therefore estimation of those two rates are critically required. Theoretically, either *in vitro* or *in silico* methods would serve the purpose of rapid parameterisation. However, the two major limitations in the currently available metabolism prediction tools are lack of (1) specificity and (2) quantitative prediction ability requiring *in vitro* experiments to generate *in vitro* metabolic constant data to extrapolate to *in vivo* by IVIVE.[3,44] For environmental chemicals, it is important to address metabolic activation in addition to incorporating total metabolic clearance of the parent chemical for safety assessment. Therefore, the estimation of metabolic parameters at the level of the individual enzyme/metabolic pathway responsible for formation of the active species is necessary in addition to the measurement or prediction of the total intrinsic clearance of the parent compound. To meet the challenge of the rapid generation of human metabolism parameters *in vitro* for chemicals, two promising pathways are worth noting. One is to increase efficiency in analytical chemistry, known as the major bottleneck in rapid metabolism characterisation[47] by adopting a similar approach developed for pharmaceutical compounds, *e.g.*, an *in silico*- and *in vitro*-combination approach known as rapid metabolite identification for drug compounds.[48] The other path forward is an application of the HT-*in vitro* screening approaches to human metabolism studies, *e.g.*, the use of human expressed enzymes for screening the major metabolic pathways for the given compound, the use of microscale 3D culture in microchips[49] *etc.* It should be noted that HT-*in vitro* screening for human metabolism will generate the parameters for biokinetic modelling that can meet the immediate needs, but at the same time, they can eventually be used to build a better training set for the development and improvement of the *in silico* prediction tools for the human metabolism of environmental chemicals.

11.3.3.3 Enzyme Ontogeny

Age-dependent development of xenobiotics metabolising enzymes can make a profound impact on chemical kinetics in humans. When combined with PBPK modelling, enzyme ontogeny data, *i.e.*, the age-dependent changes in enzyme expression levels, has been used to provide an estimate of the target tissue concentration under given exposure conditions based on *in-vitro*-based information during various life stages.[40,50] A significant amount of data have been collected resulting in a database describing the ontogeny of the majority of the key xenobiotic metabolising enzymes during growth and development in humans, in particular for hepatic phase I and

II enzymes.[51,52] The data for humans from the Hines group and collaborators have recently been compiled in a Dryad Digital Repository to provide easy access to potential users (http://dx.doi.org/10.5061/dryad.71pp6). The enzyme ontogeny database for selected enzymes including cytochromes P450 (CYPs) and UDP-glucuronosyltransferases (UGTs) has been included as a component of a HT-IVIVE module in PLETHEM to support sensitive subpopulation risk assessment especially for early ages such as infants and children. However, there remains a need to define the ontogeny of many of the enzyme systems in extra-hepatic tissues, such as the lung and kidney.[50,51] Despite this knowledge gap, combined with information on the age-dependent physiological parameters for early life stages, the database for xenobiotic metabolising enzymes ontogeny makes a significant contribution to the prediction of age-dependent kinetics of drugs and chemicals as well as inter-individual variability in chemical disposition using PBPK models.[28,53]

11.4 Conclusion

Biokinetics as a necessary tool in translating *in vitro* and *in silico* toxicity estimates to *in vivo* conditions positions itself as a key component in modern toxicity testing and safety assessment. However, in order to adapt to the new paradigm, it is essential to improve the "throughput" of biokinetic modelling. More "rapid" parameterisation capabilities are required to meet this challenge. In addition, a user-friendly modelling tool to meet the need to conduct risk assessment for a large number of chemicals would be another area to contribute to increasing the efficiency in biokinetic modelling. Leveraging big data for the development and improvement of prediction tools for human ADME parameters has significantly promoted and enabled the emergence of the "ready-to-use" and "generic" PBPK platforms dedicated to environmental chemicals in this direction. The incorporation of big data approaches in biokinetic modelling is an essential step forward to *in-vitro*-based safety assessment.

List of Abbreviations

AC50	concentration at 50% of maximum activity
ADME	absorption, distribution, metabolism and excretion
BMI	body mass index
CEM	Consumer Exposure Model
C_{ss}	steady-state concentration in the blood
f_{ub}	fraction of the compound in the blood that is unbound
GFR	glomerular filtration rate
HT-IVIVE	high-throughput IVIVE
httk	High-Throughput ToxicoKinetic
IATA	integrated approaches of toxicity assessment

IVIVE	*in vitro* to *in vivo* extrapolation
LC50	lethal concentration at which 50% of the cells die
LD50	lethal dose, 50%
LEC	lowest effective concentration
NHANES	National Health and Nutrition Examination Survey
PBBK	physiologically-based biokinetic
PBPK	physiologically-based pharmacokinetic model
PC	partition coefficient
PLETHEM	Population Lifecourse Exposure-To-Health-Effects Modelling
QSAR	Quantitative Structure Activity Relationship
QSPR	Quantitative Structure Property Relationship
SEEM	Semi Empirical Evaluation of Models
SHEDS-HT	High-Throughput Stochastic Human Exposure and Dose Simulation
TRA	Targeted Risk Assessment

Appendix: Examples of Life Stage Physiological Parameters for Human Populations

Height

The growth curve for body height (BH) in an individual male and female younger than or equal to three years old was adapted from the Jolicoeur, Pontier, Pernin, and Sempre (JPPS) model (1988)[54] and is described by eqn (11.4) and eqn (11.5):

$$H_i = h_{1,i} \cdot \left(1 - \frac{1}{1 + \left((\mathrm{AGE} + 0.75)/D1 \right)^{C1}} \right) \tag{11.4}$$

$$H_i = h_{1,i} \cdot \left(1 - \frac{1}{1 + \left((\mathrm{AGE} + 0.75)/D1 \right)^{C1}} \right) \cdot \frac{H_{3y,i}}{H_{3y,i,\mathrm{JPPS}}} \tag{11.5}$$

where H_i is BH (cm) at age AGE (years) for the ith individual; $h_{1,i}$ is the BH at adult age for the ith individual; $D1$ is a time-scale factor; $C1$ is a positive dimensionless exponent, reflecting the shape of the growth curve; 0.75 is the average age since conception in year at birth; $H_{3y,i}$ is BH at 3 years calculated for the ith individual from eqn (11.7). $H_{3y,i,JPPS}$ is BH at 3 years calculated for the ith individual from eqn (11.4).

The growth curve for BH in an individual male and female older than 3 years old was adapted from the Preece Baines model (1978)[55] and is described by eqn (11.6):

$$H_i = h_{1,i} - \frac{2\left(h_{1,i} - h_{\theta,i}\right)}{e^{S_{0,i}(\mathrm{AGE} - \theta_i)} + e^{S_{1,i}(\mathrm{AGE} - \theta_i)}} \tag{11.6}$$

$$H_{3y0,i} = h_{1,i} - \frac{2\left(h_{1,i} - h_{\theta,i}\right)}{e^{S_{0,i}(3-\theta_i)} + e^{S_{1,i}(A3-\theta_i)}} \tag{11.7}$$

where H_i is BH (cm) at age AGE (years) for the ith individual; $h_{1,i}$ is the adult height for the ith individual; θ_i is a timing parameter controlling the location of the adolescent growth spurt along the time axis for the ith individual. θ_i is highly correlated with age at peak height velocity, $h_{\theta,i}$ is the BH at age θ_i. The rate constant $S_{0,i}$ controls prepubertal growth velocity and $S_{1,i}$ controls pubertal growth velocity.

$S_{0,i}$, $S_{1,i}$, θ_i, and $h_{1,i}$ were assumed to follow a normal distribution. $h_{\theta,i}$ was defined with eqn (11.5) so that birth height (length) could vary given the same adult height and the resulting distribution of BH at early age was reasonable.

$$h_{\theta,i} = h_{1,i} \cdot \left(1 - \left(\frac{1 - \dfrac{h_\theta}{h_1}}{e^{-S_0 \cdot \theta} + e^{-S_1 \cdot \theta}} + \text{error\_}h_\theta \right) \cdot \left(e^{-S_{0,i} \cdot \theta} + e^{-S_{1,i} \cdot \theta} \right) \right) \tag{11.8}$$

Mean values of S_0, S_1, θ, h_1, h_θ, $C1$, and $D1$ were estimated based on the median BH curve from NHANES 2003–2004 data for females. The variance of the S_0, S_1, θ, h_1, and error terms were determined when the 5th, 50th and 95th percentiles of BH curves from the simulated data matched the observed corresponding NHANES data for females.

Weight

The growth curve for body weight (BW) was adapted from the O'Flaherty model (1993)[56] and is described by eqn (11.9):[56]

$$W_i = W_{\text{Birth},i} + \frac{W_{\text{Child},i} \cdot \text{AGE}}{\text{HALF}_i + \text{AGE}} + \frac{W_{\text{Adult},i}}{1 + K_i \cdot e^{-Li \cdot W_{\text{Adult},i} \cdot \text{AGE}}} \tag{11.9}$$

where W_i is BW (kg) for the ith individual at age AGE (years); $W_{\text{Birth},i}$ represents BW at birth for the ith individual; $W_{\text{Child},i}$ is the maximum weight (kg) for the ith individual during the early hyperbolic section of the growth curve; HALF_i is the midpoint age (years) for the ith individual in this hyperbolic growth segment. $W_{\text{Child},i}$ and HALF_i define the rapid growth that takes place in early childhood. A logistic growth curve, determined by the empirically determined parameters $W_{\text{Adult},i}$ (kg), K_i, and L_i, describing growth during puberty and early adulthood, is superimposed. Among these parameters, $W_{\text{Birth},i}$ was assumed to follow a normal distribution and $HALF_i$ was assumed to follow a log normal distribution.

Linking the Height and Weight Growth Curves

The age at peak height velocity and age at peak weight velocity were linked in the simulation. The age at peak height velocity was correlated with θ. The age at peak height velocity can be described as the following:

$$\text{Age\_PHV}_i = \theta_i + \ln\left(\frac{(c-b)+\sqrt{(b-c)^2-4c}}{2}\right) \Big/ (S1_i - S0_i) \qquad (11.10)$$

where

$$b = \frac{4 \cdot S_{0i}}{S_{1i}} - 1, \text{ and } c = \frac{S_{0i}^2}{S_{1i}^2}$$

The age at peak weight velocity was not directly represented by any parameters in the equation. To get the weight velocity curve, we took the first derivative of the weight equation and each term in the weight equation. The velocity of total BW change over time was parallel with the velocity of the logistic growth part. So, the age at peak weight velocity was derived by taking the second derivative of the logistic growth term in the weight equation. The derived age at peak weight velocity (APWV) is:

$$\text{APWV}_i = \frac{\ln \hat{K}2}{2 \cdot L_i \cdot W_{\text{Adult},i}} \qquad (11.11)$$

So,

$$L_i = \frac{\ln \hat{K}2}{2 \cdot \text{APWV}_i \cdot W_{\text{Adult},i}} \qquad (11.12)$$

To link the age at peak weight velocity and age at peak height velocity, we have:

Age at peak weight velocity (APWV_i) = θ_i + Diff_BH_BW + error_APWV; where Diff_BH_BW represents the lag time between age at peak height velocity and age at peak weight velocity. Mean values of W_{Birth}, W_{Child}, W_{Adult}, K, HALF and L were estimated based on the median weight curve from NHANES 2003–2004 data for females. The variance of W_{Birth}, W_{Child}, W_{Adult}, HALF and Diff_BH_BW and error terms were determined by matching the 5th, 50th and 95th percentiles of the weight curves from the simulated data to the corresponding NHANES data for females. A comparison of model-simulated BH and BW in a female as a function of age with data from NHANES study 2003–2004 is presented in Figure 11.4.

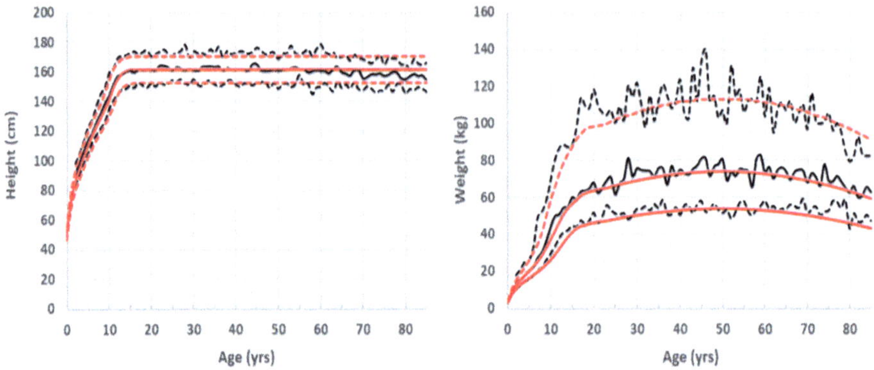

Figure 11.4 Comparison of model-simulated BH and BW in a female as a function of age with data from NHANES study 2003–2004. Red lines represent the 5th, 50th, and 95th percentiles from simulated data. Black lines represent the 5th, 50th, and 95th percentiles from observed data. (Reprinted from ref. 32 with permission from Elsevier, Copyright 2015.)

Body Fat Volume

The percentage of body fat has been linked to body measurements such as BMI in many studies. NHANES collected data on body measurements including fat volume (measured by DXA), BW, BH, and BMI in the general population. These data were obtained from the website of the Centers for Disease Control and Prevention (CDC, 2014). Models for predicting percentage of body fat were fitted to the NHANES data by performing multiple linear regression with age, and different forms of BMI including BMI, 1/BMI, BMI,[2] and combinations of these factors. The relationship between percentage of body fat, age and BMI was best described by the following equations:

$$\text{Age} \leq 25\text{ yrs: }\%\text{FAT}_i = \left(1.5334 \times e^{-0.103\times\text{AGE}} + 0.67\right) \times \text{BMI}_i + 0.6276 \times \text{AGE} + 1.0301 \quad (11.13)$$

$$\text{Age} > 25\text{yrs}: \%\text{FAT}_i = 1.9224\times\text{BMI}_i - 0.018517 \times\left(\text{BMI}_i^2 + 0.05537\times\text{AGE} - 0.794894\right) \quad (11.14)$$

where % FAT_i is percentage of body fat (%) for the *i*th individual at age AGE (years); BMI_i is the BMI for the *i*th individual at a specific age in years. A comparison of model-simulated body fat in a female as a function of age with data from NHANES study 2003–2004 is presented in Figure 11.5.

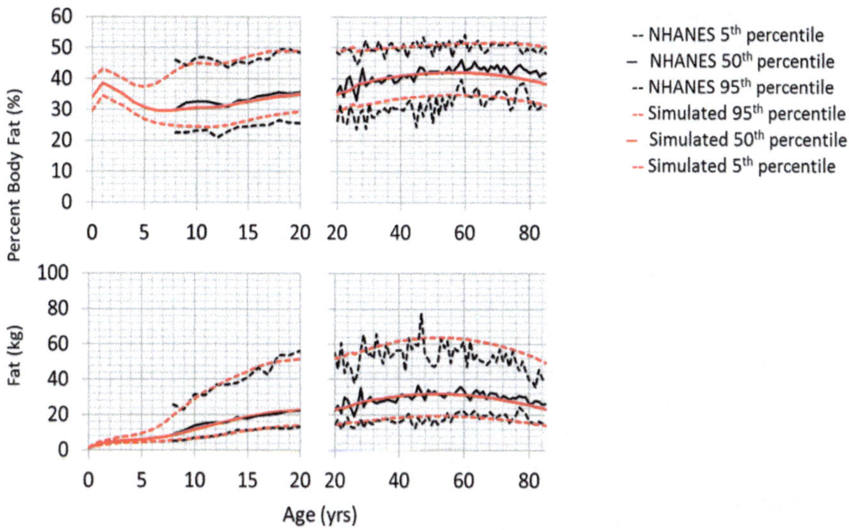

Figure 11.5 Comparison of model-simulated body fat in a female as a function of age with data from NHANES study 2003–2004. Red lines represent the 5th, 50th, and 95th percentiles from simulated data. Black lines represent the 5th, 50th, and 95th percentiles from observed data. (Reprinted from ref. 32 with permission from Elsevier, Copyright 2015.)

References

1. National Research Council, *Toxicity Testing in the 21st Century: A Vision and a Strategy*, National Academies Press, 2007.
2. B. J. Blaauboer, Biokinetic modeling and *in vitro-in vivo* extrapolations, *J. Toxicol. Environ. Health, Part B*, 2010, **13**, 242–252.
3. M. Yoon, J. L. Campbell, M. E. Andersen and H. J. Clewell, Quantitative *in vitro* to *in vivo* extrapolation of cell-based toxicity assay results, *Crit. Rev. Toxicol.*, 2012, **42**, 633–652.
4. M. Yoon, B. J. Blaauboer and H. J. Clewell, Quantitative *in vitro* to *in vivo* extrapolation (QIVIVE): An essential element for *in vitro*-based risk assessment, *Toxicology*, 2015, **332**, 1–3.
5. D. M. Rotroff, B. A. Wetmore, D. J. Dix, S. S. Ferguson, H. J. Clewell, K. A. Houck, E. L. Lecluyse, M. E. Andersen, R. S. Judson, C. M. Smith, M. A. Sochaski, R. J. Kavlock, F. Boellmann, M. T. Martin, D. M. Reif, J. F. Wambaugh and R. S. Thomas, Incorporating human dosimetry and exposure into high-throughput *in vitro* toxicity screening, *Toxicol. Sci.*, 2010, **117**, 348–358.

6. B. A. Wetmore, J. F. Wambaugh, S. S. Ferguson, M. A. Sochaski, D. M. Rotroff, K. Freeman, H. J. Clewell 3rd, D. J. Dix, M. E. Andersen, K. A. Houck, B. Allen, R. S. Judson, R. Singh, R. J. Kavlock, A. M. Richard and R. S. Thomas, Integration of dosimetry, exposure, and high-throughput screening data in chemical toxicity assessment, *Toxicol. Sci.*, 2012, **125**, 157–174.
7. G. Wilkinson and S. Schenker, Drug disposition and liver disease, *Drug Metab. Rev.*, 1975, **4**, 139–175.
8. A. D. Rule, H. M. Gussak, G. R. Pond, E. J. Bergstralh, M. D. Stegall, F. G. Cosio and T. S. Larson, Measured and estimated GFR in healthy potential kidney donors, *Am. J. Kidney Dis.*, 2004, **43**, 112–119.
9. A. Rostami-Hodjegan and G. T. Tucker, Simulation and prediction of *in vivo* drug metabolism in human populations from *in vitro* data, *Nat. Rev. Drug Discovery*, 2007, **6**, 140–148.
10. M. Yoon, A. Efremenko, B. J. Blaauboer and H. J. Clewell, Evaluation of simple *in vitro* to *in vivo* extrapolation approaches for environmental compounds, *Toxicol. In Vitro*, 2014, **28**, 164–170.
11. F. A. Groothuis, M. B. Heringa, B. Nicol, J. L. Hermens, B. J. Blaauboer and N. I. Kramer, Dose metric considerations in *in vitro* assays to improve quantitative *in vitro-in vivo* dose extrapolations, *Toxicology*, 2015, **332**, 30–40.
12. J. DeJongh, H. J. Verhaar and J. L. Hermens, A quantitative property-property relationship (QPPR) approach to estimate *in vitro* tissue-blood partition coefficients of organic chemicals in rats and humans, *Arch. Toxicol.*, 1997, **72**, 17–25.
13. S. Paterson and D. Mackay, Correlation of tissue, blood, and air partition coefficients of volatile organic chemicals, *Br. J. Ind. Med.*, 1989, **46**, 321–328.
14. W. Schmitt, General approach for the calculation of tissue to plasma partition coefficients, *Toxicol. In Vitro*, 2008, **22**, 457–467.
15. T. Peyret, P. Poulin and K. Krishnan, A unified algorithm for predicting partition coefficients for PBPK modeling of drugs and environmental chemicals, *Toxicol. Appl. Pharmacol.*, 2010, **249**, 197–207.
16. P. Poulin and S. Haddad, Toward a new paradigm for the efficient *in vitro–in vivo* extrapolation of metabolic clearance in humans from hepatocyte data, *J. Pharm. Sci.*, 2013, **102**, 3239–3251.
17. J. Dejongh, A. Forsby, J. B. Houston, M. Beckman, R. Combes and B. J. Blaauboer, An integrated approach to the prediction of systemic toxicity using computer-based biokinetic models and biological *in vitro* test methods: overview of a prevalidation study based on the ecitts project, *Toxicol. In Vitro*, 1999, **13**, 549–554.
18. B. Blaauboer, H. Clewell, R. Clothier, C. Crespi, B. Gerson, G. Hawksworth, G. Kedderis, K. Rozman and C. Willhite, *In Vitro Methods for Assessing Acute Toxicity: Biokinetic Determinations*, ICCVAM, Report of the International Workshop on *in vitro* methods for assessing acute systemic toxicity, NIEHS, NIH Publication, Research Triangle Park, NC, 2001.

19. B. J. Blaauboer, M. D. Barratt and J. B. Houston, The integrated use of alternative methods in toxicological risk evaluation - ECVAM Integrated Testing Strategies Task Force Report 1, *Altern. Lab. Anim.*, 1999, **27**, 229–237.
20. J. B. Houston and D. J. Carlile, Prediction of hepatic clearance from microsomes, hepatocytes, and liver slices, *Drug Metab. Rev.*, 1997, **29**, 891–922.
21. G. L. Kedderis, Extrapolation of *in vitro* enzyme induction data to humans *in vivo*, *Chem.-Biol. Interact.*, 1997, **107**, 109–121.
22. G. L. Kedderis, M. A. Carfagna, S. D. Held, R. Batra, J. E. Murphy and M. L. Gargas, Kinetic analysis of furan biotransformation by F-344 rats *in vivo* and *in vitro*, *Toxicol. Appl. Pharmacol.*, 1993, **123**, 274–282.
23. G. L. Kedderis and S. D. Held, Prediction of furan pharmacokinetics from hepatocyte studies: comparison of bioactivation and hepatic dosimetry in rats, mice, and humans, *Toxicol. Appl. Pharmacol.*, 1996, **140**, 124–130.
24. J. G. Teeguarden and H. A. Barton, Computational modeling of serum-binding proteins and clearance in extrapolations across life stages and species for endocrine active compounds, *Risk Anal.*, 2004, **24**, 751–770.
25. T. W. Schultz, P. Amcoff, E. Berggren, F. Gautier, M. Klaric, D. J. Knight, C. Mahony, M. Schwarz, A. White and M. T. Cronin, A strategy for structuring and reporting a read-across prediction of toxicity, *Regul. Toxicol. Pharmacol.*, 2015, **72**, 586–601.
26. T. Terada and D. Hira, Intestinal and hepatic drug transporters: pharmacokinetic, pathophysiological, and pharmacogenetic roles, *J. Gastroenterol.*, 2015, **50**, 508–519.
27. A. Wilmes, A. Limonciel, L. Aschauer, K. Moenks, C. Bielow, M. O. Leonard, J. Hamon, D. Carpi, S. Ruzek, A. Handler, O. Schmal, K. Herrgen, P. Bellwon, C. Burek, G. L. Truisi, P. Hewitt, E. Di Consiglio, E. Testai, B. J. Blaauboer, C. Guillou, C. G. Huber, A. Lukas, W. Pfaller, S. O. Mueller, F. Y. Bois, W. Dekant and P. Jennings, Application of integrated transcriptomic, proteomic and metabolomic profiling for the delineation of mechanisms of drug induced cell stress, *J. Proteomics*, 2013, **79**, 180–194.
28. M. Yoon and H. J. Clewell 3rd, Addressing early life sensitivity using physiologically based pharmacokinetic modeling and *in vitro* to *in vivo* extrapolation, *Toxicol. Res.*, 2016, **32**, 15–20.
29. H. J. Clewell 3rd, M. E. Andersen and B. J. Blaauboer, On the incorporation of chemical-specific information in risk assessment, *Toxicol. Lett.*, 2008, **180**, 100–109.
30. G. Song, C. R. Peeples, M. Yoon, H. Wu, M. A. Verner, M. E. Andersen, H. J. Clewell 3rd and M. P. Longnecker, Pharmacokinetic bias analysis of the epidemiological associations between serum polybrominated diphenyl ether (BDE-47) and timing of menarche, *Environ. Res.*, 2016, **150**, 541–548.

31. M. A. Verner, A. E. Loccisano, N. H. Morken, M. Yoon, H. Wu, R. McDougall, M. Maisonet, M. Marcus, R. Kishi, C. Miyashita, M. H. Chen, W. S. Hsieh, M. E. Andersen, H. J. Clewell 3rd and M. P. Longnecker, Associations of perfluoroalkyl substances (PFAS) with lower birth weight: An evaluation of potential confounding by glomerular filtration rate using a physiologically based pharmacokinetic model (PBPK), *Environ. Health Perspect.*, 2015, **123**, 1317–1324.
32. H. Wu, M. Yoon, M. A. Verner, J. Xue, M. Luo, M. E. Andersen, M. P. Longnecker and H. J. Clewell 3rd, Can the observed association between serum perfluoroalkyl substances and delayed menarche be explained on the basis of puberty-related changes in physiology and pharmacokinetics?, *Environ. Int.*, 2015, **82**, 61–68.
33. A. Rostami-Hodjegan, Physiologically based pharmacokinetics joined with *in vitro-in vivo* extrapolation of ADME: a marriage under the arch of systems pharmacology, *Clin. Pharmacol. Ther.*, 2012, **92**, 50–61.
34. S. N. Pendse, A. Efremenko, P. D. McMullen, M. Yoon and H. J. Clewell, PLETHEM: An interactive open-source platform for bridging the source-to-outcome continuum, *Toxicol. Sci.*, 2017, **150**, Abstract 3184.
35. R. G. Pearce, R. W. Setzer, C. L. Strope, N. S. Sipes and J. F. Wambaugh, Httk: R package for high-throughput toxicokinetics, *J. Stat. Softw.*, 2017, **79**(4), 1–26.
36. M. Moreau, J. Leonard, K. A. Phillips, J. Campbell, S. N. Pendse, C. Nicolas, M. Phillips, M. Yoon, Y. M. Tan, S. Smith, H. Pudukodu, K. Isaacs and H. Clewell, Using exposure prediction tools to link exposure and dosimetry for risk-based decisions: A case study with phthalates, *Chemosphere*, 2017, **184**, 1194–1201.
37. R. Kavlock, G. T. Ankley, T. Collette, E. Francis, K. Hammerstrom, J. Fowle, H. Tilson, G. Toth, P. Schmieder, G. D. Veith, E. Weber, D. C. Wolf and D. Young, Computational toxicology: framework, partnerships, and program development. September 29–30, 2003, Research Triangle Park, North Carolina, *Reprod. Toxicol.*, 2005, **19**, 265–280.
38. F. J. Jongeneelen and W. F. Berge, A generic, cross-chemical predictive PBTK model with multiple entry routes running as application in MS Excel; design of the model and comparison of predictions with experimental results, *Ann. Occup. Hyg.*, 2011, **55**, 841–864.
39. F. Y. Bois, M. Jamei and H. J. Clewell, PBPK modelling of inter-individual variability in the pharmacokinetics of environmental chemicals, *Toxicology*, 2010, **278**, 256–267.
40. H. J. Clewell, P. R. Gentry, T. R. Covington, R. Sarangapani and J. G. Teeguarden, Evaluation of the potential impact of age- and gender-specific pharmacokinetic differences on tissue dosimetry, *Toxicol. Sci.*, 2004, **79**, 381–393.
41. D. Hattis, G. Ginsberg, B. Sonawane, S. Smolenski, A. Russ, M. Kozlak and R. Goble, Differences in pharmacokinetics between children and

adults–II. Children's variability in drug elimination half-lives and in some parameters needed for physiologically-based pharmacokinetic modelling, *Risk Anal.*, 2003, **23**, 117–142.
42. P. S. Price, R. B. Conolly, C. F. Chaisson, E. A. Gross, J. S. Young, E. T. Mathis and D. R. Tedder, Modeling interindividual variation in physiological factors used in PBPK models of humans, *Crit. Rev. Toxicol.*, 2003, **33**, 469–503.
43. C. J. Meulenberg and H. P. Vijverberg, Empirical relations predicting human and rat tissue:air partition coefficients of volatile organic compounds, *Toxicol. Appl. Pharmacol.*, 2000, **165**, 206–216.
44. S. Coecke, H. Ahr, B. J. Blaauboer, S. Bremer, S. Casati, J. Castell, R. Combes, R. Corvi, C. L. Crespi, M. L. Cunningham, G. Elaut, B. Eletti, A. Freidig, A. Gennari, J. F. Ghersi-Egea, A. Guillouzo, T. Hartung, P. Hoet, M. Ingelman-Sundberg, S. Munn, W. Janssens, B. Ladstetter, D. Leahy, A. Long, A. Meneguz, M. Monshouwer, S. Morath, F. Nagelkerke, O. Pelkonen, J. Ponti, P. Prieto, L. Richert, E. Sabbioni, B. Schaack, W. Steiling, E. Testai, J. A. Vericat and A. Worth, Metabolism: a bottleneck in *in vitro* toxicological test development. The report and recommendations of ECVAM workshop 54, *Altern. Lab. Anim.*, 2006, **34**, 49–84.
45. J. B. Houston and A. Galetin, Methods for predicting *in vivo* pharmacokinetics using data from *in vitro* assays, *Curr. Drug Metab.*, 2008, **9**, 940–951.
46. O. Pelkonen and M. Turpeinen, *In vitro-in vivo* extrapolation of hepatic clearance: biological tools, scaling factors, model assumptions and correct concentrations, *Xenobiotica*, 2007, **37**, 1066–1089.
47. A. Tolonen and O. Pelkonen, Analytical challenges for conducting rapid metabolism characterization for QIVIVE, *Toxicology*, 2015, **332**, 20–29.
48. V. Zelesky, R. Schneider, J. Janiszewski, I. Zamora, J. Ferguson and M. Troutman, Software automation tools for increased throughput metabolic soft-spot identification in early drug discovery, *Bioanalysis*, 2013, **5**, 1165–1179.
49. S. J. Kwon, D. W. Lee, D. A. Shah, B. Ku, S. Y. Jeon, K. Solanki, J. D. Ryan, D. S. Clark, J. S. Dordick and M.-Y. Lee, High-throughput and combinatorial gene expression on a chip for metabolism-induced toxicology screening, *Nat. Commun.*, 2014, **5**, 3739.
50. X. L. Jiang, P. Zhao, J. S. Barrett, L. J. Lesko and S. Schmidt, Application of physiologically based pharmacokinetic modeling to predict acetaminophen metabolism and pharmacokinetics in children, *CPT: Pharmacometrics Syst. Pharmacol.*, 2013, **2**, e80.
51. R. N. Hines, The ontogeny of drug metabolism enzymes and implications for adverse drug events, *Pharmacol. Ther.*, 2008, **118**, 250–267.
52. S. A. Saghir, S. A. Khan and A. T. McCoy, Ontogeny of mammalian metabolizing enzymes in humans and animals used in toxicological studies, *Crit. Rev. Toxicol.*, 2012, **42**, 323–357.

53. T. N. Johnson, A. Rostami-Hodjegan and G. T. Tucker, Prediction of the clearance of eleven drugs and associated variability in neonates, infants and children, *Clin. Pharmacokinet.*, 2006, **45**, 931–956.
54. P. Jolicoeur, J. Pontier, M. O. Pernin and M. Sempe, A lifetime asymptotic growth curve for human height, *Biometrics*, 1988, **44**, 995–1003.
55. M. A. Preece and M. J. Baines, A new family of mathematical models describing the human growth curve, *Ann. Hum. Biol.*, 1978, **5**, 1–24.
56. E. J. O'Flaherty, Physiologically based models for bone-seeking elements. IV. Kinetics of lead disposition in humans, *Toxicol. Appl. Pharmacol.*, 1993, **118**, 16–29.

CHAPTER 12

Role of Toxicological Big Data to Support Read-across for the Assessment of Chemicals

MARK T. D. CRONIN*[a] AND ANDREA-NICOLE RICHARZ[b]

[a]Liverpool John Moores University, School of Pharmacy, Byrom Street, Liverpool L3 3AF, England, UK; [b]European Commission, Joint Research Centre (JRC), Ispra, Italy
*E-mail: m.t.cronin@ljmu.ac.uk

12.1 Introduction

The need to determine the hazard associated with chemicals, without resorting to animal tests, has resulted in a number of pragmatic solutions. Read-across is one such solution, relying on the premise that the biological activity, effects and properties of one molecule can be interpolated from data for other similar molecules. On the surface, it is a simple concept. Whilst it has a long history of use, there is still confusion in its application and, at times, poor practice through trivialisation and overall need for more and better-quality data to support its argumentation.

The fundamental premise of read-across is based around the similarity between molecules. There are various ways to define molecular similarity but the following are generally accepted as being the most commonly used:[1–3]

Issues in Toxicology No. 41
Big Data in Predictive Toxicology
Edited by Daniel Neagu and Andrea-Nicole Richarz
© The Royal Society of Chemistry 2020
Published by the Royal Society of Chemistry, www.rsc.org

- Structural analogues, *i.e.*, sharing the same functional group(s) but with variation in structure such as alkyl chain length.
- Mechanistic analogues, *i.e.*, sharing a similar (usually 2-D) molecular group that is associated with a mechanism of action.
- Mode of action analogues, *i.e.*, sharing a similar (usually 3-D) toxicophore that is associated with a common mode of toxic action.
- Significant degradation or metabolism to the same chemical that is responsible for toxicity.
- Chemical similarity based on a metric such as a Tanimoto score.

Each method of determining similarity has its advantages and disadvantages. As data are "read across", *i.e.*, interpolated from a compound with (preferably high quality) existing data to one or more with no data, the element of similarity must be assured to ensure that the read-across is valid. In this context the compound with data is termed the "source compound" and that to be read across to is the "target compound". Depending on the use of the read-across prediction, different levels of confidence or certainty will be required to ensure that the target compound is truly similar to the source compound.

To understand why and how toxicological data generated by new approach methodologies, increasingly available in the big data era, can support read-across, an appreciation of its context and history is required. Read-across simplifies the "structure–activity" hypothesis enabling chemicals of the same "class" to be considered together. This became popular for high tonnage, but data poor, classes of substances – something which is common for industrial chemicals.[4] Thus, in the 1980–1990s, international regulation of high production volume chemicals relied on grouping chemicals into classes, finding data (preferably of high quality) for one or more chemicals in the class, and assuming consistency in the data for all chemicals in the class.[5] Examples of this were published for groups such as the long-chain alcohols covering endpoints relating to human health[6] and environmental effects.[7] Around this time, key European legislation came to the fore, notably REACH and the Cosmetics Regulation, linking in later to initiatives such as Tox21 in the US. An increased need for toxicological information, but without the resources for animal testing, or indeed in the case of cosmetics, legal mandate, stimulated the use of read-across.[8] Allied to this were developments such as the OECD QSAR Toolbox,[9] which facilitated grouping and read-across, as well as the much greater availability of data and other tools *via* the internet.[10]

Early efforts to perform read-across, which were largely successful and at that time received regulatory acceptance, proceeded using small, or limited, data resources. However, as it became more apparent that alternatives would be required in the very near future also for "complex" endpoints such as organ-level toxicity following repeated exposure, developmental and reproductive toxicity *etc.*, but the current level of *in vitro* and *in silico* approaches were not able to provide the same level of information as the animal tests they were intended to replace, a new paradigm for hazard assessment was required.[11] Rather than relying on "problematical" quantitative structure–activity relationship (QSAR) models – even if they existed for such complex endpoints – read-across was commonly attempted for, *e.g.*, REACH submissions.[8] It was at this point that the question, and overt assumption, of chemical similarity came into focus. For regulatory decisions, especially risk assessment, similarity between molecules required proof. The reason was the well-known toxicological phenomenon whereby similar molecules, perhaps varying by only a single carbon, or even isomers (*e.g.*, thalidomide) could have very different activity. These so-called "activity cliffs" have been the ruin of many an overfit and poorly conceived model and cast doubt over the use of an *in silico* approach to model complex toxicological endpoints.[12]

Through increasing use, the problems of read-across to predict toxicity became well established.[13] The approach is, in many ways, a victim of its own simplicity. The former regulatory read-across paradigm is adequate, to some extent, with well-defined chemical classes for endpoints that are dependent on the functional group and variations in alkyl chain are either predictable or of no consequence. However, the complexity of other endpoints, and also the confirmation of chemical similarity, requires that more evidence is brought together. There are several key aspects to confirming similarity between two or more chemicals;[12] with similarity in chemical structure being only one aspect of this. There should also be similarity, or known (and quantifiable) differences, in toxicokinetics. Metabolism, or degradation, must be established and, if possible, quantified. In addition and, depending on the nature of the read-across, the most important factor is determining and confirming similarity in mechanism of action.[14] Other studies have reported the use of ADME data to support read-across for compounds with a common metabolite.[15] Some approaches to defining chemical similarity implicitly use the definition of chemical structural moieties related to the interaction of the chemical with its target, *e.g.*, DNA binding,[16] protein binding,[17] mitochondrial toxicity.[18] However, further evidence may be required for mechanistic similarity – or the absence of any specific mechanisms.

With the increasing prevalence of read-across methods to predict toxicity, and especially their complexity, guidance and examples of good practice are required. This is of great importance when predictions are intended to meet the requirements of legislatory mandate, *e.g.*, provide the same information as the animal test they replace. As such, many guidance documents have been written and tools are available; see Patlewicz *et al.*[19,20] for overviews in these areas. The European Chemicals Agency (ECHA) has developed the Read-Across Assessment Framework (RAAF)[21] to show how a read-across argumentation is evaluated and thus giving some guidance on how to structure a read-across justification and what information to include.

The need for good practice, in addition to the copious guidance, was identified and clarified by Ball *et al.*,[12] amongst others. Taking dossiers submitted to ECHA as an illustrative example, Ball *et al.*[12] determined that approximately one fifth of all disseminated compliance check decisions involved the use of read-across. Of these, there were a number of clearly identified reasons for rejection of read-across predictions. The reasons included (on the surface) trivial reasons such as a lack of definition of substance identity and read-across from inappropriate data, but also over half the rejections were due to more subjective reasons, notably the lack of scientific plausibility and insufficient evidence being provided to substantiate read-across justifications. Thus, whilst aspects of chemical similarity are required to facilitate and initiate read-across, there is a strong need to provide more evidence to support the hypothesis – especially in the context of being challenged to provide a data argument that (for REACH at least) provides the same information as the animal test it is replacing.

Other challenges apparent for read-across are for mixtures and substances of Unknown or Variable composition, Complex reaction products, and Biological materials (UVCBs). There is a desire to apply *in silico* methods to such substances, although we are limited by the lack of relevant methods to approach such a complex problem. Despite this, some read-across approaches have been applied to groups of similar compounds that may form a mixture, *e.g.*, the rosins.[22] It may also be possible to group together substances such as simple hydrocarbons of varying chain lengths. To clarify the approach and complexity of read-across for such materials, ECHA has published an additional document to the RAAF with considerations for multi-constituent substances and UVCBs.[23] Whilst progress is being made, the application of read-across to mixtures and UVCBs is currently very limited and restricted to relatively simple chemistries, the read-across justifications getting very complex for complex cases – thus there is enormous potential to extend this approach given sufficient evidence to support the similarity hypotheses.

With increasing societal desire to have a safer and cleaner environment, without recourse to animal testing to provide further information, using the best science possible to assess the safety for human health and the environment, there is a considerable challenge to innovate. Whilst on the one hand there is a real challenge to move the science of read-across forward, there are also opportunities to embrace new technologies in testing and data storage and access. With the increased use of read-across to fill toxicological data gaps for regulatory purposes, and with increasingly more toxicological and bioactivity data becoming available on a large scale, there is a tangible opportunity to use these new big data in toxicology to address problems of read-across and thus for example to increase certainty in a read-across justification.[24] This chapter outlines some of the existing examples and future potential uses of toxicological big data to support read-across. As such, it takes a rather liberal interpretation of the concept of big data – taking it to imply any data resources that go beyond what have been used previously. This can include existing data, data from chemical inventories as well as the new streams of data from *in vitro*, high-throughput and -content screening and molecular biology.

Big data for toxicology are, and increasingly will be, provided from high content screening of large numbers of chemicals. Such techniques are becoming commonplace in drug screening and provide information on cellular responses to chemical exposure. High content screening assays are used in many industries, most notably the pharmaceutical, for endpoints such as genotoxicity, cardiotoxicity, nephrotoxicity, neurotoxicity and hepatotoxicity[25] as well as significant numbers of biomarkers.[26] One of the most significant public efforts to build a significant and robust data compilation based on high-throughput and -content screening assays has been the Tox21 initiative.[27] Tox21 is the informal name given to a large collaboration, initiated in 2008, between various US agencies and centres (namely the National Institute of Environmental Health Sciences/National Toxicology Program, the US Environmental Protection Agency's (EPA's) National Center for Computational Toxicology, and the National Human Genome Research Institute/National Institutes of Health Chemical Genomics Center and the US Food and Drug Administration (FDA)). The agreed aim of the Tox 21 collaboration was to perform "high throughput screening, toxicity pathway profiling, and biological interpretation of findings" to support the concept of 21st Century Toxicology. Tox21 aimed to utilise target-specific, mechanism-based, biological observations largely obtained using *in vitro* assays.[28] Tox21 has tested thousands of chemicals in hundreds of assays, producing over 120 million data points so far, making the data freely available and constituting an immense big data resource suitable for read-across and modelling (see also Chapter 8). The vision of such an effort is to allow for the interpretation and application of a mechanistic approach to predicting and understanding toxicological responses on a mechanism of action basis.[29]

The US EPA's Toxicity Forecaster (ToxCast™) programme is a significant contributor to the overall Tox21 effort and has, in itself, developed into a significant toxicological data resource focussing on mechanistic understanding.[30] ToxCast has tested hundreds of compounds, chosen to represent not only typical chemical classes in commerce but also of mechanistic relevance in hundreds of *in vitro* and high-throughput biological assays. An aim of this initiative is to increase the knowledge and understanding of the effects of chemicals on human health and environmental species, whilst maintaining the desire to ultimately replace the traditional use of animal testing.[31] The resultant, and future, data matrix of biological and chemical information has been scrupulously quality checked and curated and placed within a robust informatics environment – a so-called "chemical landscape".[32] Access to these data is free and has been enabled through an intuitive and user-friendly interface, the "CompTox Chemistry Dashboard".[33] The output from ToxCast is, in many ways, ideal for exploration by read-across and QSAR techniques as they represent an information-rich and chemistry- and mechanism-diverse data resource.

Omics technologies constitute other large-scale data resources for toxicology and for supporting read-across, providing high-content datasets.

They include, for example, transcriptomics, metabolomics, lipidomics and proteomics (see also Chapter 7). Furthermore, high-throughput toxicokinetics (HTTK) data are becoming increasingly available (see also Chapter 11).

12.2 Role of Toxicological Big Data to Inform and Support Read-across

Before describing the opportunities for big data from toxicology to support read-across, it is worth defining the uses of big data for read-across in terms of what is frequently termed "problem formulation". The interrelationships between big data and read-across can be summarised as supporting a read-across justification, mechanistic interpretation and biological read-across. High-content omics data resources have a particular potential to contribute to the mechanistic evidence, and mining of large-scale publicly available bioactivity data resources linking to the chemical space can enhance read-across predictions. Another opportunity lies in the use of toxicological big data for read-across for hazard assessment of nanomaterials, in addressing their intrinsic complexity, variety and variability. These interrelationships between toxicological big data and read-across as well as existing and potential applications are discussed in the following.

12.2.1 Supporting a Read-across Justification

Not only relating to regulatory acceptance, there is an overwhelming need to support read-across hypotheses and justifications with further data and information. Here, bearing in mind the context and use of the read-across for data gap filling, further data can contribute to providing greater justification, in terms of biology. To illustrate this, as noted earlier, big data are taken in a rather broad context as being any "substantial" resources of information that can help support a toxicological hypothesis. In particular, these refer to large collections of bioactivity data, for example generated by high-throughput screening, or gene arrays or omics technology read-outs.

Much of the current understanding and literature on the use of read-across, especially to make assessments for regulatory purposes, has focussed on its limitations. The burden to "prove" the validity of a read-across is usually placed on the scientist undertaking read-across rather than the assessor. The need for more information to support read-across hypotheses has been noted by a number of authors, especially Patlewicz and co-workers.[12] It has also been placed in the context of the uncertainties associated with read-across. These uncertainties have been defined by a number of researchers.[34–36] Uncertainties cover all aspects of the read-across process and documentation. Schultz *et al.*[35] concluded that sources of uncertainty in read-across include a variety of elements (which are summarised by the authors) and can be assigned to two main considerations, namely the uncertainty associated with the justification of the similarity between the two or more chemicals and the completeness of the read-across argument. Of these, both are amenable to being improved by the inclusion of contributions from toxicological big data.

A number of case studies have demonstrated how read-across can be improved by the addition of further data. For example, Schultz *et al.* published four read-across examples for the assessment of repeated dose toxicity;[37–40] in each example, confidence in the read-across prediction could be improved by the inclusion of further information to confirm mechanistic understanding, toxicokinetics *etc.* Another example is Russo *et al.*[41] who extended the concept of read-across by the development of the so-called publicly available "Chemical *In vitro-In vivo* Profiling (CIIPro)" portal. This pulls on large-scale databases of biological information to form relationships between *in vitro* assay data and *in vivo* responses. The use of non-test data in this way will support the similarity hypothesis and identify analogues from optimised bioprofiles. The data required or already used to improve the read-across argumentation and similarity hypothesis can be drawn from the available large data collections of bioactivity data.

12.2.2 Investigation of Toxicological Mechanisms of Action

21st Century toxicology now has access to a range of new approaches and techniques that will assist in the identification of mechanisms of action. These can be applied to either support read-across directly in the form of supplementary evidence or the results of these techniques themselves can be used to identify similar chemicals. For instance, much is now put in the context of the Adverse Outcome Pathways (AOPs). Evidence that two chemicals are associated with the same AOP can be a crucial piece of information to support a read-across hypothesis. Data relating to the association with an AOP and, if possible, the quantitative outcome, can be found for the Key Events (KEs) and Key Event Relationships (KERs) within an AOP. Delrue *et al.*[42] described how the use of data related to an AOP could assist in the development of strategies to predict (including the use of read-across) skin sensitisation potential (qualitatively and quantitatively). They also went on to describe some of the tools that may enable this to happen including the OECD QSAR Toolbox (https://www.qsartoolbox.org) and the AOP Knowledge Base (https://aopkb.oecd.org). The concept of using such data from these technologies, within an AOP paradigm, is expounded by many other authors, *e.g.*, Rouquié *et al.*[29] who build upon concepts such as the Mode of Action framework. Whilst intuitive and useful approaches, they do not illustrate in themselves the use of big data to support read-across but more holistically how mechanistically derived data will characterise and predict adverse effects.

12.2.3 Biological Read-across

It may be hypothesised that there will be predictable commonalities in the effects of chemicals between species – albeit appreciating the differences in size and physiology. This is a fundamental tenet of experimental toxicology and the extrapolation of a surrogate species (*e.g.*, effects on a rodent) to a target species (*e.g.*, a human). The formation of such "inter-species" relationships has long been known with many and various attempts to develop quantitative models from, *e.g.*, potency data.[43] Formalisation of models for toxicity based on data for multiple chemicals, for instance as QSARs, also provides a basis to compare toxicological responses between species.[44,45] "Biological read-across" is a natural extension of interspecies comparability allowing the effects, or pattern of effects, to be extrapolated from one species to another.[46] It can also include the concept of supporting evidence, particularly if this can be placed in the context of the Molecular Initiating Event (MIE) for the AOP – for instance, data from the Ames test may assist in shedding light on the potential of a chemical to cause skin sensitisation if similarity in mechanistic chemistry can be proven.[47] Also, imaginative use of mechanistic information or even apical effects may allow for comparative toxicology to be assumed from, as an example, chronic mammalian (or even human) toxicity to environmental species – with pharmaceuticals being a good example.[48] With this in mind, there should be considerable scope within the new large-scale data compilations to forge and utilise inter-species relationships – allowing for such so-called biological read-across. The use of modern big toxicological data to support read-across has thus been proposed; Zhu *et al.*[49] demonstrated the use of ToxCast, omics and *in vitro* data to confirm biological (or toxicological) similarity between analogues. Such biological similarity will assist in the development of robust groupings, which may be anchored to high quality *in vivo* data – thus strengthening the mechanistic hypothesis and justification.

In addition to biological read-across, there is a long history of combining together information from chemical structure and biological/toxicological testing. Using large data sets, it is well established that chemical structural properties may improve inter-species relationships.[50,51] There are many reasons for this, such as descriptors accounting for differences between species, *e.g.*, uptake, distribution, metabolism or reactivity, or that there may be a strong overwhelming dependency (*i.e.*, a QSAR) between toxicity and structure – hence such relationships directly between potencies should be analysed carefully to ensure their meaning is understood. However, a data matrix derived of chemical fragments and information and the results of multiple short-term biological assays can be very powerful. Low *et al.*[52] demonstrated the use of such a data matrix with the so-called "Chemical-Biological Read-Across" (CBRA) approach to successfully identify analogues with similar adverse effects for a range of endpoints such as acute lethality hepatotoxicity, mutagenicity and hepatocarcinogenicity. The advantage here is that introducing information from chemical structure may bring in properties (*e.g.*, biokinetics or metabolism) that are not captured by *in vitro* or other non-animal tests.

A large number of analyses of the ToxCast data have been undertaken that fall under the remit of being read-across using big data. These illustrate the new directions being taken in 21st Century, animal-free, toxicology. Representative examples are described in this section. Attention has been focussed on various aspects of the toxicological response ranging from investigating relationships directly with apical effects (in a regulatory context) to searching for commonality in mechanisms. With regard to the latter, Bhhatarai *et al.*[53] used ToxCast data to probe mechanisms of mitochondrial toxicity relevant to acute toxicity. In this way, similarities and differences in responses were established for environmental species as well as for mammalian acute lethality – even the route of exposure was taken into account. An early demonstration of how the ToxCast data could inform read-across, as well as linking potential targets to molecular initiating events through the analysis of the activity data for nearly 1000 compounds in over 300 cell-free enzymatic and ligand-binding high-throughput screening (HTS) assays was provided by Sipes *et al.*[54] This clearly showed how similarity in biological response across such a data set could be the metric for toxicological read-across. Such information from ToxCast assays has also been formalised into the so-called "ToxPi™" prioritisation scheme, which has been demonstrated to be useful to determine the endocrine disrupting capability of chemicals.[55]

The data from ToxCast have also been subjected to an algorithmic, automated approach to create "bioactivity descriptors" from the output.[56] These have been utilised in a read-across approach and found to be useful when combined with chemical structure descriptors, even for *in vivo* repeated dose toxicity experimentation. The approach has been extended into the Generalised Read-Across (GenRA) analysis, with associated software and graphics by the US EPA.[57] Other studies have demonstrated the power of various multivariate analyses of the ToxCast big data resource and the combination of biological and chemical descriptors to define similarity – for instance, with reference to predicting organ-level toxicity from the ToxRefDB data.[58] In addition, the ToxCast data resource is supported by a suite of open source *in silico* models that predict physicochemical properties related to environmental fate and transport, as well as exposure potential, within a quantitative structure-property relationship (QSPR) workflow.[59] Similar analyses have been applied to the use of structural features, *i.e.*, alerts combined with *in vitro* data for the prediction of the reproductive toxicity of endocrine disrupting chemicals.[60]

As noted, in terms of significant current and future challenges for regulatory and predictive toxicology, the assessment of the hazards associated with mixtures, and/or UVCB substances, requires much attention. However, this is also an area that could be addressed by these biological/toxicological big data using biological similarity in read-across. For instance, Grimm *et al.*[61] assessed extracts of petroleum substances (mixtures/UVCBs) using bioactivity descriptors derived from *in vitro* models, *e.g.*, transcriptomic analysis of hepatocyte responses (note, the next section describes the use of molecular biological data for read-across in more detail). For the UVCBs considered, as often observed with single substances, read-across was improved when chemical and biological data were combined. In addition, an informatics structure to describe and account for the properties of UVCBs, from a purely chemical standpoint, has been developed.[62]

12.2.4 The Use of Omics Technologies for Read-across

Omics technologies in the context of supporting read-across encompass the full gamut of techniques from molecular biology including, but not limited to, transcriptomics, metabolomics, lipidomics and proteomics. These techniques have the potential to provide high-content omics datasets,[63] which in themselves constitute big data. The purpose of creating such data matrices is primarily to gain knowledge at the molecular level – which, if measured after exposure to a xenobiotic, can be translated into how that chemical affects the system at the molecular biology level. Comparison of omics effects following exposure to more than one chemical, as a means of determining commonality in function and effect, is implicitly a read-across approach and therefore can be of great assistance in data gap filling. Thus, big data from omics assays can have at least two roles in supporting read-across; namely the provision of further evidence to justify a mechanistic hypothesis as well as providing direct evidence of mechanistic similarity through comparison of omics responses.

Metabolomics is often identified as having the greatest potential to provide data and information for read-across and related approaches.[64,65] It involves the analysis of endogenous and exogenous metabolites (usually low molecular weight) in biological tissues or organs following some type of stress (for the purposes of this chapter – exposure to a chemical substance(s)). Whilst there are many applications in the biological sciences, metabolomics can provide information on the biological pathways that can be disturbed, shedding light on possible mechanisms and effects following chemical exposure[66] and this will have a potentially close application to the use of AOPs.[39] The rise in popularity of metabolomics to support toxicological investigation is in part, at least, due to the appreciation of the relevance of the outputs as well as the greater affordability of the analytical components such as mass spectroscopy.

The capability of using metabolomics to "map" modes of action has been demonstrated. For instance, Montoya *et al.*[67] studied the effects of over 500 reference compounds in rats for 28 days assessing the plasma metabolome of about 300 endogenous metabolites. This enabled the identification of specific patterns associated with more than 120 modes of action. The information is collected in the MetaMap®Tox database and the approach commercialised as the MetaMap®Tox Profiler. Whilst not a purely *in silico* approach as it relies on (limited) animal testing, the science behind the MetaMap®Tox Profiler demonstrates the power of using similarity of metabolomics outputs to identify specific modes of toxic action at a number of organ sites. Using this technology, Mattes *et al.*[68] have shown the linkage between metabolomics response, mechanisms of action and the possibility for the early identification of hepatotoxicity leading to drug-induced liver injury (DILI), with mechanisms of toxicity relevant to humans. Thus, a linkage between mechanisms of action and known apical effects is demonstrated, which can be, in the longer term, associated with AOPs for the relevant adverse effects. The MetaMap®-Tox database has subsequently been extended with data from OECD Test Guideline 407 (28-day repeated dose oral study in rat) assays to comprise information on over 550 compounds as well as comparison with inhalation studies. The possibility of using *in vitro* assays to provide data for endpoints such as hepatotoxicity is also being explored, for instance Van den Eede[69] demonstrated the use of HepaRG cells to provide biomarkers that showed the potential to identify individual and specific toxicity pathways.

The specific value of metabolomics as a read-across technique is well recognised.[70] van Ravenzwaay *et al.*[71] utilised the MetaMap®Tox database to identify metabolome patterns relating to toxicity in different organs. Whilst not an animal-free approach, it may assist in the early identification of toxicity and eliminate further testing. The possibilities for grouping of chemicals (and hence read-across) using these data are reported in terms of providing well evidenced biologically-based criteria for toxicological equivalence – the so-called development of quantitative biological–activity relationships (QBARs). The same authors[72] demonstrated the use of metabolomics in a case study using read-across to waive the need for toxicity testing within a small group of phenoxy herbicides. The evaluation of the (28-day) responses at the metabolome level demonstrated a similarity in toxicological profiles that was able to negate the requirement for a full 90-day test for all compounds in the group – specific mechanistic information was apparent after 28 days that could be related to effects seen in the liver and kidney following 90-day exposure. Whilst still reliant on animal testing to some extent, this is a potentially important use of big data derived from metabolomics to support read-across and hence eliminate the need for a full suite of testing.

Big data from metabolomics are also available for a number of environmental species (*i.e.*, non-human or rodent) following exposure to xenobiotics. This has the enormous potential to characterise modes of toxic action in such species that are not apparent, or easily determined, in humans or rodents – for instance reproductive effects in aquatic species that may result in endocrine disruption. As well as confirming known hypotheses of mechanisms of action, metabolomics has the possibility to determine information on new biochemical pathways that may be perturbed.[73] The linkage of metabolomics responses across sensitive species to AOPs was illustrated by Riedl *et al.*[74] Here, the outputs can be assessed between species enabling both chemical and biological read-across from big data, emphasising inter-species commonalities and differences.

As well as metabolomics data, there is great potential with transcriptomics to provide big data to support the similarity assessment of chemicals, thus implicitly supporting read-across. For instance, transcriptome signatures have been used to identify developmental toxicants in the migration of neural crest (MINC) assay using human cell lines.[75] Other analyses have shown the value of stem cell technology and the analysis of the transcriptome to classify a diverse selection of toxicants on the basis of developmental toxicity, as well as identifying eight genes that are sufficient to classify specific effects.[76] Another significant big data resource is the Harvard Medical School Library of Integrated Network-based Cellular Signatures (HMS LINCS or LINCS) database. LINCS provides open access to data derived from a number of biochemical and cell-based assays for over 100 cell lines and primary cell types to more than 400 perturbations across dose and time.[77] Carbonell *et al.*[78] used the LINCS database to explore the mechanistic effects of drugs at different levels, *e.g.*, pathway, cell, tissue and organ. The analysis of the data was able to link gene expression to organ toxicity forging a link from data from primary human hepatocytes to *in vivo* hepatotoxicity. There is also great potential in tools such as connectivity mapping of gene expression, with it being demonstrated to be able to use data from cell lines to group together similarly acting toxicants.[79]

12.2.5 Making Sense of Big Data Compilations to Form the Bridge With Chemical Structure

The use of read-across will be of benefit for the analysis and extraction of information from so-called compilations and inventories of toxicological big data. A number of types of data are amenable to this analysis. For instance, inventories of industry relevant substances are available and can be profiled, *e.g.*, an inventory of cosmetics ingredients[80] or pharmaceutical ingredients (*e.g.*, Drugbank[81]). The point here is that the big data are potentially being used in another manner to read-across, thus there may be many data for an individual toxicity test, related assay or *in vitro* analysis across a disparate chemical space. The ability to compile data easily, within appropriate informatics environments, has made their analysis more achievable. For instance, data compiled for the purposes of the REACH legislation have been compiled and analysed.[82] To illustrate the size and potential of such data sets, Luechtefeld *et al.*[83] obtained industry submitted (publicly available non-confidential) information from dossiers available online at the European Chemicals Agency (ECHA). A resource of over 300 000 experimental studies and other information was obtained for nearly 10 000 substances. Read-across, through the application of chemical similarity *via* well-established QSAR techniques, was attempted on endpoints such as eye irritation and skin sensitisation.[83]

Other data resources can be mined to support the development of read-across arguments. Resources such as ChEMBL[84] and PubChem[85] provide opportunities to develop similarity, grouping and read-across strategies on potential enormous sources of data. ChEMBL contains information predominantly on "small molecules" and so may be related to those that would undergo toxicological analysis, *e.g.*, in the pharmaceutical, pesticide or personal product industries. As well as structures and calculated properties, ChEMBL is a resource of various bioactivities including, in part, ADMET data as well as receptor binding and other pharmacological data. The challenge with such a large, but disparate, data compilation is finding strategies to use the data. One approach to extracting usable chemical knowledge from big data was taken by Mellor *et al.*[86] This was anchored in the use of the MIE from the AOP – in this case for nuclear receptors. Much information is publicly available within ChEMBL on the ability of (over 12 000 chemicals) to bind to a pre-identified range of nuclear receptors. Thus, information on binding to nuclear receptors associated with MIEs was utilised to develop tools to assist in mechanistically-based profiling to enable similar molecules to be identified.

An even larger publicly available resource is provided by PubChem, which is an open chemistry database focussing on collecting chemical structures and associated data and information; data have been contributed from hundreds of different sources.[85] Such a chemical structure resource (which is broader than that provided by ChEMBL) is of great use for the identification of "similar" molecules and ultimately for the classification, categorisation and stratification of diverse chemistries. PubChem also contains a number of useful and usable chemoinformatics tools that may assist in the sifting and organisation of such a large number of data. For instance, the informatics structure of PubChem was used to identify similar compounds for over 900 substances with eye irritation data. The PubChem biological data resource was also used to support read-across in the CIIPro approach[41] – this is because many *in vitro* data, some or all of which are mechanistically based, are available for compounds.

The potential for freely available tools for read-across from large chemical databases is demonstrated by the toxRead system.[87] toxRead allows the user to explore a chemical from amongst up to six similar chemicals in a chemical inventory and make a read-across prediction on the basis of the activities associated with the most similar chemicals. It is an intuitive approach, allowing visualisation of the data, to read across from a large chemical database using relatively transparent techniques to assess similarity.

In summary, the assessment of big data in terms of chemical inventories to support and undertake read-across has blossomed in the past five years, mainly due to the availability of the new resources, the improvement of informatics infrastructures and the appreciation of read-across as a viable technique to assess toxicity. There is a wide variety of approaches that can be applied to various types of data. Access to chemical inventories allows the users to superimpose toxicity data on them and, when displayed with a suitable metric to describe chemical space, can identify analogues suitable for read-across. Further, there is now the opportunity to map out areas of chemical space relevant to specific groups of substances, such as cosmetics ingredients, active pharmaceutical/pesticide ingredients, food substances and additives *etc.* as well as the even more alluring prospect of mapping the entire universe of known chemicals. Superimposition of the presence of data (such as they are) will illustrate the very limited areas where toxicological knowledge is available. As visualisation of the chemical space becomes more achievable, these approaches will certainly be more appealing. Whilst attractive, all efforts to define similarity must be made cautiously. The metric to define similarity is crucial and most are insensitive to small differences in structure, which may result in activity cliffs.[12] The phenomenon of activity cliffs is often seen for toxicity endpoints that are reliant on reactive centres in the molecule (*e.g.*, genotoxicity, skin sensitisation) or 3-D toxicophores (*e.g.*, endocrine disruption). Both of these are very sensitive to small changes in molecular structure that are often not captured with commonly applied similarity metrics. Additionally, consideration must be given to the quality and relevance of the underlying data. There could be concern over the use of large datasets of toxicological information even if from tests performed to regulatory guidelines. The use of resources such as receptor binding data can be anchored by the MIE in an AOP.[88,89]

12.2.6 Read-across for Nanomaterials

The techniques described so far in this chapter account almost exclusively for the toxicity and effects of "small" molecules. In this context, a small molecule would be thought of as having a molecular weight of less than 1000 Da and the majority of such molecules (taking pharmaceuticals as an example) would have a molecular weight somewhere in the range of 150–600 Da. The reason for this is that above this molecular weight range, molecules are likely to lack biological activity as they will either be too large to be transported or to permeate across membranes and/or they will not be soluble in aqueous solution. However, larger molecules, particularly particulates, nanomaterials and microbeads, are becoming of great concern in terms of their environmental impact and their effects on human health.[90]

The last decade has brought about considerable research into the effects of nanomaterials (defined as containing particles for 50% or more of which in the number size distribution, one or more external dimensions is in the size range 1–100 nm[91]) due to their increased use in all areas of technology. The assessment of the toxicity of nanomaterials is complicated by a number of factors, most notably their solubility, stability and problem of characterisation; also, "traditional" *in vitro* and *in vivo* tests were not designed for the assessment of nanomaterials.[92] As such, there are many recognised problems associated with their testing, which are well documented for assays such as the Ames test[90] but which may also occur in other tests.

Grouping, leading to read-across, can also be applied for the assessment of the adverse effects of nanoparticles.[93,94] A number of grouping approaches have been applied.[95,96] One approach is to consider nanoparticles with the same core, but varying dimensions (*e.g.*, carbon nanotubes of varying length) as a group. In addition, similar nanoparticles with different coatings may be considered to be a group. Forming groups in this manner will undoubtedly optimise testing and reduce the need for *in vivo* testing. It will also enable more rapid hazard assessment of nanoparticles to occur. Whilst a rational means of expanding knowledge about the hazardous effects of nanoparticles, read-across will meet the same arguments and criticism as for small molecules. However, grouping and read-across justification can be supported in the same way, *i.e.*, the definition and, where possible, quantification of uncertainties.[97,98] One means of providing more evidence for the grouping of nanomaterials is through the use of multiple lines of evidence, such as integrated testing strategies (ITS) and integrated approaches for testing and assessment (IATA).[99] Such evidence will undoubtedly benefit from including the use of toxicological/biological big data. For instance, amongst many studies in this area, Collins *et al.*[100] highlighted the use of *in vitro* approaches, and high-content data in particular to provide information that could, in theory, be used to support read-across. The techniques noted by Collins *et al.*[100] included high-throughput assays related to genotoxicity including the H2AX, comet and *in vitro* micronucleus assays. Thus, whilst not commonly applied, there is the clear potential to build evidence, based around toxicological big data, to support read-across for nanoparticles, especially considering the complexity and variety of nanomaterials. There is a particular advantage here that high-quality data used to anchor the grouping can be read across to other group members supported by evidence from assays that are tailored to the testing of nanoparticles.

12.3 The Future – Needs for the Use of Big Data for Read-across

With increasing availability and understanding of biological and toxicological big data, there is likely to be an increased application of these data to support all aspects of read-across. Whilst some examples of the use of big data exist already, there is a need clarify the potential types of usable big data as well as their applications in read-across. With this in mind, it is clear that these big data can be used to support read-across in the following scenarios:

- To support read-across justifications and similarity hypotheses
- To increase mechanistic understanding
- To develop better methods to extrapolate sets of biological data
- To evaluate chemicals databases and regulatory inventories.

In order to increase and improve the use of biological and toxicological big data for read-across and to fully benefit from the opportunities these data are offering, these scenarios need to be scoped fully to establish their potential, for their use in current legislatory and other environments, as well as their needs.

As big data use is increasing, bespoke guidance will be required – relatively new topics such as the support of read-across justifications and biological read-across bring together a disparate group of methods, thus clarity on their use would be beneficial, in particular in view of international recognition of the approaches in a regulatory context. This will require the determination of uncertainties of read-across and where and how big data can help address these uncertainties – preferably in a quantitative manner. As part of this, frameworks for read-across that explicitly incorporate toxicological big data will assist the process. In addition, guidance from regulatory agencies on how these types of new data can, and should, be used, reported or documented is required.

Upcoming areas, such as the use of metabolomics for read-across, require particular attention in the future. There is also great potential for big data to solve many of the upcoming challenges in toxicology, *e.g.*, mixtures/UVCBs, nanomaterial safety assessment *etc.* – these also need to be further identified and investigated.

As with all areas of 21st Century Toxicology, there should be no slowing in the pace of the development of methodologies generating large amounts of data, such as related to the effect mechanisms of the chemicals needed for hypothesis-driven chemical safety assessment, as well as of tools to store, distribute, analyse and process these big data. Such initiatives should build on true international and global collaboration to ensure the success of the exciting new technologies that underpin the science.

12.4 Conclusions

Read-across is an increasingly popular technique for toxicological data gap filling. Whilst it has been used widely, there are well established problems with ensuring that predictions from read-across can be accepted, especially for various regulatory purposes. *In silico* models of toxicity have benefitted recently from the availability of toxicological data and information, particularly through publicly available large-scale compilations and resources – much of this can be termed "big data". Such big data for toxicology are able to assist in read-across in many ways. They can help, particularly, with increasing confidence in read-across justifications by providing evidence to support the chemical and biological similarity hypotheses. This is seen as a vital step for the acceptance, for regulatory purposes, of read-across predictions. Big data in this context, particularly information generated with novel, mechanism-centred methodologies, will allow for a better evaluation of the mechanistic interpretation of a read-across argument, particularly when the data can be related to the MIEs or KEs of an AOP.

These toxicological and bioactivity big data will also support the so-called area of biological read-across, which is ripe for expansion. Whilst using biological data alone and in combination with information from chemistry is an attractive proposition, read-across and similarity techniques will require a full data set based on biological experimentation. As an example of what may be achieved in the future, the various analyses on Tox21 data, and ToxCast data in particular, illustrate the potential of this approach. Systems toxicology (-omics) technologies will be one of the key routes to provide big data in toxicology that will support various aspects of read-across. Metabolomics have been shown to be a particularly valuable technique, albeit possibly requiring animal testing. A number of studies have illustrated the use of metabolomics big data to identify compounds with similar mechanisms of action (hence enabling read-across) for many species. The possibilities of creating read-across predictions from large-scale chemical inventories are also well established.

Big data are being shown to provide solutions to established challenges in toxicology, particularly whether read-across can be applied to mixtures and UVCBs and whether they will provide, in combination with read-across, a solution to the hazard assessment of nanomaterials.

References

1. *Chemical Toxicity Prediction: Category Formation and Read-Across*, ed. M. T. D. Cronin, J. C. Madden, S. J. Enoch and D. W. Roberts, The Royal Society of Chemistry, Cambridge, 2013.
2. G. Patlewicz, N. Ball, P. J. Boogaard, R. A. Becker and B. Hubesch, Building scientific confidence in the development and evaluation of read-across, *Regul. Toxicol. Pharmacol.*, 2015, **72**, 117–133.
3. G. Patlewicz and J. M. Fitzpatrick, Current and future perspectives on the development, evaluation, and application of *in silico* approaches for predicting toxicity, *Chem. Res. Toxicol.*, 2016, **29**, 438–451.

4. K. Stanton and F. H. Kruszewski, Quantifying the benefits of using read-across and in silico techniques to fulfill hazard data requirements for chemical categories, *Regul. Toxicol. Pharmacol.*, 2016, **81**, 250–259.
5. C. L. Russom, R. L. Breton, J. D. Walker and S. P. Bradbury, An overview of the use of quantitative structure-activity relationships for ranking and prioritizing large chemical inventories for environmental risk assessments, *Environ. Toxicol. Chem.*, 2003, **22**, 1810–1821.
6. G. Veenstra, C. Webb, H. Sanderson, S. E. Belanger, P. Fisk, A. Nielsen, Y. Kasai, A. Willing, S. Dyer, D. Penney, H. Certa, K. Stanton and R. Sedlak, Human health risk assessment of long chain alcohols, *Ecotoxicol. Environ. Saf.*, 2009, **72**, 1016–1030.
7. S. E. Belanger, H. Sanderson, P. R. Fisk, C. Schäfers, S. M. Mudge, A. Willing, Y. Kasai, A. M. Nielsen, S. D. Dyer and R. Toy, Assessment of the environmental risk of long-chain aliphatic alcohols, *Ecotoxicol. Environ. Saf.*, 2009, **72**, 1006–1015.
8. H. Spielmann, U. G. Sauer and O. Mekenyan, A critical evaluation of the 2011 ECHA reports on compliance with the REACH and CLP regulations and on the use of alternatives to testing on animals for compliance with the REACH Regulation, *ATLA, Altern. Lab. Anim.*, 2011, **39**, 481–493.
9. https://qsartoolbox.org.
10. M. T. D. Cronin and M. Yoon, Computational Methods to Predict Toxicity, in *The History of Alternative Test Methods in Toxicology*, ed. M. Balls, R. Combes and A. Worth, The Royal Society of Chemistry, Cambridge, 2019, pp. 287–300.
11. S. Adler, D. Basketter, S. Creton, O. Pelkonen, J. van Benthem, V. Zuang, K. E. Andersen, A. Angers-Loustau, A. Aptula, A. Bal-Price, E. Benfenati, U. Bernauer, J. Bessems, F. Y. Bois, A. Boobis, E. Brandon, S. Bremer, T. Broschard, S. Casati, S. Coecke, R. Corvi, M. Cronin, G. Daston, W. Dekant, S. Felter, E. Grignard, U. Gundert-Remy, T. Heinonen, I. Kimber, J. Kleinjans, H. Komulainen, R. Kreiling, J. Kreysa, S. Batista Leite, G. Loizou, G. Maxwell, P. Mazzatorta, S. Munn, S. Pfuhler, P. Phrakonkham, A. Piersma, A. Poth, P. Prieto, G. Repetto, V. Rogiers, G. Schoeters, M. Schwarz, R. Serafimova, H. Tahti, E. Testai, J. van Delft, H. van Loveren, M. Vinken, A. Worth and J. M. Zaldivar, Alternative (non-animal) methods for cosmetics testing: current status and future prospects—2010, *Arch. Toxicol.*, 2011, **85**, 367–485.
12. G. Patlewicz, N. Ball, R. A. Becker, E. D. Booth, M. T. D. Cronin, D. Kroese, D. Steup, B. van Ravenzwaay and T. Hartung, Food for thought... read-across approaches – misconceptions, promises and challenges ahead, *ALTEX-Altern. Anim. Exp.*, 2014, **31**, 387–396.
13. N. Ball, M. T. D. Cronin, J. Shen, K. Blackburn, E. D. Booth, M. Bouhifd, E. Donley, L. Egnash, C. Hastings, D. R. Juberg, A. Kleensang, N. Kleinstreuer, E. D. Kroese, A. C. Lee, T. Luechtefeld, A. Maertens, S. Marty, J. M. Naciff, J. Palmer, D. Pamies, M. Penman, A. N. Richarz, D.

P. Russo, S. B. Stuard, G. Patlewicz, B. van Ravenzwaay, S. D. Wu, H. Zhu and T. Hartung, Toward Good Read-Across Practice (GRAP) guidance, *ALTEX-Altern. Anim. Exp.*, 2016, **33**, 149–166.
14. N. Ball, M. Bartels, R. Budinsky, J. Klapacz, S. Hays, C. Kirman and G. Patlewicz, The challenge of using read-across within the EU REACH regulatory framework; how much uncertainty is too much? Dipropylene glycol methyl ether acetate, an exemplary case study, *Regul. Toxicol. Pharmacol.*, 2014, **68**, 212–221.
15. L. H. Hand, K. Richardson, S. T. Hadfield, S. Whalley, P. Rawlinson and E. D. Booth, Use of read-across to simplify the toxicological assessment of a complex mixture of lysimeter leachate metabolites on the basis of chemical similarity and ADME behavior, *Regul. Toxicol. Pharmacol.*, 2017, **83**, 109–116.
16. S. J. Enoch and M. T. D. Cronin, A review of the electrophilic reaction chemistry involved in covalent DNA binding, *Crit. Rev. Toxicol.*, 2010, **40**, 728–748.
17. S. J. Enoch, C. M. Ellison, T. W. Schultz and M. T. D. Cronin, A review of the electrophilic reaction chemistry involved in covalent protein binding relevant to toxicity, *Crit. Rev. Toxicol.*, 2011, **41**, 783–802.
18. M. D. Nelms, C. L. Mellor, M. T. D. Cronin, J. C. Madden and S. J. Enoch, Development of an *in silico* profiler for mitochondrial toxicity, *Chem. Res. Toxicol.*, 2015, **28**, 1891–1902.
19. G. Patlewicz, G. Helman, P. Pradeep and I. Shah, Navigating through the minefield of read-across tools: A review of *in silico* tools for grouping, *Comput. Toxicol.*, 2017, **3**, 1–18.
20. G. Patlewicz, M. T. D. Cronin, G. Helman, J. C. Lambert, L. E. Lizarraga and I. Shah, Navigating through the minefield of read-across frameworks: A commentary perspective, *Compr. Toxicol.*, 2018, **6**, 39–54.
21. European Chemicals Agency, Read-Across Assessment Framework (RAAF), 2017, ECHA-17-R-01-EN, DOI: 10.2823/619212.
22. H. P. A. Illing, T. Malmfors and L. Rodenburg, Skin sensitization and possible groupings for 'read across' for rosin based substances, *Regul. Toxicol. Pharmacol.*, 2009, **54**, 234–241.
23. European Chemicals Agency, Read-Across Assessment Framework (RAAF) – considerations on multi-constituent substances and UVCBs, 2017, ECHA-17-R-04-EN, DOI: 10.2823/794394.
24. C. Mahony, R. Currie, G. Daston, N. Kleinstreuer and B. van de Water, Highlight report: 'Big data in the 3R's: outlook and recommendations', a roundtable summary, *Arch. Toxicol.*, 2018, **92**, 1015–1020.
25. M. Persson and J. J. Hornberg, Advances in predictive toxicology for discovery safety through high content screening, *Chem. Res. Toxicol.*, 2016, **29**, 1998–2007.
26. V. G. Sasseville, K. G. Mansfield and D. J. Brees, Safety biomarkers in preclinical development: translational potential, *Vet. Pathol.*, 2014, **51**, 281–291.

27. C. W. Schmidt, TOX21: New dimensions of toxicity testing, *Environ. Health Perspect.*, 2009, **117**, A348–A353.
28. R. R. Tice, C. P. Austin, R. J. Kavlock and J. R. Bucher, Improving the human hazard characterization of chemicals: A Tox21 update, *Environ. Health Perspect.*, 2013, **171**, 756–765.
29. D. Rouquie, M. Heneweer, J. Botham, H. Ketelslegers, L. Markell, T. Pfister, W. Steiling, V. Strauss and C. Hennes, Contribution of new technologies to characterization and prediction of adverse effects, *Crit. Rev. Toxicol.*, 2015, **45**, 172–183.
30. D. J. Dix, K. A. Houck, M. T. Martin, A. M. Richard, R. W. Setzer and R. J. Kavlock, The ToxCast program for prioritizing toxicity testing of environmental chemicals, *Toxicol. Sci.*, 2007, **95**, 5–12.
31. R. Kavlock, K. Chandler, K. Houck, S. Hunter, R. Judson, N. Kleinstreuer, T. Knudsen, M. Martin, S. Padilla, D. Reif, A. Richard, D. Rotroff, N. Sipes and D. Dix, Update on EPA's ToxCast Program: Providing high throughput decision support tools for chemical risk management, *Chem. Res. Toxicol.*, 2012, **25**, 1287–1302.
32. A. M. Richard, R. S. Judson, K. A. Houck, C. M. Grulke, P. Volarath, I. Thillainadarajah, C. H. Yang, J. Rathman, M. T. Martin, J. F. Wambaugh, T. B. Knudsen, J. Kancherla, K. Mansouri, G. Patlewicz, A. J. Williams, S. B. Little, K. M. Crofton and R. S. Thomas, ToxCast chemical landscape: Paving the road to 21st Century toxicology, *Chem. Res. Toxicol.*, 2016, **29**, 1225–1251.
33. A. J. Williams, C. M. Grulke, J. Edwards, A. D. McEachran, K. Mansouri, N. C. Baker, G. Patlewicz, I. Shah, J. F. Wambaugh, R. S. Judson and A. M. Richard, The CompTox Chemistry Dashboard: a community data resource for environmental chemistry, *J. Cheminf.*, 2017, **9**, 61.
34. K. Blackburn, G. Daston, J. Fisher, C. Lester, J. M. Naciff, E. S. Rufer, S. B. Stuard and K. Woeller, A strategy for safety assessment of chemicals with data gaps for developmental and/or reproductive toxicity, *Regul. Toxicol. Pharmacol.*, 2015, **72**, 202–215.
35. T. W. Schultz, P. Amcoff, E. Berggren, F. Gautier, M. Klaric, D. J. Knight, C. Mahony, M. Schwarz, A. White and M. T. D. Cronin, A strategy for structuring and reporting a read-across prediction of toxicity, *Regul. Toxicol. Pharmacol.*, 2015, **72**, 586–601.
36. T. W. Schultz, A.-N. Richarz and M. T. D. Cronin, Assessing uncertainty in read-across: Questions to evaluate toxicity predictions based on knowledge gained from case studies, *Comput. Toxicol.*, 2019, **9**, 1–11.
37. T. W. Schultz, K. R. Przybylak, A.-N. Richarz, C. L. Mellor, S. P. Bradbury and M. T. D. Cronin, Read-across of 90-day rat oral repeated-dose toxicity: A case study for selected 2-alkyl-1-alkanols, *Comput. Toxicol.*, 2017, **2**, 28–38.
38. T. W. Schultz, K. R. Przybylak, A.-N. Richarz, C. L. Mellor, S. E. Escher, S. P. Bradbury and M. T. D. Cronin, Read-across of 90-day rat oral

repeated-dose toxicity: A case study for selected n-alkanols, *Comput. Toxicol.*, 2017, **2**, 12–19.
39. C. L. Mellor, T. W. Schultz, K. R. Przybylak, A.-N. Richarz, S. P. Bradbury and M. T. D. Cronin, Read-across for rat oral gavage repeated-dose toxicity for short-chain mono-alkylphenols: A case study, *Comput. Toxicol.*, 2017, **2**, 1–11.
40. K. R. Przybylak, T. W. Schultz, A.-N. Richarz, C. L. Mellor, S. E. Escher and M. T. D. Cronin, Read-across of 90-day rat oral repeated-dose toxicity: A case study for selected β-olefinic alcohols, *Comput. Toxicol.*, 2017, **1**, 22–32.
41. D. P. Russo, M. T. Kim, W. Y. Wang, D. Pinolini, S. Shende, J. Strickland, T. Hartung and H. Zhu, CIIPro: a new read-across portal to fill data gaps using public large-scale chemical and biological data, *Bioinformatics*, 2017, **33**, 464–466.
42. N. Delrue, M. Sachana, Y. Sakuratani, A. Gourmelon, E. Leinala and R. Diderich, The Adverse Outcome Pathway concept: a basis for developing regulatory decision-making tools, *ATLA, Altern. Lab. Anim.*, 2016, **44**, 417–429.
43. M. T. D. Cronin and J. C. Dearden, QSAR in toxicology 2. Prediction of acute mammalian toxicity and interspecies relationships, *Quant. Struct.-Act. Relat.*, 1995, **14**, 117–120.
44. M. T. D. Cronin, J. C. Dearden and A. J. Dobbs, QSAR studies of comparative toxicity in aquatic organisms, *Sci. Total Environ.*, 1991, **109/110**, 431–439.
45. J. C. Dearden, M. T. D. Cronin and A. J. Dobbs, Quantitative structure-activity relationships as a tool to assess the comparative toxicity of organic chemicals, *Chemosphere*, 1995, **31**, 2521–2528.
46. M. T. D. Cronin, Biological Read-Across: Mechanistically-Based Species-Species and Endpoint-Endpoint Extrapolations, in *In Silico Toxicology: Principles and Applications*, ed. M. T. D. Cronin and J. C. Madden, The Royal Society of Chemistry, Cambridge, 2010, pp. 446–477.
47. O. Mekenyan, G. Patlewicz, G. Dimitrova, C. Kuseva, M. Todorov, S. Stoeva, S. Kotov and E. M. Donner, Use of genotoxicity information in the development of Integrated Testing Strategies (ITS) for skin sensitization, *Chem. Res. Toxicol.*, 2010, **23**, 1519–1540.
48. J. P. Berninger and B. W. Brooks, Leveraging mammalian pharmaceutical toxicology and pharmacology data to predict chronic fish responses to pharmaceuticals, *Toxicol. Lett.*, 2010, **193**, 69–78.
49. H. Zhu, M. Bouhifd, E. Donley, L. Egnash, N. Kleinstreuer, E. D. Kroese, Z. C. Liu, T. Luechtefeld, J. Palmer, D. Pamies, J. Shen, V. Strauss, S. D. Wu and T. Hartung, Supporting read-across using biological data, *ALTEX-Altern. Anim. Exp.*, 2016, **33**, 167–182.
50. C. H. Yang, A. M. Richard and K. P. Cross, The art of data mining the minefields of toxicity databases to link chemistry to biology, *Curr. Comput.-Aided Drug Des.*, 2006, **2**, 135–150.

51. I. Kahn, U. Maran, E. Benfenati, T. I. Netzeva, T. W. Schultz and M. T. D. Cronin, Comparative quantitative structure–activity–activity relationships for toxicity to *Tetrahymena pyriformis* and *Pimephales promelas*, *ATLA, Altern. Lab. Anim.*, 2007, **35**, 15–24.
52. Y. Low, A. Sedykh, D. Fourches, A. Golbraikh, M. Whelan, I. Rusyn and A. Tropsha, Integrative chemical-biological read-across approach for chemical hazard classification, *Chem. Res. Toxicol.*, 2013, **26**, 1199–1208.
53. B. Bhhatarai, D. M. Wilson, M. J. Bartels, S. Chaudhuri, P. S. Price and E. W. Carney, Acute toxicity prediction in multiple species by leveraging mechanistic ToxCast mitochondrial inhibition data and simulation of oral bioavailability, *Toxicol. Sci.*, 2015, **147**, 386–396.
54. N. S. Sipes, M. T. Martin, P. Kothiya, D. M. Reif, R. S. Judson, A. M. Richard, K. A. Houck, D. J. Dix, R. J. Kavlock and T. B. Knudsen, Profiling 976 ToxCast chemicals across 331 enzymatic and receptor signaling assays, *Chem. Res. Toxicol.*, 2013, **26**, 878–895.
55. D. Filer, H. B. Patisaul, T. Schug, D. Reif and K. Thayer, Test driving ToxCast: endocrine profiling for 1858 chemicals included in phase II, *Curr. Opin. Pharmacol.*, 2014, **19**, 145–152.
56. I. Shah, J. Liu, R. S. Judson, R. S. Thomas and G. Patlewicz, Systematically evaluating read-across prediction and performance using a local validity approach characterized by chemical structure and bioactivity information, *Regul. Toxicol. Pharmacol.*, 2016, **79**, 12–24.
57. G. Helman, I. Shah and G. Patlewicz, Extending the Generalised Read-Across approach (GenRA): A systematic analysis of the impact of physicochemical property information on read-across performance, *Comput. Toxicol.*, 2018, **8**, 34–50.
58. J. Liu, G. Patlewicz, A. J. Williams, R. S. Thomas and I. Shah, Predicting organ toxicity using *in vitro* bioactivity data and chemical structure, *Chem. Res. Toxicol.*, 2017, **30**, 2046–2059.
59. Q. Zang, K. Mansouri, A. J. Williams, R. S. Judson, D. G. Allen, W. M. Casey and N. C. Kleinstreuer, *In silico* prediction of physicochemical properties of environmental chemicals using molecular fingerprints and machine learning, *J. Chem. Inf. Model.*, 2017, **57**, 36–49.
60. G. Lewin, S. E. Escher, B. van der Burg, N. Simetska and I. Mangelsdorf, Structural features of endocrine active chemicals – A comparison of *in vivo* and *in vitro* data, *Reprod. Toxicol.*, 2015, **55**, 81–94.
61. F. A. Grimm, Y. Iwata, O. Sirenko, G. A. Chappell, F. A. Wright, D. M. Reif, J. Braisted, D. L. Gerhold, J. M. Yeakley, P. Shepard, B. Seligmann, T. Roy, P. J. Boogaard, H. B. Ketelslegers, A. M. Rohde and I. Rusyn, A chemical-biological similarity-based grouping of complex substances as a prototype approach for evaluating chemical alternatives, *Green Chem.*, 2016, **18**, 4407–4419.
62. S. D. Dimitrov, D. G. Georgieva, T. S. Pavlov, Y. H. Karakolev, P. G. Karamertzanis, M. Rasenberg and O. G. Mekenyan, UVCB substances: Methodology for structural description and application to fate and hazard assessment, *Environ. Toxicol. Chem.*, 2015, **34**, 2450–2462.

63. E. K. Brockmeier, G. Hodges, T. H. Hutchinson, E. Butler, M. Hecker, K. E. Tollefsen, N. Garcia-Reyero, P. Kille, D. Becker, K. Chipman, J. Colbourne, T. W. Collette, A. Cossins, M. Cronin, P. Graystock, S. Gutsell, D. Knapen, I. Katsiadaki, A. Lange, S. Marshall, S. F. Owen, E. J. Perkins, S. Plaistow, A. Schroeder, D. Taylor, M. Viant, G. Ankley and F. Falciani, The role of omics in the application of Adverse Outcome Pathways for chemical risk assessment, *Toxicol. Sci.*, 2017, **158**, 252–262.
64. K. B. Kell and R. Goodacre, Metabolomics and systems pharmacology: why and how to model the human metabolic network for drug discovery, *Drug Discovery Today*, 2014, **19**, 171–182.
65. M. Bouhifd, T. Hartung, H. T. Hogberg, A. Kleensang and L. Zhao, Review: Toxicometabolomics, *J. Appl. Toxicol.*, 2013, **33**, 1365–1383.
66. T. Ramirez, M. Daneshian, H. Kamp, F. Y. Bois, M. R. Clench, M. Coen, B. Donley, S. M. Fischer, D. R. Ekman, E. Fabian, C. Guillou, J. Heuer, H. T. Hogberg, H. Jungnickel, H. C. Keun, G. Krennrich, E. Krupp, A. Luch, F. Noor, E. Peter, B. Riefke, M. Seymour, N. Skinner, L. Smirnova, E. Verheij, S. Wagner, T. Hartung, B. van Ravenzwaay and M. Leist, Metabolomics in toxicology and preclinical research, *ALTEX-Altern. Anim. Exp.*, 2013, **30**, 209–225.
67. G. A. Montoya, V. Strauss, E. Fabian, H. Kamp, W. Mellert, T. Walk, R. Looser, M. Herold, G. Krennrich, E. Peter and B. van Ravenzwaay, Mechanistic analysis of metabolomics patterns in rat plasma during administration of direct thyroid hormone synthesis inhibitors or compounds increasing thyroid hormone clearance, *Toxicol. Lett.*, 2014, **225**, 240–251.
68. W. Mattes, K. Davis, E. Fabian, J. Greenhaw, M. Herold, R. Looser, W. Mellert, S. Groeters, H. Marxfeld, N. Moeller, G. Montoya-Parra, A. Prokoudine, B. van Ravenzwaay, V. Strauss, T. Walk and H. Kamp, Detection of hepatotoxicity potential with metabolite profiling (metabolomics) of rat plasma, *Toxicol. Lett.*, 2014, **230**, 467–478.
69. N. Van den Eede, M. Cuykx, R. M. Rodrigues, K. Laukens, H. Neels, A. Covaci and T. Vanhaecke, Metabolomics analysis of the toxicity pathways of triphenyl phosphate in HepaRG cells and comparison to oxidative stress mechanisms caused by acetaminophen, *Toxicol. In Vitro*, 2015, **29**, 2045–2054.
70. E. Berggren, P. Amcoff, R. Benigni, K. Blackburn, E. Carney, M. Cronin, H. Deluyker, F. Gautier, R. S. Judson, G. E. N. Kass, D. Keller, D. Knight, W. Lilienblum, C. Mahony, I. Rusyn, T. Schultz, M. Schwarz, G. Schüürmann, A. White, J. Burton, A. M. Lostia, S. Munn and A. Worth, Chemical safety assessment using read-across: assessing the use of novel testing methods to strengthen the evidence base for decision making, *Environ. Health Perspect.*, 2015, **123**, 1232–1240.
71. B. van Ravenzwaay, M. Herold, H. Kamp, M. D. Kapp, E. Fabian, R. Looser, G. Krennrich, W. Mellert, A. Prokoudine, V. Strauss, T. Walk and J. Wiemer, Metabolomics: A tool for early detection of toxicological effects and an opportunity for biology based grouping of chemicals-From

QSAR to QBAR, *Mutat. Res., Genet. Toxicol. Environ. Mutagen.*, 2012, **746**, 144–150.
72. B. van Ravenzwaay, S. Sperber, O. Lemke, E. Fabian, F. Faulhammer, H. Kamp, W. Mellert, V. Strauss, A. Strigun, E. Peter, M. Spitzer and T. Walk, Metabolomics as read-across tool: A case study with phenoxy herbicides, *Regul. Toxicol. Pharmacol.*, 2016, **81**, 288–304.
73. B. P. Lankadurai, E. G. Nagato and M. J. Simpson, Environmental metabolomics: an emerging approach to study organism responses to environmental stressors, *Environ. Rev.*, 2013, **21**, 180–205.
74. J. Riedl, R. Schreiber, M. Otto, H. Heilmeier, R. Altenburger and M. Schmitt-Jansen, Metabolic effect level index links multivariate metabolic fingerprints to ecotoxicological effect assessment, *Environ. Sci. Technol.*, 2015, **49**, 8096–8104.
75. G. Pallocca, M. Grinberg, M. Henry, T. Frickey, J. G. Hengstler, T. Waldmann, A. Sachinidis, J. Rahnenfuhrer and M. Leist, Identification of transcriptome signatures and biomarkers specific for potential developmental toxicants inhibiting human neural crest cell migration, *Arch. Toxicol.*, 2016, **90**, 159–180.
76. E. Rempel, L. Hoelting, T. Waldmann, N. V. Balmer, S. Schildknecht, M. Grinberg, J. A. Das Gaspar, V. Shinde, R. Stober, R. Marchan, C. van Thriel, J. Liebing, J. Meisig, N. Bluthgen, A. Sachinidis, J. Rahnenfuhrer, J. G. Hengstler and M. Leist, A transcriptome-based classifier to identify developmental toxicants by stem cell testing: design, validation and optimization for histone deacetylase inhibitors, *Arch. Toxicol.*, 2015, **89**, 1599–1618.
77. U. D. Vempati, C. Chung, C. Mader, A. Koleti, N. Datar, D. Vidovic, D. Wrobel, S. Erickson, J. L. Muhlich, G. Berriz, C. H. Benes, A. Subramanian, A. Pillai, C. E. Shamu and S. C. Schurer, Metadata standard and data exchange specifications to describe, model, and integrate complex and diverse high-throughput screening data from the Library of Integrated Network-based Cellular Signatures (LINCS), *J. Biomol. Screening*, 2014, **19**, 803–816.
78. P. Carbonell, O. Lopez, A. Amberg, M. Pastor and F. Sanz, Hepatotoxicity prediction by systems biology modeling of disturbed metabolic pathways using gene expression data, *ALTEX-Altern. Anim. Exp.*, 2017, **34**, 219–234.
79. K. N. De Abrew, R. M. Kainkaryam, Y. Q. K. Shan, G. J. Overmann, R. S. Settivari, X. H. Wang, J. Xu, R. L. Adams, J. P. Tiesman, E. W. Carney, J. M. Naciff and G. P. Daston, Grouping 34 chemicals based on mode of action using connectivity mapping, *Toxicol. Sci.*, 2016, **151**, 447–461.
80. C. Yang, S. M. Barlow, K. L. Muldoon Jacobs, V. Vitcheva, A. R. Boobis, S. P. Felter, K. B. Arvidson, D. Keller, M. T. D. Cronin, S. J. Enoch, A. P. Worth and H. M. Hollnagel, Thresholds of Toxicological Concern for cosmetics-related substances: new database, thresholds, and enrichment of chemical space, *Food Chem. Toxicol.*, 2017, **109**, 170–193.

81. D. S. Wishart, Y. D. Feunang, A. C. Guo, E. J. Lo, A. Marcu, J. R. Grant, T. Sajed, D. Johnson, C. Li, Z. Sayeeda, N. Assempour, I. Iynkkaran, Y. Liu, A. Maciejewski, N. Gale, A. Wilson, L. Chin, R. Cummings, D. Le, A. Pon, C. Knox and M. Wilson, DrugBank 5.0: a major update to the DrugBank database for 2018, *Nucleic Acids Res.*, 2018, **46**, D1074–D1082.
82. T. Hartung, Making big sense from big data in toxicology by read-across, *ALTEX-Altern. Anim. Exp.*, 2016, **33**, 83–93.
83. T. Luechtefeld, A. Maertens, D. P. Russo, C. Rovida, H. Zhu and T. Hartung, Analysis of Draize eye irritation testing and its prediction by mining publicly available 2008-2014 REACH data, *ALTEX-Altern. Anim. Exp.*, 2016, **33**, 123–134.
84. A. P. Bento, A. Gaulton, A. Hersey, L. J. Bellis, J. Chambers, M. Davies, F. A. Krüger, Y. Light, L. Mak, S. McGlinchey, M. Nowotka, G. Papadatos, R. Santos and J. P. Overington, The ChEMBL bioactivity database: an update, *Nucleic Acids Res.*, 2014, **42**, 1083–1090.
85. S. Kim, P. A. Thiessen, E. E. Bolton, J. Chen, G. Fu, A. Gindulyte, L. Han, J. He, S. He, B. A. Shoemaker, J. Wang, B. Yu, J. Zhang and S. H. Bryant, PubChem substance and compound databases, *Nucleic Acids Res.*, 2016, **44**, D1202–D1213.
86. C. L. Mellor, F. P. Steinmetz and M. T. D. Cronin, Using molecular initiating events to develop a structural alert based screening workflow for nuclear receptor ligands associated with hepatic steatosis, *Chem. Res. Toxicol.*, 2016, **29**, 203–212.
87. G. Gini, A. M. Franchi, A. Manganaro, A. Golbamaki and E. Benfenati, ToxRead: A tool to assist in read across and its use to assess mutagenicity of chemicals, *SAR QSAR Environ. Res.*, 2014, **25**, 999–1011.
88. M. T. D. Cronin, S. J. Enoch, C. L. Mellor, K. R. Przybylak, A.-N. Richarz and J. C. Madden, *In silico* prediction of organ level toxicity: linking chemistry to adverse effects, *Toxicol. Res.*, 2017, **33**, 173–182.
89. M. T. D. Cronin and A. N. Richarz, Relationship between Adverse Outcome Pathways and chemistry-cased in silico models to predict toxicity, *Appl. In Vitro Toxicol.*, 2017, **3**, 286–297.
90. N. Golbamaki, B. Rasulev, A. Cassano, R. L. Marchese Robinson, E. Benfenati, J. Leszczynski and M. T. D. Cronin, Genotoxicity of metal oxide nanomaterials: review of recent data and discussion of possible mechanisms, *Nanoscale*, 2015, **7**, 2154–2198.
91. European Commission Recommendation of 18 October 2011 on the definition of nanomaterial, Official Journal of the European Union L 275, 2011, 275, 38–40.
92. L. Lubinski, P. Urbaszek, A. Gajewicz, M. T. D. Cronin, S. J. Enoch, J. C. Madden, D. Leszczynska, J. Leszczynski and T. Puzyn, Evaluation criteria for the quality of published experimental data on nanomaterials and their usefulness for QSAR modelling, *SAR QSAR Environ. Res.*, 2013, **24**, 995–1008.

93. A. Gajewicz, K. Jagiello, M. T. D. Cronin, J. Leszczynski and T. Puzyn, Addressing a bottle neck for regulation of nanomaterials: quantitative read-across (Nano-QRA) algorithm for cases when only limited data is available, *Environ. Sci.: Nano*, 2017, **4**, 346–358.
94. A. Gajewicz, M. T. D. Cronin, B. Rasulev, J. Leszczynski and T. Puzyn, Novel approach for efficient predictions properties of large pool of nanomaterials based on limited set of species: nano-read-across, *Nanotechnology*, 2015, **26**, 015701.
95. M. T. D. Cronin, S. J. Enoch, J. C. Madden and A.-N. Richarz, Read-Across to fill toxicological data gaps: good practice to ensure success with nanoparticles, in *Computational Nanotoxicology: Challenges, Pitfalls, and Perspectives*, ed. A. Gajewicz and T. Puzyn, Jenny Stanford Publishing Pte. Ltd., 2020, pp. 381–400.
96. L. Lamon, K. Aschberger, D. Asturiol, A. Richarz and A. Worth, Grouping of nanomaterials to read-across hazard endpoints: a review, *Nanotoxicology*, 2019, **13**(1), 100–118.
97. L. Lamon, D. Asturiol, A. Richarz, E. Joossens, R. Graepel, K. Aschberger and A. Worth, Grouping of nanomaterials to read-across hazard endpoints: from data collection to assessment of the grouping hypothesis by application of chemoinformatic techniques, *Part. Fibre Toxicol.*, 2018, **15**, 37.
98. K. Aschberger, D. Asturiol, L. Lamon, A. Richarz, K. Gerloff and A. Worth, Grouping of multi-walled carbon nanotubes to read-across genotoxicity: A case study to evaluate the applicability of regulatory guidance, *Comput. Toxicol.*, 2019, **9**, 22–35.
99. R. Landsiedel, Concern-driven integrated approaches for the grouping, testing and assessment of nanomaterials, *Environ. Pollut.*, 2016, **218**, 1376–1380.
100. A. R. Collins, B. Annangi, L. Rubio, R. Marcos, M. Dorn, C. Merker, I. Estrela-Lopis, M. R. Cimpan, M. Ibrahim, E. Cimpan, M. Ostermann, A. Sauter, N. El Yamani, S. Shaposhnikov, S. Chevillard, V. Paget, R. Grall, J. Delic, F. Goni-de-Cerio, B. Suarez-Merino, V. Fessard, K. N. Hogeveen, L. M. Fjellsbo, E. R. Pran, T. Brzicova, J. Topinka, M. J. Silva, P. E. Leite, A. R. Ribeiro, J. M. Granjeiro, R. Grafstrom, A. Prina-Mello and M. Dusinska, High throughput toxicity screening and intracellular detection of nanomaterials, *Wiley Interdiscip. Rev.: Nanomed. Nanobiotechnol.*, 2017, **9**, e1413.

Subject Index

Acute Toxic Class Method, 41
acute toxicity testing
  limitations, 41
  test guidelines, 40–41
  use of, 41–42
adverse outcome pathways (AOPs), 39, 52–54
  framework, 15
  networks
    steatosis, 289–290
    sufficient key events, 287–289
  predicted and putative AOP generation, 283–284
    computational AOP generation, 284–287
    from computationally predicted AOPs to putative AOPs, 287
Aggregated Computational Toxicology Resource (ACToR), 13
Ames test, 49
AOPs. *See* adverse outcome pathways (AOPs)
Application Programming Interfaces (APIs), 12
aromatase inhibition model development, 292–295
ArrayAnalysis, 223
ArrayExpress, 220
artificial neural networks (ANN), 191
Babelomics, 223
bacterial reverse mutation test (Ames test), 49
BAO. *See* BioAssay Ontology (BAO)
big ('omics) data, 229–232
Big Data for Better Outcomes (BD4BO), 2
Big Data to Knowledge (BD2K) initiative, 13
BioAssay Ontology (BAO), 137
bio-descriptors, 320–324
biokinetics, 331–333
  approach, 340–341
  chemical grouping, 339–340
  chemical specific parameters
    enzyme ontogeny, 347–348
    metabolism parameters, 346–347
    partition coefficients (PCs), 346
  high-throughput testing, 333–339
  minimalist biokinetics approach, 340–341
  risk assessment, 342
    in silico humans, 344–345
    PLETHEM stands, 343–344
biological significance, 62
biomarkers, 324–326

Canada's Existing Substances Assessment Repository (CESAR), 12
cancer effects, 58, 286
carcinogenicity, 42, 43, 51, 58, 130, 228
Carcinogenic Potency Database (CPDB), 56
CAS. *See* Chemical Abstracts Service (CAS)
CCRIS. *See* Chemical Carcinogenesis Research Information System (CCRIS)
CEBS. *See* Chemical Effects in Biological Systems (CEBS)
cell transformation assays (CTAs), 51
Center for Expanded Data Annotation and Retrieval (CEDAR), 13
ChemAgora portal. *See* JRC ChemAgora portal
ChEMBL, 223, 231
Chemical Abstracts Service (CAS), 186
chemical–bioactivity relationships
    HTS assays, 274–275
    HTS data, 278–279
    in vivo toxicity, 278–279
    publicly available databases, 275–278
Chemical Carcinogenesis Research Information System (CCRIS), 55–56
chemical descriptors, 313–315
Chemical Effects in Biological Systems (CEBS), 138, 221
chemical substances, 101–104
chemoinformatics approaches
    for chemical structure representation, 69–73
    chemical substances, 101–104
    constitutional representations, 77–78
    fingerprints, 91–95
    hash codes, 91–95
    linear notations, 81–82
        conformational search, 89–91
        4D structure representations, 85–87
        InChI Linear Notation, 84
        SMARTS linear notation, 84–85
        SMILES linear notation, 82–83
        3D structure representations, 85–89
    molecular descriptors, 91–95
    structure representation
        challenging cases for, 95–97
        handling aromaticity, 97–98
        miscellaneous challenges, 100–101
        tautomerism, 99
        unique structure representations, 98
    structure representation characteristics, 74–76
    topological representations, 78–81
CIBEX, 220
Clinical Data Interchange Standards Consortium (CDISC), 120
cloud technology, 172
Comparative Toxicogenomics Database (CTD), 56, 221
connection table (CT), 71
connectivity mapping (CMap), 227–228
ConsensusPathDB (CPDB), 228
Core Information for Metabolomics Reporting (CIMR), 220

Data Infrastructure for Chemical Safety Assessment (diXa) resource, 13
data integration

AOP networks
steatosis, 289–290
sufficient key events, 287–289
and application, 265–268
approaches, 168–171
chemical–bioactivity relationships
HTS assays, 274–275
HTS data, 278–279
in vivo toxicity, 278–279
publicly available databases, 275–278
de novo network inference, 279–280
dynamic networks, 281
networks and inference, 282–283
steady-state networks, 280–281
globally accessible knowledge base, 268–269
high-throughput datasets, 290–291
aromatase inhibition model development, 292–295
mechanistic modelling methods, 291–292
predicted and putative AOP generation, 283–284
computational AOP generation, 284–287
from computationally predicted AOPs to putative AOPs, 287
supporting decision-making, 273–274
use of structured knowledge, 269–273
data interpretation, 61–62
data set representation, 191–192
data sharing infrastructure, 174–175
decision trees (DT), 191
dendritic cells (DC), 52
de novo network inference, 279–280
dynamic networks, 281
networks and inference, 282–283
steady-state networks, 280–281
Distributed Structure-Searchable Toxicity (DSSTox) Database, 10
diXa (Data Infrastructure for Chemical Safety), 221
DrugMatrix database, 221

ebKB, 221
electrotopological (E-state) descriptors, 314
Environmental Protection Agency (EPA), 48
enzyme ontogeny, 347–348
Estimation Programme Interface (EPI) Suite™, 10
EU Integrated Training Network BIGCHEM project, 2
European Chemicals Agency (ECHA), 23, 57, 59, 169
European Food Safety Authority (EFSA), 59
European Grid Infrastructure (EGI) federation, 175
European Nucleotide Archive (ENA), 221
European Union Reference Laboratory for alternatives to animal testing (EURL ECVAM), 48

fingerprints, 91–95
fixed dose procedure, 40–41
fragments as descriptors, 314

GenePattern, 223
genetic algorithm (GA), 313, 317–320
GENE-TOX (Genetic Toxicology Data Bank), 56
(epi)genomics, 217–218
genotoxicity, 47, 48–51, 130
GEO, 220

Global Alliance for Genomics and Health (GA4GH), 173
Good *In vitro* Method Practices (GIVIMP), 5
good laboratory practice (GLP), 5, 109

H atoms' information, 71
Hazardous Substances Data Bank (HSDB), 56
high throughput/content screening (HTS/HCS) assay, 3
high-throughput screening (HTS) assays, 48
HTS data analyses tools (HDAT), 24

immunotoxicity, 46–47
InChI Linear Notation, 84
Innovative Medicines Initiative (IMI), 2
in silico humans, 344–345
Integrated Risk Information System (IRIS), 58
integrative data analysis, 174
interactive Chemical Safety for Sustainability (iSSC), 187
in vitro mammalian cell gene mutation test, 50
in vitro mammalian cell micronucleus test, 50–51
in vitro mammalian chromosomal aberration test, 49–50
in vitro toxicity testing, data generated by
    carcinogenicity, 51
    genotoxicity/mutagenicity testing, 48–51
    skin sensitisation, 51–52
in vivo micronucleus study record, 145–159
in vivo toxicity testing
    data generated by
        acute toxicity testing, 40–42
        repeated dose administration, 42–43
    genotoxicity, 47
    immunotoxicity, 46–47
    neurotoxicity, 46
    reproductive and developmental toxicity tests, 44–46
ISAcreator configurator, 124

JRC ChemAgora portal, 223
JRC QSAR Model Database, 205

Kyoto Encyclopedia of Genes and Genomes (KEGG), 56–57

Library of Integrated Network Based Cellular Signature (LINCS), 221
locality sensitive hash codes (LSH), 95
local lymph node assay (LLNA), 47
Long-Range Research Initiative (LRI) AMBIT software, 169

mammalian cell gene mutation test. *See in vitro* mammalian cell gene mutation test
mathematical representation, 190–191
Medical Subject Headings (MeSH), 251
MetaboLights, 221
metabolomics, 219–220
MicroArray Quality Control Consortium (MAQC), 220
minimum information about a microarray experiment (MIAME), 220
minimum information about a proteomics experiment (MIAPE), 220
modes of action (MoA), 39
Molecular Design Limited (MDL), 111
molecular structures, 314

multiple linear regression
(MLR), 191
mutagenicity testing, 48–51

Nanotechnology Informatics
Working Group, 173
National Academies of Sciences
(NAS), 3
National Cancer Informatics
Program (NCIP), 173
National Cancer Institute (NCI), 173
National Center for Environmental
Assessment (NCEA), 58
National Chemical Genomics Center
(NCGC), 137
National Institute of Environmental
Health Sciences (NIEHS), 137,
138
National Institutes of Health (NIH),
13, 137
National Toxicology Program (NTP),
58, 137
neurotoxicity, 45–46
next-generation sequencing
(NGS), 216
non-cancer effects, 58
Nucleic Acids Research (NAR)
database, 167, 172

OECD Harmonised Templates
(OHTs), 9
Office of Research and Development
(ORD), 58
omics data
comparative assessments, 226
connectivity mapping (CMap),
227–228
mechanistic analysis, 228–229
unsupervised and super-
vised pattern recognition,
226–227
'omics technologies
(epi)genomics, 217–218
metabolomics, 219–220
proteomics, 218–219
transcriptomics, 216–217
one-dimensional (constitutional)
descriptors, 314
Online Chemical Modelling
Environment (OCHEM), 13
ontologies, 171–172
Open TG-GATEs, 221
Organisation for Economic
Co-operation and Development
(OECD), 9–10, 39, 40, 42, 44, 59,
122, 268–269

pancreatic ductal carcinoma cells,
315
partition coefficients (PCs), 346
peroxisome-proliferator-treated
hepatocytes, 315
physico–chemical descriptors,
314–315
physiological parameters, human
populations
body fat volume, 352–353
height, 349–350
height and weight growth
curves, 351–352
weight, 350–351
physiology-based pharmacokinetic
(PBPK), 307
PLETHEM stands,
343–344
point of departure (POD), 60
Predictive Model Markup Language
(PMML), 199–201
predictive models
anatomy of
data set representation,
191–192
mathematical
representation,
190–191
metadata, 192–193
model (and data)
provenance, 192–193
data and file formats, 195–196
applications, 204–205

predictive models (*continued*)
integrated modelling environment approaches, 206–208
Predictive Model Markup Language (PMML), 199–201
QsarDB Archive Format, 201–204
QSAR-ML, 198–199
QSAR model reporting format, 196–198
repository approaches, 205–206
storage solutions, 204–205
model governance, 193–195
predictive toxicology, 2–3
biological data for, 62–63
challenges, 6–7
adequate infrastructure, 7–9
data, making sense of, 14–15
data curation, 9–11
data integration infrastructures, 11–14
relevant data, 11
standardisation, 9–11
omics data in, 226–229
opportunities, 16–17
applications, 19–25
big picture, 18–19
chemical safety assessment, 23
emerging risks, 20–21
exposure to multiple chemicals, 21–22
population variability, 22–23
predictive nanotoxicology, 24–25
specific susceptibility, 22–23
validity, 4, 6
value, 4, 6
variability, 4, 5–6
variety, 4, 5
veracity, 4, 5
visibility, 4, 6
visualisation, 4, 6
volatility, 4, 6
volume, 3–5
principal component analysis (PCA), 226
proteomics, 218–219, 307–310
application
bio-descriptors, 320–324
biomarkers, 324–326
clustering and classification, 315–317
genetic algorithm, 317–320
chemometrical methods
bio-descriptors, 313–315
chemical descriptors, 313–315
clustering and classification, 311–313
genetic algorithm (GA), 313
similarity relationship, 310–311
PRoteomics IDEntifications (PRIDE), 221
PubChem, 231

QsarDB Archive Format, 201–204
QsarDB Repository, 205–206
QSAR-ML, 198–199
QSAR Model Database. *See* JRC QSAR Model Database
QSAR Model Report Format (QMRF), 197

QSAR model reporting format, 196–198
quality of data, 59–61

read-across, 359–375
biological, 366–368
chemical structure, 370–372
justification, 364–365
for nanomaterials, 372–373
omics technologies, 368–370
toxicological big data, 364
investigation of mechanisms of action, 365
receiver operating characteristic (ROC) curve, 252
Registry of Toxic Effects of Chemical Substances (RTECS), 56
Resource Description Framework (RDF), 12

Scientific Committee of Consumer Products (SCCP), 59
Scientific Committee on Consumer Safety (SCCS), 59
Screening Information Dataset (SIDS), 59
Secure Hash Algorithm (SHA), 95
self-organising map (SOM), 251, 252
self-organising map (SOM)-based clustering analysis, 24
Semantic Enrichment of the Scientific Literature (SESL) project, 12
Sequencing Quality Control (SEQC), 220
shared vocabulary, 171–172
skin sensitisation, 51–52
SMARTS linear notation, 84–85
SMILES linear notation, 82–83
standalone software, 206–207
Standard Data Tabulation Model (SDTM), 120
statistical significance, 62
support vector machines (SVM), 191
surface-enhanced laser desorption ionisation time of flight (SELDI-TOF) mass spectrometry, 325

tautomerism, 99
Test Guidelines. *See* Organisation for Economic Co-operation and Development (OECD)
three-dimensional descriptors, 314
Tox21 process
big data collection
data modelling, 257–259
in vivo toxicity prediction, 252–256
toxicity, compound mechanism of, 251–252
compound collection, 243–246
data quality control, 248–251
robotic platform, 247
screening process, 247–248
ToxFX, 221
toxicity testing
acute
limitations, 41
test guidelines, 40–41
use of, 41–42
after repeated dose administration
carcinogenicity test data, 43
sub-chronic toxicity test data, 43
test guidelines, 42–43
use of chronic toxicity data, 43
use of short-term data, 43
genotoxicity, 47
immunotoxicity, 46–47
in vitro

toxicity testing (*continued*)
carcinogenicity, 51
genotoxicity/mutagenicity testing, 48–51
skin sensitisation, 51–52
in vivo
acute toxicity testing, 40–42
genotoxicity, 47
immunotoxicity, 46–47
neurotoxicity, 46
repeated dose administration, 42–43
reproductive and developmental toxicity tests, 44–46
neurotoxicity, 46
reproductive and developmental toxicity tests, 44–46
toxicogenomics, 214–216
challenges, 232–234
opportunities, 232–234
perspectives, 232–234
toxicoinformatics, 214–216
annotation, 223–225
omics data, 220–223
standardisation, 220–223
storage, 220–223
toxicokinetic data, 54–55
toxicological big data
databases and data sets, 125–132
exchange and organising
ISA-Tab, 123–125
OECD harmonised templates, 122
SEND, 120–121
Structure Data (SD) file, 111–112
ToxML (DB), 113–120
inform and support read-across
biological read-across, 366–368
chemical structure, 370–372
omics technologies, 368–370
read-across for nanomaterials, 372–373
read-across justification, 364–365
toxicological mechanisms of action, 365
integrated approaches, 138–143
NIEHS/CEBS, 138
organisation of, 108–111
process management, 133–134
PubChem, 138
regulatory compliance databases, 133–134
sources of
Aggregated Computational Toxicology Resource (ACToR), 57
COSMOS Next Generation, 57–58
data from assessment reports, 58–59
European Chemicals Agency (ECHA) Data, 57
Kyoto Encyclopedia of Genes and Genomes (KEGG), 56–57
Registry of Toxic Effects of Chemical Substances (RTECS), 56
TOXicology Data NETwork (TOXNET), 55–56
ToxBank, 135–136
ToxCast/DSSTox/Tox21, 137–138

toxicological significance, 62
TOXicology Data NETwork
 (TOXNET), 55–56
TOXLINE, 55
TOXNET, 55
transcriptomics, 216–217
Transcriptomics Reporting
 Framework (TRF), 10

two-dimensional descriptors, 314

up-and-down procedure, 41

web services, 172
web solutions, 207–208